羌塘地区晚古生代—早中生代放射虫及地层

Late Palaeozoic and Early Mesozoic Radiolarian Palaeontology and Stratigraphy of Qiangtang Region

王玉净 沙金庚 朱同兴 郭通珍 解超明 著

资 助 项 目

中国科学院先导性项目B（XDB03010101）
国家自然科学基金资助项目（41730317）

中国科学技术大学出版社

内 容 简 介

我国青藏高原北部羌塘地区龙木错-双湖缝合带和可可西里龙木错-玉树缝合带内保存着晚古生代—早中生代含放射虫的蛇绿岩和放射虫硅质岩,是探讨特提斯演化的难得的地区。本书总结了羌塘和可可西里地区晚古生代—早中生代地层;系统描述了龙木错-双湖和龙木错-玉树缝合带内晚泥盆世—中三叠世放射虫化石,计25科、53属和155种(25新种),并建立了14个放射虫化石带及其区内、国内和全球对比格架;证明了古特提斯洋最晚闭合于二叠纪末,发现新特提斯洋最早形成于中三叠世安尼期而不是二叠纪。

这一综合性的放射虫古生物专著可供科研机构、大专院校、博物馆、地质矿产部门和石油系统从事地层学、古地理学、构造地质学和青藏高原或特提斯演化研究、教育、学习和地质生产的人员参考。

图书在版编目(CIP)数据

羌塘地区晚古生代—早中生代放射虫及地层:中文、英文/王玉净等著. —合肥:中国科学技术大学出版社,2024.5
ISBN 978-7-312-05769-4

Ⅰ. 羌… Ⅱ. 王… Ⅲ. ①羌塘高原—晚古生代—古动物—放射虫目—研究—汉、英 ②羌塘高原—中生代—古动物—放射虫目—研究—汉、英 ③羌塘高原—晚古生代—地层—研究—汉、英 ④羌塘高原—中生代—地层—研究—汉、英 Ⅳ. ①Q915.811 ②P535.245

中国国家版本馆 CIP 数据核字(2023)第 168448 号

审图号:GS(2024)0247 号

羌塘地区晚古生代—早中生代放射虫及地层
QIANGTANG DIQU WAN-GUSHENGDAI—ZAO-ZHONGSHENGDAI FANGSHECHONG JI DICENG

出版	中国科学技术大学出版社
	安徽省合肥市金寨路96号,230026
	http://press.ustc.edu.cn
	https://zgkxjsdxcbs.tmall.com
印刷	合肥华苑印刷包装有限公司
发行	中国科学技术大学出版社
开本	880 mm×1230 mm 1/16
印张	12
插页	15
字数	411千
版次	2024年5月第1版
印次	2024年5月第1次印刷
定价	99.00元

前　言

古特提斯洋何时闭合？新特提斯洋什么时候打开？国际地学界对这两个涉及特提斯演化的重要科学问题关注和争论已久。我国青藏高原是揭示东特提斯演化，特别是澄清以上重大古地理事件的关键地区。

采用地层名称术语，并按照大洋存在的时代，特提斯洋分为古（生代）特提斯、中（生代）特提斯和新（生代）特提斯（黄汲清，陈炳蔚，1987），但按照大洋形成的先后顺序，先形成的特提斯洋称古特提斯洋，后形成的特提斯洋称新特提斯洋（如：沙金庚，2018）。

早在 20 世纪 80 年代，黄汲清和陈炳蔚（1987）根据构造、地层、古生物等综合研究，总结了中国及邻区特提斯演化，认为古特提斯洋在接近晚二叠世末就已关闭，此前南欧、伊朗、阿富汗、巴基斯坦、印度北部、藏南和藏北（包括拉萨和羌塘）、中缅地块、马来西亚、苏门答腊及印度尼西亚的大部地区位于冈瓦纳大陆的北缘。中、新特提斯洋（相当于本书的新特提斯洋）开始于早三叠世。

然而，可能由于研究的手段、依据、地区甚至对特提斯、特提斯洋和特提斯海的理解不同，后来的特提斯研究者对古特提斯洋关闭时代和新特提斯洋打开时代的认识产生了很大分歧。

特提斯的研究者普遍认为古特提斯洋闭合于中生代晚三叠世甚至更晚（如：刘本培等，1993；Glonka，2004；王玉净等，2005；Muttoni et al.，2009；Zanchi，Gaetani，2011；Metcalfe，2013；Angiolini et al.，2013，2015；Jolivet，2017），新特提斯洋打开于晚古生代中二叠世甚至早二叠世或更早（如：Muttoni et al.，2009；Zanchi，Gaetani，2011；Angiolini et al.，2013，2015；Metcalfe，2013；Berra，Angiolini，2014；张以春，王玥，2019）。但也有研究者认为古特提斯洋闭合于中二叠世末或晚二叠世初（如：张以弢，1991；沙金庚等，1992；边千韬等，1993，1996，1997；青海可可西里综合科学考察队，1994；张以弢，郑健康，1994；王玉净，1995；张俊明，1995；沙金庚，1998；Sha，Fürsich，1999；Sha et al.，2004；沙金庚，2009，2018）或晚二叠世（如：张以弢，1993），新特提斯洋形成于中生代中三叠世晚期拉丁期（Ladinian）—晚三叠世早期卡尼期（Carnian）（如：沙金庚，2018）。

以上关于古特提斯洋闭合和新特提斯洋打开时代讨论的依据分别是北主缝合带（或古特提斯缝合带）喀喇昆仑山口-龙木错/湖-玉树-金沙江-昌宁-双江-孟连古特提斯缝合带和南主缝合带（或中特提斯缝合带）印度斯/河-雅鲁缝合带及其分支班公错-怒江缝合带（黄汲清，陈炳蔚，1987）。

我们认为含放射虫的蛇绿岩和放射虫硅质岩是判别深水大洋或洋盆和裂谷的最可靠的标志，放射虫化石是判别洋盆或裂谷的必不可少的依据。

近 30 年来，青藏高原地质工作者已在我国青藏高原北部羌塘地区可可西里北主缝合带龙木错-玉树古特提斯缝合带和龙木错-双湖缝合带内蛇绿岩和放射虫硅质岩建造中发现了晚古生代（如：边千韬，郑祥身，1991；李红生，1993；王玉净，1995；朱同兴等，2010）和三叠纪（如：苟金，邢国忠，1990；王玉净等，2005；朱同兴等，2010）放射虫化石记录，暗示龙木错-玉树和龙木错-双湖缝合带内可能记录了晚古生代—早中生代特提斯的演化过程，从中可以找到古特提斯洋闭合时代和新特提斯洋打开时代的证据。

为了更全面合理地理解我国青藏高原北部北特提斯演化，揭开古、新特提斯洋的闭合和打开时代之谜，本书作者王玉净教授等经过对羌塘和可可西里等地区近30年的调查、研究、积累和总结，取得了以下4项成果：

（1）发现岗齐曲北侧康特金、西金乌兰湖移山湖、西金乌兰湖北、汉台山南蛇形沟、西金乌兰湖构造混杂岩带西北、羊湖东南、才多茶卡北岸、雅曲乡、题玛茶卡北、绒玛北东、角木日、才玛尔错南和才玛尔错北地区（图1）的晚古生代—早中生代蛇绿岩和放射虫硅质岩建造中均含丰富的放射虫化石。

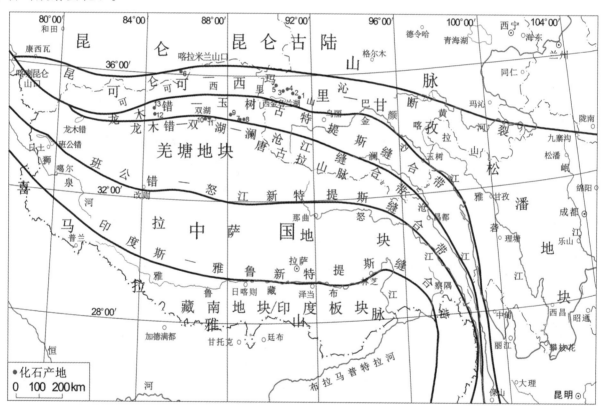

图1 羌塘地区晚古生代—早中生代放射虫化石分布图

1. 岗齐曲北侧康特金；2. 西金乌兰湖移山湖；3. 西金乌兰湖北部；4. 汉台山南蛇形沟；5. 西金乌兰湖构造混杂岩带西北；6. 羊湖东南；7. 才多茶卡北岸；8. 雅曲乡；9. 题玛茶卡北；10. 绒玛北东；11. 角木日；12. 才玛尔错南；13. 才玛尔错北

Fig. 1 Map showing the late Palaeozoic and Early Mesozoic radiolarian localities of Qiangtang region

1. Kangtejin of the north side of Gangqiqu; 2. Yishanco (co=lake) of Xijir Ulan Hu (Hu=lake) (Xijinwulanco); 3. north side of Xijir Ulan Hu (Xijinwulanco); 4. Shexinggou of the south of Hantaishan; 5. northwest of the Xijir Ulan Hu (Xijinwulanco) tectonic mélange zone; 6. southeast of Yanghu; 7. north shore of Cêdo Caka (Caiduochaka); 8. Yaqu village; 9. north of Timachaka; 10. northeast of Rongma; 11. Jiaomuri; 12. south of Caimaerco; 13. north of Caimaerco

（2）总结、描述和对比了羌塘地区晚古生代—早中生代地层系统（见第1章）及西藏双湖和青海可可西里地区晚古生代—早中生代放射虫地层（见第2章）。

（3）对积累了近30年的放射虫样品进行了分析处理、系统研究和详细描述，在龙木错-双湖缝合带和龙木错-玉树缝合带内识别出了25科、53属和155种晚泥盆世-中三叠世放射虫化石，其中包括25新种（见第4章和第6章）；建立了西藏双湖和青海可可西里地区晚古生代（9个）和早中生代（5个）放射虫化石带，并对它们进行了区内、国内和全球对比（见第3章和第5章）。

（4）讨论了龙木错-双湖缝合带的古地理意义，证明了我国古特提斯洋开始于晚泥盆世甚至更早，最晚闭合于二叠纪末（黄汲清，陈炳蔚，1987），发现我国新特提斯洋最早形成于中三叠世早期安尼期（Anisian）而不是二叠纪（见第7章和第8章）。

本书作者有：中国科学院南京地质古生物研究所现代古生物学和地层学国家重点实验室的王玉净、沙金庚，中国地质调查局成都地质调查中心的朱同兴，青海省地质调查院的郭通珍，吉林大学地球科学学院的解超明。本书的写作分工如下：前言，沙金庚；第1章，朱同兴、沙金庚、王玉净；第2章，郭通珍、朱同兴、沙金庚；第3—6章、图版说明和图版，王玉净；第7章，王玉净、沙金庚、解超明；第8章及英文目录和摘要，王玉净、沙金庚；全书统稿，沙金庚、王玉净。

本书中研究的放射虫样品由曾繁年、史世福分析处理，化石扫描照相由茅永强、樊晓羿完成，陈雪儿和张小弘协助图、表的绘制和负责文字的打印。作者对以上鼎力相助的同仁表示由衷的感谢！由于藏北的野外工作环境艰险和地质背景复杂，此项研究涉及的地层时代长，放射虫化石的研究工作量特别大，不当和错误之处，敬请地质古生物学界的同仁和读者批评指正。

羌塘双湖和可可西里无人区的高寒、缺氧、无交通等恶劣环境导致了那里化石标本采集的艰难，我们向那些冒着生命危险的样品采集者表示由衷的感谢和深深的敬意。

目 录

前言 ……………………………………………………………………………………………………（ⅰ）

第1章　羌塘地区晚古生代—早中生代地层系统 ……………………………………………（ 1 ）
  1.1　南羌塘（昂达尔错）地层分区 ………………………………………………………（ 2 ）
  1.2　中羌塘（西雅尔岗）构造-地层分区 …………………………………………………（ 5 ）
  1.3　北羌塘（多格错仁）地层分区 ………………………………………………………（ 5 ）

第2章　西藏双湖和青海可可西里地区晚古生代—早中生代放射虫地层剖面 ………………（ 10 ）
  2.1　青海可可西里地区移山湖泥盆系实测剖面 …………………………………………（ 10 ）
  2.2　青海可可西里地区西金乌兰湖西北部构造混杂岩带实测剖面 ……………………（ 11 ）
  2.3　青海可可西里地区蛇形沟蛇绿混杂岩剖面 …………………………………………（ 14 ）
  2.4　青海可可西里地区岗齐曲康特金剖面 ………………………………………………（ 15 ）
  2.5　青海可可西里地区西金乌兰湖北移山湖剖面 ………………………………………（ 15 ）
  2.6　西藏双湖地区才多茶卡北岸晚古生代放射虫硅质岩实测剖面 ……………………（ 16 ）
  2.7　西藏双湖地区雅曲乡硅质岩 …………………………………………………………（ 17 ）
  2.8　四川甘孜-理塘构造带沉积混杂岩 …………………………………………………（ 17 ）
  2.9　西藏角木日地区玄武岩剖面 …………………………………………………………（ 17 ）

第3章　西藏双湖和青海可可西里地区晚古生代放射虫动物群及其时代和对比 ……………（ 19 ）
  3.1　西藏双湖和青海可可西里地区晚古生代放射虫动物群及其时代 …………………（ 19 ）
  3.2　西藏双湖和青海可可西里地区晚古生代放射虫动物群的对比 ……………………（ 30 ）

第4章　晚古生代放射虫分类描述 …………………………………………………………（ 40 ）

第5章　西藏双湖和青海可可西里地区早中生代放射虫动物群及其时代和对比 ……………（ 81 ）
  5.1　西藏双湖和青海可可西里地区早中生代放射虫动物群及其时代 …………………（ 81 ）
  5.2　西藏双湖和青海可可西里地区早中生代放射虫化石带及其对比 …………………（ 88 ）

第6章　早中生代放射虫分类描述 …………………………………………………………（ 92 ）

第7章　龙木错-双湖缝合带的古地理意义 …………………………………………………（117）
  7.1　蛇绿岩和放射虫动物群 ………………………………………………………………（117）
  7.2　古地理意义 ……………………………………………………………………………（119）

第8章　结束语 ………………………………………………………………………………（121）

参考文献（REFERENCES） ……………………………………………………………………（123）

索引（INDEX OF GENERA AND SPECISE） ……………………………………………………… (136)
1. 拉-汉属种名称 ……………………………………………………………………………………… (136)
2. 汉-拉属种名称 ……………………………………………………………………………………… (140)

英文摘要（ENGLISH SUMMARY） ……………………………………………………………… (147)
  Introduction ………………………………………………………………………………………… (148)
  1 Late Palaeozoic and Early Mesozoic stratigraphic system in Qiangtang region ……………… (148)
  2 Late Palaeozoic and Early Mesozoic stratigraphic sections containing
     radiolarian fossils in Shuanghu of north Tibet and Hoh Xil of Qinghai ………………………… (149)
  3 Composition, age and correlation of Late Palaeozoic radiolarian faunas in Shuanghu of
     north Tibet and Hoh Xil of Qinghai ……………………………………………………………… (150)
  4 Systematic classification and description of new species of Late Palaeozoic radiolarians ……… (153)
  5 Composition, age and correlation of Early Mesozoic radiolarian faunas in Shuanghu of
     north Tibet and Hoh Xil of Qinghai ……………………………………………………………… (162)
  6 Systematic classification and description of new species of Early Mesozoic radiolarians ……… (164)
  7 Palaeogeographic implications of the Longmuco-Shuanghu suture zone ……………………… (171)
  8 Conclusions ………………………………………………………………………………………… (175)

图版说明（PLATE EXPLANATIONS） …………………………………………………………… (177)
图版（PLATES） …………………………………………………………………………………… (187)

第1章 羌塘地区晚古生代—早中生代地层系统

以羌塘中部龙木错-双湖缝合带北缘长蛇山断裂和南缘查桑断裂为界，羌塘地区大地构造单元自北向南依次划分为北羌塘地块、龙木错-双湖缝合带、南羌塘地块（图1.1）。在大地构造单元划分基础上，羌塘地层区自北向南也依次划分为3个地层分区：北羌塘（多格错仁）地层分区、中羌塘（西雅尔岗）构造-地层分区（龙木错-双湖缝合带）和南羌塘（昂达尔错）地层分区。各地层分区的泥盆系—三叠系地层序列见表1.1。

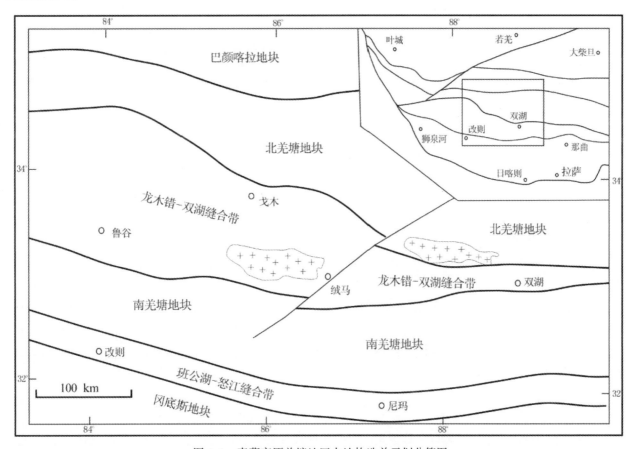

图 1.1 青藏高原羌塘地区大地构造单元划分简图

Fig. 1.1 Sketch of geotectonic unit division of Qiangtang region, Tibetan Plateau

表 1.1 青藏高原羌塘地层区划及泥盆系—三叠系地层划分

Table 1.1 Stratigraphic division and Devonian—Triassic sequences of Qiangtang region, Tibetan Plateau

地层系统		北羌塘（多格错仁）地层分区			中羌塘（西雅尔岗）构造-地层分区			南羌塘（昂达尔错）地层分区		
三叠系	上统	藏夏河组	菊花山组	土门格拉群				日干配错群	肖茶卡群	扎那组 / 角木茶卡组
										肖切保组
	中统	康南组							欧拉组	
	下统	硬水泉组			红脊山蛇绿混杂岩（P—T）	角木日蛇绿混杂岩（P—T）	双湖蛇绿混杂岩（D—T）	?	孜狮桑组	?
		康鲁组								
二叠系	上统	热觉茶卡组						吉普日阿组		
	中统	先遣组						龙格组		
								吞龙共巴组		
	下统	冈玛错组						霍尔巴错群	曲地组	
									展金组	
石炭系	上统							擦蒙组		
	下统	日湾茶卡组			?	?		?		
泥盆系	上统	拉竹龙组								
	中统	查桑组				?				
	下统	兽形湖组								

1.1 南羌塘（昂达尔错）地层分区

南羌塘地层分区在构造单元划分上属于南羌塘地块。区内大面积地分布着晚古生代和中生代地层。日土县多玛地区上古生界发育齐全，自下而上依次为石炭系—二叠系霍尔巴错群（包括上石炭统擦蒙组、下二叠统展金组和曲地组）、中二叠统吞龙共巴组和龙格组、上二叠统吉普日阿组、三叠系孜师桑组、欧拉组、日干配错群（改则地区）、肖茶卡群（双湖地区）等。

1.1.1 擦蒙组

上石炭统擦蒙组主要为一套以含砾板岩、含砾粉砂质板岩为特征的冰水杂砾岩沉积组合。在区域上，擦蒙组下部为灰色含砾粉砂质板岩与含砾泥质粉砂岩互层；中部为灰色薄—中层状含砾粉砂质板岩、灰绿色泥质粉砂岩与粉砂质板岩互层；上部为灰黄色中层状细粒砂岩夹细砾岩、含砾细粒砂岩。在双湖孔孔茶卡剖面采获牙形刺：*Adetognathus lautus*（Gunneu），*A. paralautus* Orchara，*Lonchodina* sp.（朱同兴等，2010）。

区内擦蒙组均未见底，顶部与下二叠统展金组整合接触（日土地区），或与上三叠统肖茶卡群假整合接触（双湖地区）。擦蒙组厚度变化大：东部双湖孔孔茶卡地区大于 96.82 m，西部日土多玛地区大于 470 m。擦蒙组板岩、粉砂质板岩中广泛分布含量不等的复成分砾石。砾石成分包括灰岩、变砂岩、千枚岩、硅质岩、脉石英、玄武岩、花岗岩、石英斑岩等。砾石无分选，部分具压裂面结构。结合前人研究资料（刘本培等，1983；尹集祥，1997），南羌塘地区上石炭统擦蒙组应为冰海相沉积（朱同兴等，2010）。

1.1.2 展金组

下二叠统展金组为整合于下伏擦蒙组和上覆曲地组之间的一套凝灰质砂板岩夹多层基性玄武岩、灰岩地层。南羌塘地区展金组东西向岩性、厚度变化大，由东向西，火山岩和灰岩层减少，砂板岩增多。

西部日土地区展金组产冈瓦纳相冷水型双壳类 *Eurydesma* 动物群 *Eurydesma perversum* Reed, *E. playfordi* Dickins, *E. mytiloides* Reed, *E.* sp., *Nuculopsis* sp.（刘本培等，1983；梁定益等，1983）；冷水型珊瑚化石 *Amplexocaninia*－*Cyathaxonia* 组合（范影年鉴定，2005）和腕足类化石 *Bandoproductus*－*Brachythyrina* 组合（张以春等鉴定，2010）。

在西部日土多玛剖面，展金组粉砂质板岩中发育不同型式的鲍马序列组合，沉积构造发育正粒序、平行层理、沙纹层理、水平纹层以及滑塌角砾、滑塌变形构造，产薄壳小腕足类化石，说明其沉积相为大陆斜坡远缘浊积岩。

在东部双湖玛日保索古来-江额立果日一带地质路线调查表明，展金组岩性主要为致密状玄武岩、杏仁状玄武岩夹火山角砾岩、凝灰岩、凝灰质砂岩等，厚度大于 800 m，产腕足类 *Liraplecta richthoferi* Chao 等（朱同兴等，2010）。

在东部双湖知塞剖面，展金组岩性为一套凝灰质砂泥岩夹玄武岩、泥晶灰岩等，厚度大于 648.8 m。其中产䗴 *Schwagerina shuanghuensis* Wang et Zhang, *Sphaeroschwagerina glomerosa* Schwag, *Robustoschwagerina obesa* Yang, *R. guangxiensis* Yang, *Pseudofusulina vulgaris* Schellwie et Dyhrenfurth, *Pseudoschwagerina* sp., *Ozawainella guangxiensis* Chen et Wang；腕足类 *Streptorhynchus tibetanus* Chang, *Costiferina obesa* Fang；海百合类 *Cyclocticus* cf. *lubricus* Li, *Pentagonocyclicus* sp.；苔藓虫 *Fenestella elusa* Reed, *Ogbinopora sinopermiana* Fan, *Protoretepora* sp. 等（朱同兴等，2010）。

在东部双湖肖茶卡地区采获早二叠世皱纹珊瑚：*Kepingophyllum shuanghuense* Yu（范影年鉴定，2005）。

1.1.3 曲地组

下二叠统曲地组为整合于下伏展金组和上覆吞龙共巴组之间的一套粗碎屑岩地层，厚度大于 2100 m，产腕足类、珊瑚和双壳类化石，但保存不好。

1.1.4 吞龙共巴组

中二叠统吞龙共巴组仅局限分布于西部日土多玛地区，是指整合于下伏曲地组碎屑岩与上覆龙格组灰岩之间的一套细碎屑岩与灰岩互层为特征的地层，厚度大于 1360 m。产䗴 *Schwagerina shuanghuensis* Wang et Zhang, *Misellina* sp., *Nankinella* sp.；腕足类 *Costiferina obesa* Fang, *Streptorhynchus tibetanus* Chang；苔藓虫 *Fenestella elusa* Reed, *Ogbinopora sinopermiana* Fan（夏代祥，刘世坤，1997；赵政璋等，2001）及皱纹珊瑚 *Waagenophyllum* sp.（范影年鉴定，2005）等化石。

1.1.5 龙格组

羌塘地区广泛分布的中二叠统龙格组为一套碳酸盐岩地层，主要岩性为重结晶灰岩、生物礁灰岩、鲕状灰岩等，产丰富的䗴、珊瑚、腕足类等化石，厚度普遍大于 1000 m。

双湖知塞剖面产䗴 *Verbeekina spheara* Ozawa, *Parafusulina splendens* Dunbar, *Parafusulina elliptica* Sheng, *Nankinella* cf. *quasihunnaensis* Sheng, *Eopolydiexodina* sp., *Sumatrina* sp.,

Afghannella sp.（朱同兴等，2010）及皱纹珊瑚 *Wentzelellites zhisaiensis* Zhang et Zheng（sp. nov，M. S.）（朱同兴等，2010）等化石。

角木茶卡剖面产蜓 *Sumatrina fusiformis* Sheng，*S. annae* Voly，*Afghannella schencki* Thompson，*Neoschwagerina craticulifera*（Schwager），*Pseudofusulina* sp.（朱同兴等，2010）。

查桑剖面产蜓 *Parafusulina jiangsuensis* Chen，*Parafusulina rothi* Dunbar et Skinner，*P. lineate* Dunbar et Skinner，*P. splendens* Dunbar et Skinner（朱同兴等，2010）以及珊瑚 *Multimurinus biformis* Fontaine，*Szechuanophylum polygonale* Yu，*Szechuanophylum paraszechuanense* Yu，*Shuanghuphyllia typical* Yu，*S. xiaochakaensis* Yu，*S. simplex* Yu，*S. stenostabulata* Zhang（sp. nov，M. S.），*Sakamotosawenella intermedia* Yu 等（朱同兴等，2010）。

1.1.6 吉普日阿组

上二叠统吉普日阿组仅局限分布于日土地区，为一套含煤陆源碎屑岩夹灰岩地层，含蜓、腕足类等化石（赵政璋等，2001）。吉普日阿组底部与中二叠统龙格组呈整合或假整合接触；顶部与上三叠统日干配错群呈假整合接触，厚度大于1520 m。

1.1.7 孜狮桑组

下三叠统孜狮桑组仅局限分布于尼玛县维多乡孜狮桑加波日一带，岩性为灰黑色、灰白色砂屑灰岩、白云质细晶灰岩、泥晶灰岩，产牙形刺 *Neospathodus triangularis*，*N. homeri*，*Enantiognathus delicatulus*，*Hibbardella tricornigera*，*Ellisonia* sp.，*Prioniodella ctenoides*，*Hindeodella* sp.，*Ozarkodina* sp. 等（王永胜等，2012）。剖面未见底，厚度大于239.33 m。

1.1.8 欧拉组

下—中三叠统欧拉组仅局限分布于日土多玛、加措等地，为一套碳酸盐岩夹泥岩地层，产双壳类 *Claraia*、*Eumorphotis inaequicostata*（Benedke）（赵政璋等，2001）。欧拉组顶、底界出露不全，厚度大于600 m。

1.1.9 日干配错群

上三叠统日干配错群主要出露于南羌塘中、南部的改则县日干配错地区，主要由碳酸盐岩夹细碎屑岩组成，产双壳类、海绵、海百合茎 *Pentagonocyclicus* sp.，*Cyclocyclicus* sp. 及珊瑚 *Stylophyllopsis* sp. 等化石（曾庆高等，2011）。在区域上，日干配错群多未见顶、底（或顶、底接触关系不清楚），厚度大于1591.49 m。

1.1.10 肖茶卡群

上三叠统肖茶卡群主要出露于南羌塘北缘的肖茶卡地区。根据岩性组合，此组自下而上分为肖切保组火山岩、角木茶卡组灰岩和扎那组碎屑岩（冯心涛等，2005）。

肖切保组岩性主要由致密块状玄武岩、安山质玄武岩夹多层火山角砾岩、灰岩透镜体，熔岩中枕状构造发育。此组的 K-Ar 年龄为 223 Ma±5 Ma，厚度大于670.4 m。

角木茶卡组总厚大于551.75 m，分为上、下两段：下段发育厚块状砾岩，局部夹海绵点礁体。砾石成分以基性火山岩为主，少量灰岩、硅质岩；上段由中层状砂砾屑灰岩、厚块状海绵礁灰岩、角砾状灰岩等组成，产珊瑚、双壳类、腕足类、海绵、海百合茎等化石。

扎那组总厚度大于1350.61 m，三段结构明显：下段为薄—中层状细粒岩屑砂岩、凝灰质砂岩、粉砂岩和薄层状凝灰质泥岩互层；中段为中—厚层状含砾粗砂岩、中—细粒长石岩屑砂岩、钙质粉砂岩夹粉砂质泥岩、泥质粉砂岩；上段为中层状细粒岩屑长石砂岩、粉砂岩与薄—中层状钙质粉砂质泥岩、钙质泥岩互层，夹少量泥质灰岩，产孢粉和少量植物、双壳类化石。

在区域上，肖茶卡群未见顶，底部多为整合或假整合接触。

1.2 中羌塘（西雅尔岗）构造-地层分区

中羌塘构造-地层分区在大地构造单元划分上相当于龙木错-双湖缝合带范围，出露前奥陶系至新生界地层，但多不连续，主要呈构造接触，并发生不同程度的变质变形。区内地层层序表现为局部有序、整体无序格局。

区内广泛出露晚古生代—三叠纪蛇绿混杂岩群或构造混杂岩，自西向东分别命名为红脊山蛇绿混杂岩群（P—T）、角木日蛇绿混杂岩群（P—T）、双湖蛇绿混杂岩群（D—T）。其中角木日地区具有相对完整的蛇绿岩套层序，包括底部变质超基性橄榄岩；中部基性堆晶岩和辉长岩、辉长岩席状岩墙；上部枕状玄武岩、放射虫硅质岩以及远洋砂板岩复理石等。区内普遍发育榴辉岩、多硅白云母片岩、蓝片岩等高压变质岩构造透镜体。

最新研究资料表明，在上述蛇绿混杂岩群硅质岩岩块中，共发现5个层位放射虫动物群：才多茶卡北岸 *Helenifore robustum* 动物群［晚泥盆世弗拉期（Frasnian）］和 *Neoalbaillella ornithoformis* 动物群［晚二叠世长兴期（Changhsingian）］；雅曲乡 *Pseudoalbaillella sakmarensis-P. lomentaria* 动物群［早二叠世隆林期（Longlinian）］和 *Albaillella ishigai* 动物群［中二叠世栖霞期（Chihsian）］；角木日 *Eptingium nakasekoi* 动物群［（中三叠世安尼期（Anisian）］（朱同兴等，2010）。

1.3 北羌塘（多格错仁）地层分区

北羌塘地层分区在构造单元划分上属于北羌塘地块。被中生代侏罗系大面积覆盖，晚古生代—三叠纪地层发育齐全，但出露零星。需要指出的是，北羌塘地层分区上，三叠统土门格拉群、菊花山组、藏夏河组为同时异相岩石地层单位。

1.3.1 兽形湖组

下泥盆统兽形湖组主要分布于西部兽形湖一带，岩性主要为灰黄绿色薄层状泥质粉砂岩、灰色中-薄层细粒长石石英砂岩，生物遗迹（爬行迹、觅食迹）发育，厚度大于204 m，产腕足类化石 *Meristella* sp.，*Atrypa* cf. *gurjerskensis* Alekseeva，*Levenea* sp.，*Nucleospina* sp.，*Strophome nids*，*Megastrophia* sp.，*Howellella* sp.，*Howellella* cf. *papaoensis*（Grabau）等（夏代祥等，1997）。

1.3.2 查桑组

中泥盆统查桑组主要分布于双湖查桑地区，岩性为浅灰、浅紫红色中-厚层状生物碎屑细晶灰岩、浅紫红色中层状粉晶—细晶灰岩、（含）生物碎屑灰岩、砂屑灰岩，以富含腕足类、珊瑚化石为特征，厚度大于530.53 m。查桑组未见底，顶部与拉竹龙组整合接触。产腕足类化石 *Desquamatia*（*D.*）*hunanensis* Grabau，*Spinatrypa kuangsiensis*（Grabau），*S. subkuangsiensis*（Tien）（朱同兴等，2010）；珊瑚化石 *Cyathophyllum* sp.，*Favosites* sp.（朱同兴等，2010）。

1.3.3 拉竹龙组

上泥盆统拉竹龙组主要分布于羌塘北部拉竹龙地区，岩性主要由上、下两部分组成：下部为浅紫红色中层状含生物碎屑砂砾屑中晶灰岩夹灰色薄-中层状泥质灰岩和泥晶灰岩；上部浅紫、浅灰色中-厚层状（含）生物碎屑结晶灰岩夹棘屑灰岩、浅紫红色生物碎屑球粒灰岩、层孔虫礁灰岩等。拉竹龙组厚度大于631.41 m，岩石普遍重结晶为细晶-中晶方解石结构，含珊瑚、腕足类、层孔虫、海百合、苔藓虫、腹足类等化石。其中珊瑚化石包括 *Disphyllum cylindricum* Sun，*D. longiseptatum* Yoh，*D.* sp.，*Hunanophrentis* sp.，*Sinodisphyllum* cf. *simplex* Sun，*Temnopyllum* sp.（朱同兴等，2010）。

1.3.4 日湾茶卡组

下石炭统日湾茶卡组主要分布于日土县月牙湖-改则县日湾茶卡-双湖区查桑一带。日湾茶卡组系由谢义木（1983）命名的日湾茶卡群演变而来，其岩性组合以一套碳酸盐岩夹细碎屑岩为特征，下部为（含）生物碎屑灰岩夹粉砂质灰岩、粉砂质泥岩，上部为鲕粒灰岩、灰质白云岩、白云岩夹泥质灰岩，含丰富的早石炭世腕足类和珊瑚化石：腕足类 *Gigantoproductus-Striatifera* 组合（赵政璋等，2001）；暖水型珊瑚化石 *Yuanophyllum kansuense*，*Kueichouphyllum sinense*，*Arachnolasma* sp.（赵政璋等，2001），*Acrocyathus* cf. *simplex*，*Prosmilia jiaomuchakaensis* Zhang et Zhang（sp. nov.），*Chaetetella filiformis* Sokolov（朱同兴等，2010）；冷水型珊瑚化石 *Groenlandophyllum pulchrun* Fan，*Amolexum* cf. *coralloidea* Sowerby（朱同兴等，2010）。

日湾茶卡组底部与拉竹龙组，顶部与冈玛错组均为整合接触，厚度东西向变化不大，210—417 m。日湾茶卡组沉积相为碳酸盐开阔台地相，含台内点礁相、台内浅滩相等。

1.3.5 冈玛错组

上石炭统-下二叠统冈玛错组零星分布于改则查布冈玛错地区，岩性为浅灰-灰黄色长石石英细粒砂岩、粉砂岩夹粉砂质灰岩、泥晶灰岩等，底部以长石石英砂岩与日湾茶卡组灰岩分界，厚度大于150 m，产腕足类 *Chonetes* sp.，*Phricodothyris ovate*；珊瑚 *Campophyllum kiaeri*；䗴 *Robustoschwagerina obesa* Yang，*Sphaeroschwagerina glomerosa* Schwager（赵政璋等，2001）；牙形刺 *Adetognathus lautus* Gunnen，*A. paralautus* Orchara，*Lonchodina* sp.，*Lithostrotion* sp. 化石（朱同兴等，2010）。上述腕足类、珊瑚、牙形刺门类中的代表性分子在我国南方和西北等地均有分布，时代主要为晚石炭世，个别分子延续至早二叠世。

1.3.6 先遣组

中二叠统先遣组零星分布于改则查布先遣-鲁谷地区，岩性为泥晶灰岩夹鲕粒灰岩、碎屑岩，厚度大于 2000 m，产䗴 *Schwagerina shuanghuensis* Wang，*Misellina* sp.，*Nankinella* sp.，*N.* cf. *quaslhunnensis* Sheng，*Neoschwagerina craticulifera* Schwager，*Sumatrina fusiformis* Sheng，*S. annae* Voly，*Verbeekina spheara* Ozawa，*Parafusulina splendens* Dunbar et Skinner，*P. elliptica* Sheng，*P. kiangsuensis* Chen，*P. rothi* Dunbar et Skinner，*Afghannella schencki* Thompson，*Eopolydiexodina* sp.；腕足类 *Costiferina obesa* Fang，*Streptorhynchus tibetanus* Chang；海绵 *Amblysiphonella randuiensis* Deng，*A. radicifera* Waagen et Wentzel，*A. markamensis* Deng；苔藓虫 *Fenestella elusa* Reed，*Ogbinopora sinopermiana* Fan；皱纹珊瑚 *Waagenophyllum* sp.，*Wentzelellites zhisaiensis* Zhang et Zhang（sp. nov.），*Shuanghuphyllia typical* Yu，*Sh. xiaochakaensis* Yu，*Sh. simplex* Yu，*Sh. stenostabulata* Zhang et Zhang（sp. nov.），*Multimurinus biformis* Fontaine，*Szechuanophylum polygonale*

Yu，*Sze. paraszechuanense* Yu，*Sakamotosawenella intermedia* Yu 等化石（赵政璋等，2001）。

1.3.7 热觉茶卡组

上二叠统热觉茶卡组零星分布于玛依岗日北部热觉茶卡一带，岩性三分明显：下部为灰色薄层细粒砂岩、粉砂岩夹炭质页岩；中部为灰色薄层粉砂质灰岩、泥质灰岩与粉砂岩互层；上部为灰色、深灰色中层细粒砂岩、粉砂岩夹黑色炭质页岩、煤线（层）。热觉茶卡组未见底，顶部与三叠系康鲁组呈整合或假整合接触，厚度大于 510 m，含䗴、腕足类、植物等化石。其中䗴包括 *Palaeofusulina nana*，*P. sinensis*，*P. fusiformis*；腕足类为 *Peltichia zigzag*，*Squamularia waagani*，*Cathaysia chontoidas*，*Leptodus* sp.；植物化石包括 *Gigantonoclea* sp.，*Pecopteris* sp.，*Lobatanularia* sp.，*Rajahia* sp.（赵政璋等，2001）。

1.3.8 康鲁组

下三叠统康鲁组下部为灰绿色中-厚层状砂质砾岩、含砾岩屑粗砂岩、中细粒长石岩屑砂岩、粉砂岩；上部为钙质粉砂岩、粉砂质泥岩夹泥晶灰岩，组成明显的向上变细的沉积序列，厚度为 99—449 m，产丰富的双壳类（*Eumorphotis*）动物群：*Eumorphotis* cf. *multiformis* Bittner，*E. teilhardi* Patte，*E.* sp.，*Unionites* cf. *canalensis* Catullo，*U.* cf. *spicatus* Chen，*Myophoria* (*Laviconcha*) cf. *praeorbicularis* (Bittner)，*Entolium* sp.（朱同兴等，2010）。康鲁组底部与上二叠统热觉茶卡组含煤碎屑岩、顶部与下三叠统硬水泉组灰岩均呈整合接触关系。

区域地质调查表明，区内康鲁组分布局限，仅出露于热觉茶卡和江爱达日那等地，但其岩石组合、生物类型和沉积相变化不大，地层厚度由东往西增厚，江爱达日那地区厚度仅为 99 m，至热觉茶卡地区厚度达 449 m。

1.3.9 硬水泉组

下三叠统硬水泉组岩性有灰色、灰绿色薄-中层状泥晶灰岩、泥质灰岩夹生物扰动灰岩、鲕粒灰岩、豆粒灰岩、粉砂质泥岩、泥岩、泥质粉砂岩，厚度为 298—453 m，产丰富的双壳类和牙形刺等。硬水泉组与下伏康鲁组、上覆康南组皆为整合接触关系。

硬水泉组产丰富的双壳类（*Eumorphotis*）动物群：*Eumorphotis rugosa* Chen，*E. multiformis* Bittner，*E. inaequicostata* (Benecke)，*E.* aff. *venetiana* (Hauer)，*E.* (*Asoella*) *paradoxical* Chen，*E.* (*A.*) *subillyrica* (Hsu)，*E.* (*A.*) cf. *subillyrica* (Hsu)，*E.* sp.，*Unionites fassaensis* (Wissmann)，*U.* cf. *fassaensis* (Wissmann)，*U. spicatus* Chen，*U. albertii* (Assmann)，*U.* sp.，*Plagiostoma* aff. *striatum* (Schl.)，*Myophoria laevigata* (Ziethen)，*Leptochondria albertii* (Goldfuss)，*L.* cf. *michaeli* (Assmann)，*Donaldina* sp.，*Nuculana subperlonga* Chen（朱同兴等，2010）；牙形刺化石 *Pachycladina inclinata* Staesche，*P. bidentata* Wang et Cao，*P. multidentata* Wang et Cao，*P. trigona* Wang et Cao，*P.* sp.，*Parachirognathus geiseri* Clark，*P.* sp.，*Cypridodella muelleri* Tatge，*Hindeodella suevica* Tatge，*H. navadensis* Muller，*Diplododella triassica* Muller，*D. magnidentata* Tatge（朱同兴等，2010）。

区域地质调查资料表明，硬水泉组局限出露于热觉茶卡和江爱达日那等地，地层厚度由东往西加大，江爱达日那一带仅 298 m，至康鲁山一带为 453 m，热觉茶卡地区大于 767 m。在区域上，硬水泉组岩性组合、生物组合和沉积相变化不大，都是以一套生物扰动灰岩、鲕粒灰岩、豆粒灰岩，产丰富双壳类和牙形刺化石的碳酸盐台地相为特征。

1.3.10 康南组

中三叠统康南组零星分布于热觉茶卡康鲁山南和江爱达日那等地，岩性为灰色细碎屑岩夹泥灰岩、双壳生物灰岩、菊石生物灰岩。康南组最大特征是产丰富的双壳和菊石，厚度为100—311 m。康南组与下伏硬水泉组呈整合接触，与上覆土门格拉群呈整合或假整合接触关系。

康南组产丰富的双壳类化石 *Myophoria minor* Chen, *Posidonia* cf. *baiyushiensis* Lu, *Enantiostreon* sp., *Unionites* sp., *Murchisonia* sp., *Pinna* sp., *Eumorphotis multiformis reticulata* Chen, *E.* (*Asoella*) cf. *illyrica* Bittner, *Mytilus eduliformis praecursor* (Frech), *M.* sp.；丰富的菊石 *Hollandites voiti* Oppel, *H. hidimba* Diener, *H. truncus* Oppel, *H. vyasa* Diener, *H. visvakarma* Diener, *H.* sp., *Paraceratites* cf. *binodosus* Hauer, *Cuccoceras taramellii* Mojs（朱同兴等，2010）；牙形刺化石 *Hindeodella suevica* Tatge, *H. multinamata* Huckriede, *Cypridodella* sp.（朱同兴等，2010）。

1.3.11 土门格拉群

土门格拉群最早由青海省区域地质测量队（1970）创名于土门格拉地区，指一套不见顶底的三叠系含煤碎屑岩系，产双壳类和植物化石，厚度大于873 m。西藏自治区地质矿产局（1997）建议废除土门格拉群，而改称结扎群。朱同兴等（2005，2010）认为土门格拉群含煤地层和结扎群细碎屑岩地层应分属于不同的地层小区或沉积相带，它们是不同沉积环境相的产物，因此建议恢复土门格拉群，并将其地层年代归属晚三叠世。

上三叠统土门格拉群呈近东西向广泛分布于羌塘北部的江爱达日那-土门格拉等地，主要岩性为灰色含煤碎屑岩系，在江爱达日那地区三分岩性组合非常明显：

下段为灰、深灰色薄-中层状细粒岩屑长石砂岩、炭质页岩、粉砂质页岩、泥质粉砂岩夹薄层状泥灰岩，局部夹煤线，厚度为129 m，底部发育1 m厚的复成分细砾岩。产双壳类化石 *Myophoria* (*Costatoria*) *minor* Chen, *Halobia* sp., *Myophoricardium tulongense* Wen et Lan, *Entolium quotidianum* Healey, *Prototrigonia seranensis* Krumbeek, *Heminajas fissidentata* Wohrmana, *Pachycardia* sp., *Trigonia* (*Kumatrigonia*) *jingguensis* Chen, *Entolium* cf. *tenuistriatum rotundum* Chen, *E.* sp., *Nuculana* cf. *miaocunensis* Chen, *Gervillia lanpingensis* Chen, *Meleagrinella* sp., *Posidonia* sp., *Amonotis* cf. *salinaria* (Bronn), *Ostrea* sp.（朱同兴等，2005；2010）。

中段为灰色薄-中层泥晶灰岩、泥灰岩夹泥页岩、钙质泥岩，厚度为211 m。产双壳类化石 *Entolium quotidianum* Healey, *Schafhaeutlia gigantean* Krumbeck, *Amonotis* sp., *Pinna* sp.；菊石 *Juvavites* cf. *xizangensis* Wang et He（朱同兴等，2005；2010）；珊瑚化石 *Margarophyllum decora* Wu, *Paromphalophyllia sparsa* Deng et Zhang（朱同兴等，2005；2010）。

上段为灰-深灰色薄-中层状粉砂质泥岩、粉砂岩、中细粒岩屑砂岩夹深灰色炭质页岩、煤层煤线。地层层序总体具有下细上粗的逆韵律结构，出露厚度大于210 m。产植物碎片和双壳类化石 *Amonotis* cf. *rothpletzi yushensis* Sha et Chen, *Halobia* sp.（朱同兴等，2005；2010）。

在区域上，土门格拉群底部与下伏康南组为整合接触，顶部与中侏罗统雀莫错组不整合覆盖。沉积相主要表现为浅水陆棚—碳酸盐台地—海陆过渡三角洲向上变浅的沉积序列。

1.3.12 菊花山组

上三叠统菊花山组零星分布于甜水河、照沙山、菊花山等地，主要岩性为碳酸盐岩，包括灰色泥晶灰岩、泥质灰岩、生物碎屑灰岩、核形石灰岩、介壳灰岩、砂屑灰岩、鲕粒灰岩等，厚度为479—669 m。在区域上，菊花山组多未见底，顶部与那底岗日组呈假整合接触，产双壳类 *Halobia* sp., *Entolium* cf. *quotidianum*，*Burmesia* sp.，*Indopecten* cf. *seinanmensis*，*I. himalayensis*，*I. serraticosta*，*I. globraminor*，*Palaeocardita langnongensis*，*P.* cf. *globiformis*，*Unionites griesbachi*，*Chlamys jingdingensis*，*C. yadongensis*，*C.* cf. *biformatus*，*C. dingriensis*，*Monotis* sp.；腕足类 *Amphiclina intermedia*，*A.* cf. *taurica*，*Caucasorhynchia* cf. *trigonalis*，*Rhaetinopsis pentagonalis*，*R. zadoensis*，*Timorhynchia nimaensis*，*T. sulcata*，*Yidunella* cf. *pentagona*（赵政璋等，2001）；珊瑚 *Margarosmilia zieteni*，*Conophyllopsis qamdoensis*，*Pachythecophora* sp.，*Pachytheca* sp.，*Stylophyllopsis* sp.（朱同兴等，2010）；牙形刺 *Epigdolella postera*，*E. abneptis spatulatus*，*Neohindeod triassica*，*N. kobayashii*，*Neogondolella* sp.，*Xaniognathus abstractus*（Clark et Ethingtol），*X. deflectenst*（朱同兴等，2010）。

1.3.13 藏夏河组

上三叠统藏夏河组由朱同兴等（2005）新建，代表北羌塘地层分区北部晚三叠世一套砂泥质深水复理石盆地相沉积地层，呈东西向分布于北羌塘北部藏夏河-多色梁子-冈盖日一带，主要岩性为灰—深灰色薄至中厚层状含砾砂岩、细粒岩屑长石砂岩、长石岩屑砂岩、粉砂岩、粉砂质泥岩和页岩组成夹层或互层，产孢粉 *Cadargasporites grranulatus*，*Aratrisporites* sp.，*Chasmetooporites* sp.，*Biretisporites* sp.；双壳类 *Halobia plicasa*，*H. superbescens*；植物 *Neocalamites* sp. 等化石（朱同兴等，2005，2012），厚度为 627—1063 m。在区域上，藏夏河组多未见底，顶部与那底岗日组呈不整合接触。

综上所述，羌塘地层区南羌塘地层分区晚古生代地层擦蒙组发育有以含砾板岩为特征的冰水杂砾岩沉积；展金组发育有冈瓦纳相冷水型双壳类 *Eurydesma* 动物群和冷水型珊瑚 *Amplexocaninia-Cyathaxonia* 组合。从古气候和古生物面貌方面显示，南羌塘地块在晚石炭世—早二叠世与冈瓦纳陆块亲缘关系密切。

北羌塘地层分区晚二叠世地层热觉茶卡组灰岩中发育 *Paleofusulina* 动物群，这是我国华南地区二叠纪长兴阶的最高的一个蜓带，并广泛分布于古特提斯海相地层中；热觉茶卡组碎屑岩中发育 *Gigantonoclea* 植物群，这是目前为止在我国大陆境内分布最西端的华夏植物群。古生物面貌显示，北羌塘地块在晚二叠世与华南扬子陆块亲缘关系密切，应属于相同的一级古地理单元。

作为龙木错-双湖缝合带的核心区域，中羌塘构造-地层分区发育了一系列蛇绿混杂岩群。在硅质岩岩块中发现，晚古生代—三叠纪地层至少发育 5 个放射虫动物群，完全可以与北邻的可可西里蛇绿混杂岩带、昌宁-双江-孟连蛇绿混杂岩带及我国华南地区对比（详见表 7.1）。

第 2 章　西藏双湖和青海可可西里地区晚古生代—早中生代放射虫地层剖面

羌塘地区晚古生代和早中生代含放射虫硅质岩地层剖面和地质点，来自 3 个地区，即青海可可西里地区、西藏双湖-角木日地区和四川甘孜-理塘地区。

其中编号为 8P$_2$W16-2、8P$_2$W22-2、W5031、W5032、W5418-2、W7008、Bb8311-1、8PFS6、8PFS7、8PFS5014 的含放射虫动物群的样品由郭通珍等进行青海可可西里湖幅 1:25 万区域地质调查（2000—2002）时采集；编号为 CDP14WF1、D9002WF1、D9002WF3-2、D7052WF1、D7052WF3、D1339WF1 的样品由朱同兴等（2006）在西藏双湖-角木日地区进行野外考察时采集；ZD305-90b1 和 Z 顶部 Z144b1 这 2 块样品由原四川省第三地质大队采集；b90-125、b90-129、b90-45、b90-64、b90-69 和 FKGK1、FKXG1 样品由沙金庚等（1990）在参加青海可可西里地区综合科学考察时采集。

2.1　青海可可西里地区移山湖泥盆系实测剖面（图 2.1）

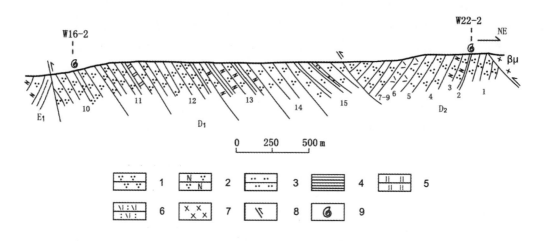

图 2.1　青海省治多县可可西里地区移山湖泥盆纪地层实测剖面图（8P$_2$）
1. 石英砂岩；2. 长石砂岩；3. 粉砂岩；4. 页岩；5. 白云岩；6. 凝灰岩；7. 辉绿岩；8. 逆冲断裂；9. 化石层位

Fig. 2.1　Measured section sketch of the Devonian in Yishanco of Zhidoi County, Hoh Xil of Qinghai Province (8P$_2$)
1. Quartz sandstone; 2. Feldspathic sandstone; 3. Siltstone; 4. Shale; 5. Dolomite; 6. Tuff; 7. Diabase; 8. Thrust fault; 9. Fossil beds

拉竹龙组（D$_3^2$）

15. 灰色厚层状变质细中粒石英砂岩夹灰绿色粉砂质板岩（未见顶）
14. 灰色厚层状变质细中粒石英砂岩
13. 灰色中厚层状变质中细粒长石石英砂岩夹灰色板岩
12. 灰白色中厚层状变质中细粒石英砂岩

11. 灰黑色中层状中粗粒石英砂岩夹灰黑色碳质板岩
10. 灰白色厚层状细粒石英砂岩夹灰黑色硅质岩，在硅质岩中发现放射虫 *Holoeciscus foremanae* 动物群（8P$_2$W16-2）

======== 断层 ========

9. 灰白色中厚层状变质细中粒石英砂岩
8. 土黄色厚层状变质中粗粒石英砂岩
7. 青灰色中厚层状变质细中粒石英砂岩
6. 土黄色厚层状变质凝灰岩
5. 灰白色中-厚层状变质细粒石英砂岩
4. 灰白色厚层状变质中-粗粒石英砂岩
3. 灰白色中层状变质中细粒长石石英砂岩
2. 深灰色硅质岩，含有放射虫 *Holoeciscus foremanae* 动物群（8P$_2$W22-2）
1. 灰白色厚层状变质细-中粒石英砂岩（未见底）

2.2　青海可可西里地区西金乌兰湖西北部构造混杂岩带实测剖面（图2.2）

本书采用西金乌兰群代表石炭纪至中二叠世地层，汉台山群代表晚二叠世至中三叠世地层。西金乌兰群进一步可分为3个非正式岩石地层单位：碎屑岩组、碳酸盐岩组和火山岩组（表2.1）。由于断裂破坏，地层连续性差。

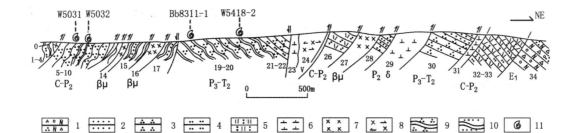

图2.2　青海省治多县可可西里地区西金乌兰湖西北部构造混杂岩带实测剖面图（7P$_3$）

1. 角砾状、砾状长石砂岩；2. 砂岩；3. 石英砂岩；4. 粉砂岩；5. 凝灰岩；6. 闪长岩；
7. 辉绿岩；8. 辉长岩；9. 石英片岩；10. 变质砂板岩；11. 化石层位

Fig. 2.2　Measured section sketch of the tectonic mélange belt of northwest Xijir Ulan Hu
(Xijinwulamco) Zhidoi County, Hoh Xil, Qinghai Province（7P$_3$）

1. Breccia. rudaceous feldspathic sandstone; 2. Sandstone; 3. Quartz sandstone; 4. Siltstone; 5. Tuff; 6. Diorite;
7. Diabase; 8. Gabbro; 9. Quartz schist; 10. Metamorphic sandy slate; 11. Fossil beds

(0) 第四系洪积砂砾石层
覆盖
碎屑岩组（C-P$_2$Xa）
灰色中-薄层状变质粉砂岩夹深灰色黏土硅质岩，发育水平纹理
1. 深灰色蚀变辉绿岩（蛇绿构造混杂岩岩块）

表 2.1 青海可可西里地区蛇绿岩混杂岩带地层及化石带
Table 2.1 Strata and fossil zonation of the ophiolitic mélange belt of Hoh Xil, Qinghai Province

系	地层	化 石 带
三叠系	汉台山群 (P_3—T_2)	*Spongoserrula rarauana* *Annulotriassocampe multisegmantatum* *Tiborella florida*
二叠系	西金乌兰群 (C—P_2)	*Pseudoalbaillella globosa*
二叠系	西金乌兰群 (C—P_2)	*Albaillella xiaodongensis*
二叠系	西金乌兰群 (C—P_2)	*Ps. rhombothoracata*
二叠系	西金乌兰群 (C—P_2)	*Ps. sakmarensis–Ps. lomentaria*
石炭系	西金乌兰群 (C—P_2)	*Albaillella indensis* *Cyrtisphaeractenium* aff. *crassum–* *Astroentactinia multispinosa* *Polyentactinia* sp. A
泥盆系	拉竹笼组 (D_3^2)	*Holocciscus foremanae*

2. 浅灰绿色片理化中-薄层变质黏土质粉砂岩，发育包卷层理及水平纹理
3. 深灰色蚀变辉绿玢岩（蛇绿构造混杂岩岩块）
======== 断层 ========
碎屑岩组（C-P_2Xa）
4. 浅灰色变质黏土质细粉砂岩
5. 灰色中厚层状变质细中粒石英砂岩
6. 深灰色放射虫硅质岩，含放射虫 *Albaillella xiaodongensis* 动物群（W5031）和 *Pseudoalbaillella globosa* 动物群（W5032）
7. 灰色中厚层变质凝灰质不等粒石英砂岩夹灰色中层状变质凝灰质含砾不等粒石英砂岩
8. 灰绿色片理化薄层石英细-粉砂岩
9. 浅灰色中-厚层状变质中细粒石英砂岩
======== 断层 ========
10. 灰白色蚀变二长岩
11. 深灰绿色蚀变角闪辉长苏长岩
======== 断层 ========
碎屑岩组（C-P_2Xa）
12. 浅灰色含粉砂黑云斑点板岩
13. 深灰色蚀变角闪辉绿玢岩（岩块）

14. 浅灰色含粉砂黑云斑点板岩夹灰色中层状变质石英粉砂岩

15. 深灰色蚀变角闪辉绿玢岩（岩块）

========断层========

17. 灰色含粉砂黑云斑点板岩夹灰色中层状变质石英粉砂岩

========断层========

汉台山群下段（$P_3-T_2h^1$）

19. 灰色中厚层状变质粗中粒石英砂岩，夹灰白色放射虫硅质岩，在硅质岩中含放射虫 *Annulotriassocampe multisegmantatum* 动物群（Bb8311-1）和 *Spongoserrula rarauana* 动物群（W5418-2）

20. 浅灰白色角岩化石英岩

21. 灰色中厚层状熔岩角砾岩夹灰色粗粒石英砂岩

22. 灰色中层状角岩化夹灰色粗中粒石英砂岩

========断层========

23. 灰色蚀变闪长岩

侵入

24. 灰色蚀变角闪苏长辉长岩

========断层========

火山岩组（$C-P_2Xc$）

26. 灰绿色中厚层状变质沉凝灰岩

========断层========

27. 深灰色蚀变角闪苏长辉长辉绿岩

========断层========

汉台山群下段（$P_3-T_2h^1$）

28. 浅白灰色中厚层状变质粗中粒石英砂岩

========断层========

29. 深灰色中细粒闪长岩

========断层========

汉台山群下段（$P_3-T_2h^1$）

30. 浅灰色中层状角岩化中粗粒石英砂岩

========断层========

31. 深灰色蚀变闪长岩

========断层========

火山岩组（$C-P^2Xc$）

32. 浅灰绿色蚀变基性晶屑岩屑凝灰岩夹片理化纤闪石化玄武岩（岩块）

33. 浅灰绿色片理化流纹质凝灰岩夹灰色透镜状中细粒砂岩及灰色条带状硅质岩

========断层========

沱沱河组（E_1）

34. 紫红色厚层状含砾细中粒岩屑石英砂岩

其后的 3 个剖面采用了边千韬等（1997）的材料，并进行少量的修改。这些剖面也采用西金乌兰群（$C—P_2$）和汉台山群（$P_3—T_2$）2 个地层单位。

2.3 青海可可西里地区蛇形沟蛇绿混杂岩剖面（图 2.3，BB′）

汉台山群（$P_3-T_2h^1$）
1. 石英砂岩
2. 灰岩

第四系覆盖

3. 大理岩

======== 断层 ========

西金乌兰群（$C-P_2$）

4. 枕状玄武岩

第四系覆盖

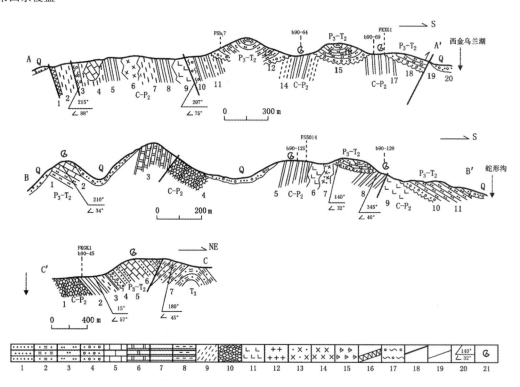

图 2.3 可可西里地区蛇绿混杂岩剖面图（来自边千韬等，1997，图 2）

AA′. 西金乌兰湖北综合剖面；BB′. 蛇形沟信手剖面；CC′. 康特金综合剖面

1. 砂岩；2. 杂砂岩；3. 石英砂岩；4. 砾岩及含砾砂岩；5. 灰岩；6. 大理岩；7. 硅质岩；8. 泥质硅质岩；9. 千枚岩；
10. 枕状玄武岩；11. 块状玄武岩；12. 花岗岩；13. 辉绿岩；14. 辉长岩；15. 变质橄榄岩；16. 石英脉；17. 糜棱岩；
18. 主要断裂；19. 一般断裂；20. 地层产状或流劈理产状；21. 化石层位

Fig. 2.3 Sketch of sections of ophiolitic mélanges in different outcrops in the Hoh Xil region (after Bian et al., 1997, fig. 2)

AA′. Composite profile of the Xijir Ulan Hu (Xijinwulamco); BB′. Hand profile sketch Shexinggou; CC′. Composite profile of Kangtejin

1. Sandstone; 2. Graywacke; 3. Quartz sandstone; 4. Conglomerate and pebbly sandstone; 5. Limestone; 6. Marble; 7. Silicalite; 8. Pelitic silicalite; 9. Phyllite; 10. Pillow basalt; 11. Massive basalt; 12. Granite; 13. Diabase; 14. Gabbro; 15. Metamorphic peridotite; 16. Quartz vein; 17. Mylonite; 18. Major fault; 19. Fault; 20. Stratigraphic or flow cleavage attitude; 21. Fossil beds

5. 硅质岩，含放射虫 *Pseudoalbaillella sakmarensis* - *P. lomentaria* 动物群（b90-125）和 *P. rhombothoracata* 动物群（8PFS5014）

========= 断层 =========

6. 块状玄武岩

========= 断层 =========

7. 辉绿岩

8. 硅质岩，含放射虫 *Pseudoalbaillella sakmarensis* - *P. lomentaria* 动物群（b90-129）

========= 断层 =========

9. 块状玄武岩

～～～～ 不整合接触 ～～～～

汉台山群（$P_3 - T_2h^1$）

10. 砾岩和含砾砂岩
11. 砂岩

第四系覆盖

2.4　青海可可西里地区岗齐曲康特金剖面（图 2.3，CC′）

西金乌兰群（$C - P_2$）

1. 枕状玄武岩
2. 硅质岩含放射虫 *Pseudoalbaillella rhombothoracata* 动物群（b90-45）和 *P. sakmarensis* - *P. lomentaria* 动物群（FKGK1）
3. 泥质硅质岩

========= 断层 =========

汉台山群（$P_3 - T_2h^1$）

4. 砾岩和含砾砂岩
5. 砂岩
6. 灰岩夹砂岩

========= 断层 =========

巴颜喀拉群（T_2）

7. 杂砂岩

2.5　青海可可西里地区西金乌兰湖北移山湖剖面（图 2.3，AA′）

西金乌兰群（$C - P_2$）

1. 石英脉
2. 千枚岩
3. 糜棱岩
4. 千枚岩含灰岩团块
5. 硅质岩
6. 辉绿岩
7. 千枚岩
8. 硅质岩
9. 块状玄武岩

========= 断层 =========

10. 糜棱岩

11. 硅质岩和千枚岩互层

~~~~~~ 不整合接触断层 ~~~~~~

汉台山群（$P_3 - T_2h^1$）

12. 砾岩和含砾砂岩

13. 砂岩含少量硅质岩，在硅质岩中发现放射虫 *Tiborella florida* 动物群（8PFS6，8PFS7）

~~~~~~ 不整合接触 ~~~~~~

西金乌兰群（$C - P_2$）

14. 千枚岩夹硅质岩，在硅质岩中含放射虫 *Astroentactinia multispinosa* 动物群（b90-64）

~~~~~~ 不整合接触 ~~~~~~

汉台山群（$P_3 - T_2h^1$）

15. 砾岩和含砾砂岩

16. 砂岩

~~~~~~ 不整合接触 ~~~~~~

西金乌兰群（$C - P_2$）

17. 硅质岩，含放射虫 *Albaillella indensis* 动物群（FKXG1）和 *Cyrtisphaeractenium* aff. *crassum* 动物群（b90-69）

~~~~~~ 不整合接触 ~~~~~~

汉台山群（$P_3 - T_2h^1$）

18. 砾岩和含砾砂岩

19. 砂岩

========= 断层 =========

20. 砾岩和含砾砂岩

第四系覆盖

2.6 西藏双湖地区才多茶卡北岸晚古生代放射虫硅质岩实测剖面（图2.4）

14. 黑色中-薄层含放射虫硅质岩夹黑色页岩（未见顶），在硅质岩中含放射虫 *Helenifore robustum* 动物群（CDP14WF1）

13. 灰绿-灰白色中-细粒长石石英砂岩

12. 黑色薄层硅质岩夹页岩（内部有断层通过），硅质岩中未分离出放射虫化石

11. 灰绿色细粒长石石英砂岩

10. 杂色页岩

9. 灰-黑色薄层放射虫硅质岩夹页岩。在硅质岩中含放射虫 *Neoalbaillella ornithoformis* 动物群（D9002WF3-2，D9002WF1）

8. 灰色页岩

7. 第四系掩盖，推测岩性为灰色页岩

6. 灰绿色蓝片岩

5. 灰绿色含钠长石大理岩夹灰绿色变基性岩

4. 灰绿色变基性岩

3. 糜棱岩化构造碎裂岩带

2. 灰绿色绿帘钠长片岩夹杏仁状变基性火山岩

1. 灰绿色块状-杏仁状变基性火山岩（未见底）

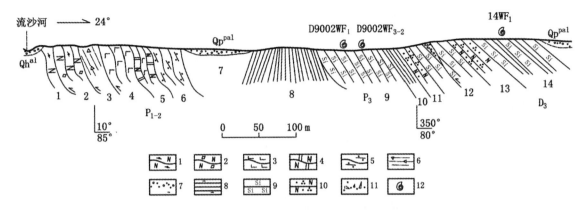

图 2.4 西藏双湖地区才多茶卡北岸构造混杂岩晚古生代放射虫硅质岩实测剖面

1. 杏仁状变基性火山岩；2. 绿帘钠长片岩；3. 灰绿色变基性岩；4. 含钠长石大理岩；5. 蓝片岩；6. 糜棱岩化构造碎裂岩带；7. 第四系掩盖；8. 页岩；9. 薄层硅质岩；10. 长石石英砂岩；11. 断层角砾岩；12. 放射虫化石产出层位；Qp^{pal}. 更新世冲、洪积物；Qh^{al}. 全新世冲积物；P_{1-2}. 早—中二叠世地层；P_3. 晚二叠世地层；D_3. 晚泥盆世地层

Fig. 2.4　Measured section sketch of the Late Palaeozoic radiolarian cherts in the north shore of Cêdo Caka (Caiduochaka) tectonic mélange of Shuanghu, northern Tibet

1. Amygdaloidal metamorphic basic volcanic rock; 2. Helsinkite slate; 3. Greyish-green metamorphic basic rock; 4. Marble with albite; 5. Blue achist; 6. Mylonitized tectonic cataclasite belt; 7. Quaternary; 8. Shale; 9. Thin-bedded silicalite; 10. Feldspathic quartz sandstone; 11. Fault breccia; 12. Radiolarian fossil beds; Qp^{pal}. Pleistocene alluvium and proluvium; Qh^{al}. Holocene alluvium; P_{1-2}, Early-Middle Permian strata; P_3, Late Permian strata; D_3, Late Devonian strata

2.7　西藏双湖地区雅曲乡硅质岩

这一硅质岩含有 2 个放射虫动物群：*Pseudoalbaillella ishigai* 动物群 (D7052WF1) 和 *Pseudoalbaillella sakmarensis* – *P. lomentaria* 动物群 (D7052WF3)。

2.8　四川甘孜-理塘构造带沉积混杂岩

四川甘孜-理塘构造带内广泛分布着晚古生代—中生代沉积混杂岩。其中上三叠统哈工组上部第三段含有许多时代从中泥盆世至中晚三叠世的放射虫硅质岩岩块，本书采用的 2 块硅质岩均采自中甸地区，一块含有放射虫 *Pseudoalbaillella globosa* 动物群 (ZD305-90b1)，另一块含有放射虫 *Muelleritortis cochleata* 动物群 (Z144b1)。

2.9　西藏角木日地区玄武岩剖面（图 2.5）

1. 玄武质角砾岩
2. 枕状玄武岩
3. 钙质粉砂岩和块状玄武岩
4. 枕状玄武岩
5. 火山砾岩

6. 枕状玄武岩夹硅质岩，在硅质岩中含有放射虫 *Eptingium nakasekoi* 动物群（D1339WF1）
7. 玄武质角砾岩
8. 粉砂岩
9. 块状玄武岩
10. 辉长岩岩墙
11. 块状玄武岩
12. 辉石岩墙
13. 辉石橄榄岩
14. 块状玄武岩夹辉长岩岩墙

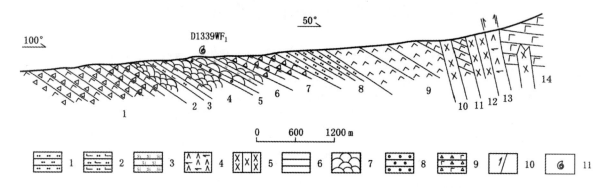

图 2.5 西藏角木日玄武岩剖面 [据翟庆国等（2006），略有修改]

1. 粉砂岩；2. 钙质粉砂岩；3. 硅质岩；4. 辉石橄榄岩；5. 辉长岩岩墙；6. 块状玄武岩；7. 枕状玄武岩；8. 火山砾岩；9. 玄武质角砾岩；10. 逆断层；11. 化石层位

Fig. 2.5 Section of Jiaomuri basalt at Jiaomuri, Tibet (after Zhai et al., 2006)

1. Siltstone; 2. Calcareous siltstone; 3. Silicalite; 4. Pyroxene peridotite; 5. Gabbro dike; 6. Massive basalt; 7. Pillow basalt; 8. Volcanic conglomerate; 9. Basaltic breccia; 10. Reverse fault; 11. Fossil beds

第3章 西藏双湖和青海可可西里地区晚古生代放射虫动物群及其时代和对比

3.1 西藏双湖和青海可可西里地区晚古生代放射虫动物群及其时代讨论

3.1.1 青海可可西里标本 8P$_2$W22-2 放射虫动物群及其时代（表 3.1）

这个放射虫动物群包括 9 属 22 种，其中包括 1 个新种（*Stigmosphaerostylus spiciosus* Wang）。它们归属于 4 目（阿尔拜虫目、内射虫目、球虫目和罩笼虫目）、6 超科（古蓬虫超科、角舍虫超科、内射虫超科、多内射虫超科、海绵内射虫超科和门孔虫超科）、8 科（古蓬虫科、石鱼虫科、星内射虫科、内射虫科、多内射虫科、单内射虫科、海绵内射虫科和门孔虫科），其中以 *Astroentactinia*，*Stigmosphaerostylus* 和 *Archocyrtium* 3 属的种数最多，约占这个动物群的 60% 以上。*Asrtoentactinia paronae*，*Holoeciscus elongatus*，*Spongentactinia exilispina*，*S. spongites* 和 *Tetrentactinia spongacea* 5 种在美国俄亥俄、德国北巴伐利亚、马来西亚、俄罗斯南乌拉尔地区和新西伯利亚鲁德内（Rudny）阿尔泰地区、澳大利亚、泰国和中国华南的晚泥盆世法门期（Famennian）地层中都已找到。*Astroentactinia stellata*，*Spongentactinia indisserta*，*Stigmosphaerostylus diversitus* 和 *S. proceraspina* 4 种在俄罗斯南乌拉尔地区和新西伯利亚鲁德内阿尔泰地区、澳大利亚、泰国、德国、中国华南和新疆出现的时代为晚泥盆世弗拉期—法门期。*Archocyrtium diductum*，*A. castuligerum*，*A. ludicrum*，*A. strictum*，*Polyentactinia aranea*，*Pylentonema*? sp. A，*Stigmosphaerostylus pantotolma* 7 种常见于澳大利亚、法国、德国、土耳其、美国何克拉荷马、中国华南和新疆晚泥盆世法门期至早石炭世维宪期（Visean）早期地层中。*Astroentactinia mirousi*，*A. multispinosa*，*Archocyrtium ferreum*，*A. wonae*，*Stigmosphaerostylus micula*，*S. variospina*，*Triaenosphaera hebes*，*T. sicarius* 8 种在美国俄亥俄和何克拉荷马、德国北巴伐利亚、法国、马来西亚、泰国、澳大利亚、中国华南发现的时代是晚泥盆世弗拉期至早石炭世维宪期。还有两个种，即 *Astroentactinia biaciculata* 和 *Palaeoscenidium cladophorum*，它们的时限较长，在澳大利亚、俄罗斯南乌拉尔地区、马来西亚、泰国、德国北巴伐利亚、法国、美国俄亥俄、日本、中国华南出现于中泥盆世艾菲尔期（Eifelian）至早石炭世维宪期地层中。在时代上，这个放射虫动物群属种主要出现于晚泥盆世法门期至早石炭世杜内期（Tournaisian）。虽然在当前样品中没有发现晚泥盆世弗拉期的标准带化石 *Helenifore laticlavium*，*H. robustum* 和法门期的标准带化石 *Holoeciscus foremanae*，也没有发现早石炭世以 *Albaillella* 动物群为主的带化石，但是，这个动物群中所出现的主要属种，如 *Holoeciscus elongatus*，*Spongentactinia exilispina*，*S. spongites*，*Tetrentactinia spongacea*，*Astroentactinia paronae*，*A. stellata*，*Polyentactinia aranea*，*Archocyrtium diductum*，*A. strictum*，*Triaenosphaera sicarius* 等都是 *Holoeciscus foremanae* 带中的常见分子，因此，我们把这个放射虫动物群归于 *Holoeciscus foremanae* 带，时代为晚泥盆世法门期。

表 3.1 青海可可西里标本 8P₂W22-2 放射虫动物群属种时代分布表

Table 3.1 Time ranges of the radiolarians of sample 8P₂W22-2 from Hoh Xil of Qinghai

属种名称 \ 分布 时代	泥盆纪			石炭纪	
	中泥盆世	晚泥盆世		早石炭世	
		弗拉期	法门期	杜内期	维宪期
Astroentactinia biaciculata		■■■■■■■■■■■■■■■■■■■■■■■■■■■■■■			
A. paronae			■■■■■■■■■■■■■■		
A. stellata		■■■■■■■■■■■■■■			
Archocyrtium castuligerum			■■■■■■■■■■■■■■■■■■■■■■		
A. diductum			■■■■■■■■■■■■■■■■■■■■■■		
A. ferreum			■■■■■■■■■■■■■■■■■■■■■■		
A. ludicrum				■■■■■■■■■■■■■■	
A. strictum			■■■■■■■		
Holoeciscus elongatus			■■■■■■■		
Palaeoscenidium cladophorum	■■				
Polyentactinia aranea			■■■■■■■■■■■■■■		
Pylentonema? sp. A			■■■■■■■■■■■■■■		
Spongentactinia exilispina			■■■■■■■		
S. spongites			■■■■■■■		
Stigmosphaerostylus diversitus		■■■■■■■			
S. micula		■■■■■■■■■■■■■■■■■■■■■■■■■■■■■■			
S. proceraspina		■■■■■■■■■■■■■■■■■■■■■■			
S. variospina		■■■■■■■■■■■■■■■■■■■■■■			
Tetrentactinia spongacea			■■■■■■■		
Triaenosphaera hebes		■■■■■■■■■■■■■■■■■■■■■■■■■■■■■■			
T. sicarius		■■■■■■■■■■■■■■■■■■■■■■■■■■■■■■			

3.1.2 青海可可西里标本 8P₂W16-2 放射虫动物群及其时代（表 3.2）

这个放射虫动物群包括 6 属 10 种，其中包括 1 个新种（*Tetrentactinia gigantia* Wang）。它们归属于 3 目（罩笼虫目、内射虫目和球虫目）、3 超科（门孔虫超科、内射虫超科和海绵内射虫超科）、5 科（门孔虫科、内射虫科、星内射虫科、单内射虫科和海绵内射虫科）。在这个放射虫动物群中，除了新种和 *Astroentactinia multispinosa* 2 种外，其余 8 种都是在标本 8P₂W22-2 放射虫动物群中出现过的分子，因此，我们把这个放射虫动物群也归属于 *Holoeciscus foremanae* 带，时代为晚泥盆世法门期。

表 3.2 青海可可西里标本 8P₂W16-2 放射虫动物群属种时代分布表
Table 3.2 Time ranges of the radiolarians of sample 8P₂W16-2 from Hoh Xil of Qinghai

分布 时代 属种名称	泥盆纪			石炭纪	
	中泥盆世	晚泥盆世		中石炭世	
		弗拉期	法门期	杜内期	维宪期
Archocyrtium diductum			━━━	━━━	━━━
A. wonae		━━━━━━			
Astroentactinia multispinosa		━━━━━━			
Spongentactinia indisserta		━━━━━━			
S. spongites		━━━━━━			
Stigmosphaerostylus micula		━━━━━━━━━━━━━━━			
S. pantotolma			━━━━━━━━━━		
S. variospina		━━━━━━			
Triaenosphaera hebes		━━━━━━			

3.1.3 青海可可西里标本 W5032 放射虫动物群及其时代（表 3.3）

这个放射虫动物群的属种异常丰富，主要有三大类，即假阿尔拜虫类、隐管虫类和内射虫类。本文共描述 11 属 21 种，其中包括 5 个新种（*Archaeospongoprunum sinisterispinosum* Wang，*Stigmosphaerostylus cruciformis* Wang，*S. gracilentus* Wang，*S. vetulus* Wang，*Raciditor oblatum* Wang）。在这个放射虫动物群中，最重要的一个化石属种为 *Pseudoalbaillella globosa* Ishiga et Imoto。它是一个世界性分布的属种，常见于我国广西、云南西部、苏皖南部（孤峰组）以及日本西南部、美国西海岸、加拿大、俄罗斯远东地区、新西兰北岛、马来西亚、泰国、约旦安曼、希腊克里特岛等地。与这个种常共存的放射虫较多，如 *Pseudoalbaillella fusiformis*，*P. longtanensis*，*Hegleria mammilla*，*Latentifistula texana*，*Stigmosphaerostylus itsukaichiensis*，*Copicyntra akikawaensis*，*Pseudotormentus kamigoriensis* 等。它的生存时间较短，仅限于茅口期（Maokouan）早期，现在已成为瓜德鲁普期（Guadalupian）早期的一个标准带化石。值得注意的是，在当前这个放射虫动物群中，还发现一些过去只在晚二叠世才发育的属种，如 *Raciditor gracilis*，*R. phlogidea*，*Stigmosphaerostylus ichikawai* 等，我们把这个放射虫动物群称为 *Pseudoalbaillella globosa* 带，时代为中二叠世茅口期早期，与美国瓜德鲁普期早期时代相当。

表 3.3 青海可可西里标本 W5032 放射虫动物群属种时代分布表

Table 3.3 Time ranges of the radiolarians of sample W5032 from Hoh Xil of Qinghai

属种名称 \ 时代	晚石炭世	早二叠世		中二叠世					晚二叠世					
	马平期	隆林期		栖霞期			茅口期		吴家坪期		长兴期			
化石带*	P. nodosa	P. bulbosa	P. u – forma – P. elegans	P. lomentaria – P. sakmarensis	P. rhombothoracata	A. xiaodongensis	A. sinuata	P. ishigai	P. globosa	P. monacantha	F. scholasticus – F. ventricosus	F. bipartitus – F. charveti – F. orthogonus	Foremanhelena triangularis	A. protolevis	A. levis – A. excelsa	N. ornithoformis	N. optima
Copiellintra diploacantha							━━━━━━━━━━━							
Copicyntra akikawaensis								━━━━━━━━━━━━━━━					
Hegleria mammilla								━━━━━━━━━━━━					
Latentibifistula triacanthophora							━━━━━━━							
Pseudoalbaillella fusiformis							━ ━							
P. globosa							━ ━							
P. longtanensis							━ ━							
P. scalprata							━━━━							
Pseudotormentus kamigoriensis						━ ━ ━ ━ ━ ━ ━ ━ ━ ━							
Quadricaulis femoris							━━━━━━━━━━━━━					
Quinquermis robusta							━━━━━━━━━━━━━━━					
Raciditor gracilis							━━━━━━━━━━━━━━━					
R. phlogidea							━━━━━━━━━━━━━━━					
Stigmosphaerostylus ichikawai							━━━━━━━━━━━━━━━					
S. itsukaichiensis							━━━━━━━━━━━━━━━					
S. modesta							━━━━━━━━━━━━━━━					

*A: Albaillella; F: Folliculullus; N: Neoalbaillella; P: Pseudoalbaillella.

3.1.4 青海可可西里标本 W5031 放射虫动物群及其时代（表 3.4）

这个放射虫动物群包括 9 属 11 种，其中含有 2 个新种（*Stigmosphaerostylus cruciformis* Wang，*Raciditor oblatum* Wang），归属于 4 目（阿尔拜虫目、隐管虫目、内射虫目和球虫目）、5 超科（阿尔拜虫超科、丑巾虫超科、隐管虫超科、内射虫超科和海绵星虫超科）、8 科（阿尔拜虫科、假阿尔拜虫科、隐管虫科、四桨虫科、鲁仁采夫海绵虫科、奥米斯顿虫科、内射虫科和奥特尔海绵虫科）。在这个放射虫动物群中，最重要的一个化石种是 *Albaillella xiaodongensis* Wang。它主要发现于中国广西钦防地区的板城、小董、钦北地区及日本的西南部，这次在青海可可西里地区又一次被发现。这一化石放射虫种位于 *Albaillella sinuata* 带之下，与它共存的放射虫主要有 *Albaillella sinuata*，*Copicyntra cuspidata*，

Hegleria mammilla，*Latentifistula patagilaterala*，*L. texana*，*Pseudoalbaillella rhombothoracata*，*Quadriremis minima*，*Pseudotormentus kamigoriensis*，*Stigmosphaerostylus itsukaichiensis* 等，其中 *A. sinuata* 在日本、美国德克萨斯（Texas）和阿拉斯加、马来西亚、泰国、俄罗斯远东地区和中国华南都产于伦纳德期（Leonardian）地层中。*C. cuspidata* 在俄罗斯南乌拉尔地区、美国德克萨斯也见于伦纳德阶或阿丁斯克阶（Artinskian）。*Latentifistula patagilaterala* 这个种在美国德克萨斯、日本、马来西亚、泰国、中国华南和西藏双湖地区常见于伦纳德阶。*Pseudoalbaillella rhombothoracata* 在日本、马来西亚、俄罗斯远东地区、中国华南常见于下二叠统狼营阶（Wolfcampian）上部，是此阶最上部的一个带化石，其残余分子可以延续到伦纳德阶下部。其余的一些共存属种，时限都比较长，在美国、日本、菲律宾、马来西亚、泰国、希腊克里特岛、俄罗斯远东地区、中国华南和西藏双湖地区可以从狼营期延续到长兴期。虽然这个放射虫化石目前分布还不广，生存时限极短，但我们仍把它作为一个化石带，即 *Albaillella xiadongensis* 带，时代为中二叠世栖霞早期，与美国伦纳德期早期或俄罗斯阿丁斯克期早期时代相当。

表 3.4 青海可可西里标本 W5031 放射虫动物群属种时代分布表
Table 3.4 Time ranges of the radiolarians of sample W5031 from Hoh Xil of Qinghai

时代 / 化石带* / 属种名称	晚石炭世	早二叠世				中二叠世						晚二叠世				
		马平期	隆林期			栖霞期			茅口期			吴家坪期		长兴期		
	P. nodosa	*P. bulbosa*	*P. u – forma – P. elegans*	*P. lomentaria – P. sakmarensis*	*P. rhombothoracata*	*A. xiadongensis*	*A. sinuata*	*P. ishigai*	*P. globosa*	*P. monacantha*	*F. scholasticus – F. ventricosus*	*F. bipartitus – F. charveti – F. orthogomus*	*Foremanhelena triangularis*	*A. protolevis*	*A. levis – A. excelsa*	*N. ornithoformis*	*N. optima*
Albaillella sinuata							—	—	—								
A. xiadongensis							—										
Copicyntra cuspidata							—	- -	- -								
Hegleria mammilla							—	—	—								
Latentifistula patagilaterala							—	—	—								
L. texana							—	—	—								
P. rhombothoracata							—	—									
Pseudotormentus kamigoriensis							—	—	—								
Quadriremis minima							—	- -	- -								

*A: *Albaillella*; F: *Follicucullus*; N: *Neoalbaillella*; P: *Pseudoalbaillella*.

3.1.5 青海可可西里标本 P8FS5014 放射虫动物群及其时代（表3.5）

这个放射虫动物群包括4属7种，归属于2目（阿尔拜虫目和隐管虫目）、3超科（丑巾虫超科、隐管虫超科、鲁仁采夫海绵虫超科）、4科（假阿尔拜虫科、隐管虫科、鲁仁采夫海绵虫科和奥米斯顿虫科）。在这个放射虫动物群中，最重要的2种是 *Pseudoalbaillella rhombothoracata* 和 *P. elongata*，它们在日本、马来西亚、俄罗斯远东地区、意大利西西里岛、中国华南仅见于早二叠世狼营期最晚期地层中，

P. rhombothoracata 已成为这个时期的一个标准带化石。其余一些与其共存的放射虫，如 *Pseudoalbaillella sakmarensis*，在日本、俄罗斯南乌拉尔地区和远东地区、马来西亚、中国华南发现于狼营阶或萨克马尔阶（Sakmarian）；*P. simplex* 在日本、智利、泰国、马来西亚、中国华南出现于狼营期地层中；*Quinqueremis arundinea* 在俄罗斯南乌拉尔地区仅见于阿丁斯克阶。*Pseudotormentus kamigoriensis*，*Latentibifistula asperspongiosa*，*Ormistonella robusta* 3 种时限较长，在日本、菲律宾、泰国、美国、俄罗斯远东地区、马来西亚以及中国华南、内蒙古和西藏双湖地区，从狼营期或阿塞尔期（Asselian）延续至长兴期地层。我们把这个放射虫动物群称为 *Pseudoalbaillella rhombothoracata* 带，时代为早二叠世隆林期最晚期，与美国狼营期最晚期或俄罗斯萨克马尔期最晚期相当。

表 3.5　青海可可西里标本 8PFS5014 放射虫动物群属种时代分布表
Table 3.5　Time ranges of the radiolarians of sample 8PFS5014 from Hoh Xil of Qinghai

时代	晚石炭世	早二叠世				中二叠世					晚二叠世					
		马平期		隆林期		栖霞期			茅口期		吴家坪期			长兴期		
化石带* 属种名称	*P. nodosa*	*P. bulbosa*	*P. u – forma – P. elegans*	*P. lomentaria – P. sakmarensis*	*P. rhombothoracata*	*A. xiadongensis*	*A. sinuata*	*P. ishigai*	*P. globosa*	*P. monacantha*	*F. scholasticus – F. ventricosus*	*F. bipartitus – F. charveti F. orthogonus*	*Foremanhelena triangularis*	*A. protolevis*	*A. levis – A. excelsa*	*N. ornithoformis*	*N. optima*
Latentibifistula asperspongiosa				───	───	───	───	───	───	───	───	───	───	───			
Pseudoalbaillella elongata				───													
P. rhombothoracata				───	───												
P. sakmarensis				───													
P. simplex			───	───													
Pseudotormentus kamigoriensis				───	───	───	───	───	───	───	───	───	───	───	───	───	───
Quinqueremis arundinea				───	───												

*A: *Albaillella*; F: *Folliculullus*; N: *Neoalbaillella*; P: *Pseudoalbaillella*.

3.1.6　西藏双湖才多茶卡标本 CDP14WF1 放射虫动物群及其时代（表 3.6）

这个放射虫动物群共包括 4 属 10 种，其中包括 1 个新种（*Triaenosphaera robustispina* Wang sp. nov.），归属于 2 目（内射虫目和球虫目）、3 超科（内射虫超科、多内射虫超科和海绵内射虫超科）、3 科（内射虫科、单内射虫科和海绵内射虫科）。在这个放射虫动物群中最重要的化石是 *Trilonche* 属，从中共识别出 6 个种，其中 *T. davidi*，*T. echinata*，*T. elegans*，*T. minax*，*T. pittmani* 5 个种在澳大利亚、美国俄亥俄、俄罗斯南乌拉尔地区和新西伯利亚鲁德内阿尔泰地区、哈萨克斯坦、德国、法国、泰国、马来西亚以及中国云南、贵州、广西和新疆主要见于晚泥盆世弗拉期地层，但这些分子从中泥盆世地层已经开始出现，还可延续到法门期地层。*T. tretactinia* 和 *Tetrentactinia spongacea* 2 种在美国俄亥俄、中国华南和青海可可西里地区均发现于晚泥盆世法门期地层中。另外 2 个共存种放射虫 *Stigmospha-*

erostylus variospina 和 *Triaenosphaera sicarius* 时限比较长，在德国、法国、泰国、马来西亚、美国阿拉斯加以及中国华南、新疆和青海可可西里地区都见于晚泥盆世弗拉期至早石炭世维宪期地层中。在这个动物群中没有发现弗拉期的标准带化石 *Helenifore laticlavium* 和 *H. robustum*，也没有发现法门期的标准带化石 *Holoeciscus foremanae*，更没有发现早石炭世才开始出现的 *Albaillella* 动物群，因此，我们把这个放射虫动物群暂时归入 *Helenifore robustum* 带，时代为晚泥盆世弗拉期晚期，比法门期要古老一些。

表 3.6　西藏双湖才多茶卡标本 CDP14WF1 放射虫动物群属种时代分布表

Table 3.6　Time ranges of the radiolarians of Cêdo Caka sample CDP14WF1 from Shuanghu of Tibet

分布　　　时代		泥盆纪		石炭纪	
	中泥盆世	晚泥盆世		早石炭世	
属种名称		弗拉期	法门期	杜内期	维宪期
Stigmosphaerostylus variospina		■■■■	■■■■		
Tetrentactinia spongacea			■■■■		
Triaenosphaera sicarius		■■■■	■■■■		
Trilonche davidi	■■■■	■■■■			
T. echinata	■■■■	■■■■			
T. elegans	■■■■	■■■■			
T. minax	■■■■	■■■■			
T. pittmani	■■■■	■■■■			
T. tretactinia		■■■■			

3.1.7　西藏双湖雅曲乡标本 D7052WF1 放射虫动物群及其时代（表 3.7）

这个放射虫动物群包括 5 属 9 种，其中包括 2 个新种（*Pseudoalbaillella monopteryla* Wang 和 *P. nonpteryla* Wang）。在这个放射虫动物群中，其中最重要的也是最丰富的一个化石种是 *Pseudoalbaillella ishigai* Wang，这个种在日本、俄罗斯远东地区、中国华南是栖霞阶上部的一个标准带化石，它的伴生放射虫 *Albaillella sinuata* 和 *Latentifistula patagilaterala* 在日本、美国德克萨斯和阿拉斯加、马来西亚、泰国、俄罗斯远东地区、中国华南和青海可可西里地区常见于伦纳德阶，*Latentifistula texana* 在日本、美国德克萨斯、菲律宾、中国华南和青海可可西里地区常见于伦纳德期至瓜德鲁普期地层中。其余 3 个伴生放射虫 *Raciditor gracilis*，*R. inflata* 和 *Pseudotormentus kamigoriensis* 的时限较长，在日本、美国、菲律宾、泰国、马来西亚以及中国华南、内蒙古和青海可可西里地区都见于中晚二叠世伦纳德期至长兴期地层中。因此，我们把这个放射虫动物群称为 *Pseudoalbaillella ishigai* 带，时代为中二叠世栖霞期晚期，在时代上与美国伦纳德期晚期或俄罗斯阿丁斯克期晚期相当。

3.1.8　西藏双湖雅曲乡标本 D7052WF3 放射虫动物群及其时代

这个放射虫动物群比较单调，只含 2 属 3 种，其中包括 1 个新种（*Latentifistula conica* Wang）。它们归属于 2 目（阿尔拜虫目和隐管虫目）、2 超科（丑巾虫超科和隐管虫超科）、2 科（假阿尔拜虫科和隐管虫科）。其中 *Pseudoalbaillella lomentaria* 和 *P. sakmarensis* 2 种个体占据了这个放射虫动物群的 95%

以上，它们在日本、美国俄勒冈（Oregon）、马来西亚、泰国、俄罗斯南乌拉尔地区和远东地区、中国华南和青海可可西里地区常见于狼营期早期地层中。*P. lomentaria* 经常被日本学者（Ishiga，1986，1990；Yao and Kuwahara，2004；Shimakawa and Yao，2006）当作早二叠世萨卡莫托瓦扎期（Sakamotozawan）中期一个标准带化石。由于这2个种经常同时出现，我们把这个放射虫动物群称为 *Pseudoalbaillella sakmarensis - P. lomentaria* 带，时代为早二叠世隆林期早期，在时代上与美国狼营期早期或俄罗斯萨克马尔期早期相当。

表 3.7　西藏双湖雅曲乡标本 D7052WF1 放射虫动物群属种时代分布表
Table 3.7　Time ranges of the radiolarians of Yaqu Village sample D7052WF1 from Shuanghu of Tibet

时代 / 化石带* / 分布 / 属种名称	晚石炭世	早二叠世				中二叠世					晚二叠世				
		马平期		隆林期		栖霞期			茅口期		吴家坪期	长兴期			
	P. nodosa	*P. bulbosa*	*P. u - forma - P. elegans*	*P. lomentaria - P. sakmarensis*	*P. rhombothoracata*	*A. xiadongensis*	*A. sinuata*	*P. ishigai*	*P. globosa*	*P. monacantha*	*F. scholasticus - F. ventricosus*	*F. bipartitus - F. charveti - F. orthogonus*	*Foremanhelena triangularis*	*A. protolevis*	*A. levis - A. excelsa*	*N. ornithoformis*	*N. optima*
Albaillella sinuata							━━										
Latentifistula patagilaterala				┄┄	┄┄	┄┄	━━	━━									
L. texana						━━	━━	━━									
Pseudoalbaillella ishigai						━━	━━										
Pseudotormentus kamigoriensis						━━	━━	━━	━━	━━	━━	━━	━━	━━	━━	━━	━━
Raciditor gracilis							━━	━━	━━	━━	━━	━━	━━	━━	━━	━━	━━
R. inflata							━━	━━	━━	━━	━━	━━	━━	━━	━━	━━	━━

*A: *Albaillella*；F: *Folliculcullus*；N: *Neoalbaillella*；P: *Pseudoalbaillella*.

3.1.9　西藏双湖才多茶卡标本 D9002WF3-2 放射虫动物群及其时代（表3.8）

这个放射虫动物群属种异常丰富，丰度和分异度都比较高，共包括9属16种，归属于4目（阿尔拜虫目、隐管虫目、内射虫目和球虫目）、4超科（阿尔拜虫超科、鲁仁采夫海绵虫超科、内射虫超科和海绵尾虫超科）、5科（阿尔拜虫科、新阿尔拜虫科、奥米斯顿虫科、内射虫科和古海绵梅虫科）。这个放射虫动物群中最重要的化石是 *Neoalbaillella* 和 *Albaillella* 2属，前一个属包括4个种，后一个属含有3个种，其中 *N. ornithoformis* 种的标本最多，成为这个放射虫动物群的主导化石，在日本、菲律宾、马来西亚、泰国、俄罗斯远东地区、中国华南等地这个种已经成为晚二叠世晚期一个标准的带化石种。*N. optima* 在这个放射虫动物群中只是零星出现，可能是 *N. optima* 带的先遣种。*N. optima* 带位于 *N. ornithoformis* 带之上，是二叠纪放射虫最高的化石带。它曾在日本、菲律宾、泰国、俄罗斯远东地区、中国华南发现。*N. gracilis*，*Albaillella lauta* 和 *Ishigaum craticula* 3种在日本和中国华南见于晚二叠世晚期 *N. ornithoformis* 带和 *N. optima* 带中。*N. pseudogrypa* 在日本、美国俄勒冈、俄罗斯远东地区、中国华南也是发现于晚二叠世长兴期晚期的 *Neoalbaillella* 2个化石带中。*A. levis* 和 *A. triangularis* 2种在日本、美国加利福尼亚及内华达和俄勒冈、菲律宾、马来西亚、泰国、俄罗斯远东地区、中国华南常见于晚二叠世长兴期中晚期 *Albaillella levls – A. excelsa* 带至 *N. optima* 带地层中。其余的共存放射虫，如 *Triplanospongos musashiensis*，*Stigmosphaerostylus itsukaichiensis*，*Trilonche pseudocimelia*，*Ormistonella robusta*，*Raciditor gracilis*，*Copicyntra akikawaensis* 等，时限比较长，在日本、美国俄勒冈和加利福尼亚、菲律宾、马来西亚、泰国、俄罗斯远东地区以及中国华南、内蒙古和青海可可西里地区出现于中晚二叠世地层中。从上述动物群分析，我们把这个放射虫动物群称为 *Neoalbaillella ornithoformis* 带，时代为晚二叠世长兴期晚期，这是世界上目前分布最靠西北的放射虫动物群。

表 3.8　西藏双湖才多茶卡标本 D9002WF3-2 放射虫动物群属种时代分布表

Table 3.8　Time ranges of the radiolarians of Cêdo Caka sample D9002WF3-2 from Shuanghu of Tibet

时代	晚石炭世	早二叠世				中二叠世						晚二叠世				
		马平期		隆林期		栖霞期			茅口期			吴家坪期			长兴期	
化石带* / 分布 / 属种名称	*P. nodosa*	*P. bulbosa*	*P. u – forma – P. elegans*	*P. lomentaria – P. sakmarensis*	*P. rhombothoracata*	*A. xiadongensis*	*A. sinuata*	*P. ishigai*	*P. globosa*	*P. monacantha*	*F. scholasticus – F. ventricosus*	*F. bipartitus – F. orthogonus / F. charveti*	*Foremanhelena triangularis*	*A. protolevis*	*A. levis – A. excelsa*	*N. ornithoformis*	*N. optima*
Albaillella lauta																─	
A. levis															──	─	
A. triangularis																─	
Copicyntra akikawaensis									─────	──	──	──	──	──	──	─	
Ishigaum craticula													──	──	──	─	
I. obesum													──	──	──	─	
I. trifistis													──	──	──	─	
Ormistonella robusta				───	──	──	──	──	──	──	──	──	──	──	──	─	
Raciditor gracilis								──	──	──	──	──	──	──	──	─	
Neoalbaillella gracilis																─	
N. optima																	─
N. ornithoformis																─	
N. pseudogrypa																─	
Stigmosphaerostylus itsukaichiensis							──	──	──	──	──	──	──	──	──	─	
Trilonche pseudocimilia								──	──	──	──	──	──	──	──	─	
Triplanospongos musashiensis							──	──	──	──	──	──	──	──	──	─	

*A: *Albaillella*；F: *Follicucullus*；N: *Neoalbaillella*；P: *Pseudoalbaillella*.

3.1.10　西藏双湖才多茶卡标本 D9002WF1 放射虫动物群及其时代（表 3.9）

这个放射虫动物群包括 6 属 6 种，它们归属于 4 目（阿尔拜虫目、隐管虫目、内射虫目和球虫目）、5 超科（丑巾虫超科、隐管虫超科、鲁仁采夫海绵虫超科、内射虫超科和海绵尾虫超科）、5 科（新阿尔拜虫科、考勒特虫科、奥米斯顿虫科、内射虫科和古海绵梅虫科）。在这个放射虫动物群中，最重要的化石是 *Neoalbaillella ornithoformis* 和 *Archaeospongoprunum chiangdaoensis*。这 2 个种在日本、菲律宾、马来西亚、泰国、俄罗斯远东地区、中国华南都出现于长兴期晚期 Neoalbaillella ornithoformis 带和 N. optima 带中。其他 4 个伴存放射虫 Ishigaum trifistis，Raciditor gracilis，Trilonche pseudocimelia 和 Triplanospongos musashiensis 在世界各地的分布时限较长，可以从中二叠世伦纳德期（Leonardian）晚期 Pseudoalbaillella ishigai 带延续到长兴期晚期 N. optima 带。由于在这个放射虫动物群中出现了 Neoalbaillella ornithoformis，所以我们称它为 Neoalbaillella ornithoformis 带，时代为晚二叠世长兴期晚期。

表 3.9　西藏双湖才多茶卡标本 D9002WF1 放射虫动物群属种时代分布表

Table 3.9　Time ranges of the radiolarians of Cêdo Caka sample D9002WF1 from Shuanghu of Tibet

属种名称 \ 时代	晚石炭世	早二叠世				中二叠世					晚二叠世				
	马平期	隆林期		栖霞期		茅口期			吴家坪期		长兴期				
化石带*	P. nodosa	P. bulbosa	P. u–forma – P. elegans	P. lomentaria – P. sakmarensis	P. rhombothoracata	A. xiadongensis	A. sinuata	P. ishigai	P. globosa	P. monacantha	F. scholasticus – F. ventricosus	F. bipartitus – F. charveti / F. orthogonus	Foremanhelena triangularis	A. protolevis	A. levis – A. excelsa	N. ornithoformis	N. optima
Archaeospongoprunum chiangdaoensis																██	██
Ishigaum trifistis												██	██	██	██	██	██
Neoalbaillella ornithoformis																██	██
Raciditor gracilis								██	██	██	██	██	██	██	██	██	██
Triplanospongos musashiensis												██	██	██	██	██	██
Trilonche pseudocimelia								██	██	██	██	██	██	██	██	██	██

*A: Albaillella; F: Folliculullus; N: Neoalbaillella; P: Pseudoalbaillella.

3.1.11　青海可可西里岗齐曲北部康特金标本 FKGK1 放射虫动物群及其时代（表 3.10）

这个放射虫动物群包括 5 属 10 种，其中包括 2 新种（Latentifistula conica Wang 和 Raciditor oblatum Wang），归属于 2 目（阿尔拜虫目和隐管虫目）、2 超科（丑巾虫超科和隐管虫超科）、3 科（假阿尔拜虫科、隐管虫科和奥米斯顿虫科）。在这个放射虫动物群中，最重要的化石是 Pseudoalbaillella 属，包括 5 种，占这个放射虫动物群种级组成的 1/2。其中 Pseudoalbaillella lomentaria 和 Pseudoalbaillella sakmarensis 2 种经常伴存，在日本、美国俄勒冈、马来西亚、泰国、俄罗斯南乌拉尔地区和远东地区以及中国华南和西藏双湖地区常见于狼营期早期地层中。其他伴存放射虫，如 Pseudoalbaillella elongata 和 Pseudoalbaillella postscalprata 2 种在日本、马来西亚、意大利西西里岛、中国华南和西藏双湖地区也发现于狼营阶。Latentifistula texana，Pseudoalbaillella scalprata，Quadricaulis femoris 和 Quinquiremis robusta 的时限较长，在日本、美国、菲律宾、泰国、马来西亚以及中国华南和西藏双湖地区出现于狼营期早期至瓜德鲁普期晚期地层中。这个放射虫动物群与西藏双湖雅曲乡标本 D7052WF1 放射虫动物群在主要属种上十分相似。我们把这个放射虫动物群称为 Pseudoalbaillella lomentaria-Pseudoalbaillella sakmarensis 带，时代为早二叠世隆林期（Longlinan）早期，与美国狼营期早期或俄罗斯萨克马尔期早期时代相当。

表 3.10 青海可可西里岗齐曲北部康特金标本 FKGK1 放射虫动物群属种时代分布表

Table 3.10 Time ranges of the radiolarians of sample FKGK1 from Kantejin, northern Gangqiqu, Hoh Xil of Qinghai

时代 / 化石带* / 分布 / 属种名称	晚石炭世	早二叠世			中二叠世				晚二叠世					
		马平期	隆林期		栖霞期			茅口期	吴家坪期		长兴期			
	P. nodosa	P. bulbosa	P. u – forma – P. elegans	P. lomentaria – P. sakmarensis	P. rhombothoracata	A. xiadongensis	A. sinuata	P. ishigai	P. globosa	P. monacantha	F. scholasticus – F. ventricosus	F. bipartitus – F. charveti – F. orthogonus	Foremanhelena triangularis	A. protolevis	A. levis – A. excelsa	N. ornithoformis	N. optima
Latentifistula texana				----	━━━												
Pseudoalbaillella elongata				━━━													
P. lomentaria				━━━													
P. postscalprata				━━━━━━━━━━━━━━													
P. scalprata				━━ ----													
P. sakmarensis				━													
Quadricaulis femoris				━━━━━━━━━━━━━━━━━━━━━━━━━━━━━━━━													
Quinquiremis robusta				━━━━━━━━━━━━━━━━━━													

*A: *Albaillella*; F: *Folliculucullus*; N: *Neoalbaillella*; P: *Pseudoalbaillella*.

3.1.12 青海可可西里移山湖西北部标本 FKXG1 放射虫动物群及其时代（表 3.11）

本书对这个放射虫动物群共描述了 5 属 7 种，归属于 3 目（阿尔拜虫目、内射虫目和罩笼虫目）、3 超科（阿尔拜虫超科、内射虫超科和门孔虫超科）、4 科（阿尔拜虫科、内射虫科、星内射虫科和门孔虫科）。在这个放射虫动物群中，最重要的化石是 *Albaillella indensis*，这个种在德国、法国、澳大利亚、中国华南常见于早石炭世杜内期晚期—维宪期早期地层中，因此，它被认为是这些地区的一个标准带化石。与其伴存的放射虫 *Albaillella undulata* 在美国、法国、德国、土耳其、泰国、中国华南在中上杜内阶和下维宪阶中找到。*Astroentactinia biaciculata*，*Pylentonema mira*，*Stigmosphaerostylus variospina* 和 *Stigmosphaerostylus vulgaris* 这些种的时限较长，在澳大利亚、马来西亚、泰国、德国、法国、俄罗斯南乌拉尔地区、美国何克拉荷马州、中国华南通常发现于中晚泥盆世—早石炭世地层中。因此，我们把这个放射虫动物群称为 *Albaillella indensis* 带，时代为早石炭世杜内期晚期—维宪期早期。

3.1.13 四川理塘标本 ZD305-90b1 放射虫动物群及其时代

这个放射虫动物群属种组成极其单调，仅含 1 属 3 种，其中有 1 个新种（*Pseudoalbaillella litangensis* Wang），归属于 1 目（阿尔拜虫目）、1 超科（丑巾虫超科）、1 科（假阿尔拜虫科）。在这个动物群中，最重要的化石是 *Pseudoalbaillella globosa* Ishiga et Imoto。这是一个世界级分布的种，它主要发现于我国广西、云南西部、广东、苏皖南部以及日本西南部、美国西海岸、加拿大、俄罗斯远东地区、新西兰北岛、马来西亚、泰国、约旦安曼、希腊克里特岛等地。由于它生存时间较短，仅限于中二叠世茅口期早期，现在已成为瓜德鲁普期早期的一个标准化石带。*Pseudoalbaillella scalprata* Holdsworth et Jones 种也是广布于世界各地，但时限比较长，出现于早中二叠世地层中。我们把这个放射虫动物群称为 *Pseudoalbaillella globosa* 带，时代为中二叠世茅口期早期，与美国的瓜德鲁普期早期时代相当。

表 3.11　青海可可西里移山湖西北部标本 FKXG1 放射虫动物群属种时代分布表
Table 3.11　Time ranges of the radiolarians of sample FKXG1 from northwestern Yishanhu, Hoh Xil of Qinghai

时代 化石带 分布 属种名称	晚泥盆世		早石炭世					
	弗拉期	法门期	杜内期			维宪期		
	Helenifore laticlavium	*H. robustum*	*Holoeciscus foremanae*	*A. pseudoparadoxa*	*A. paradoxa*	*A. indensis*	*Eosyringodictya rota*	*A. cartalla*	*Latentifistula impella-L. turgida*
Albaillella indensis						──		
A. undulata					────			
Archocyrtium lagabriellei					────			
Astroentactinia biaciculata	────────							
Pylentonema mira		────						
Stigmosphaerostylus variospina	────────							
S. vulgaris	────────							

3.2　西藏双湖和青海可可西里地区晚古生代放射虫动物群的对比

西藏双湖、青海可可西里和四川理塘地区共发现 9 个晚古生代放射虫动物群（或称带），自老至新分别为：（1）*Helenifore robustum* 带（弗拉期晚期），（2）*Holoeciscus foremanae* 带（法门期），（3）*Albaillella indensis* 带（杜内期晚期—维宪期早期），（4）*Pseudoalbaillella lomentaria - Pseudoalbaillella sakmarensis* 带（隆林期早期），（5）*Pseudoalbaillella rhombothoracata* 带（隆林期晚期），（6）*Albaillella xiadongensis* 带（栖霞期早期），（7）*Pseudoalbaillella ishigai* 带（栖霞期晚期），（8）*Pseudoalbaillella globosa* 带（茅口期早期），（9）*Neoalbaillella ornithoformis* 带（长兴期晚期）（表 3.12）。这 9 个放射虫化石带的时代从晚泥盆世弗拉期晚期至晚二叠世长兴期晚期，尽管由于标本采集不系统或由于板块碰撞而导致地层缺失等原因，而造成了放射虫化石带的不连续，但这些放射虫化石带完全可以同世界各地已经描述过的同时代的古特提斯洋中放射虫化石带进行对比。

表 3.12 青海可可西里地区蛇绿岩混杂岩带地层及放射虫化石带

Table 3.12 Stratigraphy of ophiolite mélange and radiolarians zones, Hoh Xil of Qinghai

系	地层	化 石 带
三叠系	汉台山群 (P_3—T_2)	*Spongoserrula rarauana* *Annulotriassocampe multisegmantatum* *Tiborella florida*
二叠系	西金乌兰群 (C—P_2)	*Pseudoalbaillella globosa*
		Albaillella ishigai
		Ps. rhombothoracata
		Ps. sakmarensis—Ps. lomentaria
石炭系		*Albaillella indensis*
		Cyrtisphaeractenium aff. *crassum-* *Astroentactinia multispinosa*
		Polyentactinia sp. A
泥盆系	拉竹笼组 (D_3^2)	*Holoeciscus foremanae*

这些放射虫动物群中前 3 个动物群，即 *Helenifore robustum* 带、*Holoeciscus foremanae* 带、*Albaillella indensis* 带可以同我国广西钦防地体（Wang et al.，1994，1998，2012）和云南西部昌宁-孟连地体（Wang et al.，2000；Feng and Ye，2006；Feng et al.，2007）3 个同名带对比。前 2 个带也可同我国新疆北部晚泥盆世的 2 个同名带 *Helenifore robustum* 带（Wang et al.，2013）和 *Holoeciscus foremanae* 带（Wang，1997）对比。在国外，澳大利亚东南部新英格兰造山带也具有这 3 个同名放射虫带（Aitchison，1993）。在美国西海岸，也产 *Helenifore robustum* 带（Boundy-Sanders et al.，1999）、*Holoeciscus foremanae* 带（Won et al.，1999）、*Albaillella indensis* 带（Holdsworth and Murchey，1988；Murchey，1990）。

在广西钦防地区板城石梯水库石梯水库组层状硅质岩中发现的 *Helenifore robustum* 带还含有放射虫 *Haplentactinia rhinophyusa*，*Ceratoikiscum planistellare*，*C. elmensis*，*Archocyrtium delicata*，*A. ludicrum*，*Trilonche pittmani*，*T. parapalimbola*，*Pylentonema mira*，*Spongentactinia corynacantha* 等；*Holoeciscus foremanae* 带还含有放射虫 *Holoeciscus elongatum*，*Archocyrtium ormistoni*，*A. wonae*，*A. delicatum*，*A. validum*，*Tetrentactinia spongacea*，*Polyentactinia leptosphaera*，*Astroentactinia stellata*，*Trilonche tretactinia*，*T. elegans*，*T. Vetustum*，*T. echinatum*，*T. minax*，*Triaenosphaera sicarius*，*Palaeoscenidium cladophorum*，*Stigmosphaerostylus vulgaris*，*S. variospina* 等；下石炭统石夹组中的 *Albaillella indensis* 带还含有放射虫 *Albaillella ferreum*，*Pylentonema mendax*，*Astroentactinia multispinosa* 等（Wang et al.，1998，2012）。

云南西部祥云晒经坡长育村组上部含有 *Helenifore laticlavium*（= *Helenifore robustum*）动物群，

其中还包括放射虫 *Palaeoscenidium cladophorum*，*Stigmosphaerostylus variospina*，*Spongentactinella* cf. *corynacantha*；在澜沧里拉 38 道班房公路南侧里拉组上部发现 *Helenifore laticlavium*（= *Helenifore robustum*）动物群，其中还包括放射虫 *Trilonche grandis*，*T. menax*，*Stigmosphaerostylus* cf. *dimidiata*，*S.* sp. B 等；在澜沧太布尔山南公路北侧里拉组中发现 *Helenifore laticlavium*（= *Helenifore robustum*）动物群还含有放射虫 *Palaeoscenidium cladophorum*，*Stigmosphaerostylus variospina*，*S.* cf. *dimidiata*，*S.* sp. B，*T. trilonche davidi*，*T. vetusta*，*T.* sp. B，*Triaenosphaera* cf. *sicarius*，牙形类 *Palanatolepis hasii*，*P. rhenana rhenana*，*P. rhenana nasuta* 等；在孟连回库里拉组硅质岩中发现 *Helenifore laticlavium*（= *Helenifore robustum*）动物群，其中还包括放射虫 *Ceratoikiscum planistellare*，*Trilonche vetusta*，*Archocyrtium* cf. *typicum*，*A.* cf. *dilatipes*，*Popofskyellum* sp. 等。在澜沧阿里地区"曼信组"硅质岩中发现 *Holoeciscus foremanae* 动物群还包含放射虫 *Archocyrtium wonae*，*A.* cf. *ormistoni*，*A. calidum*，*A.* cf *typicum*，*Popofskyellum* sp. B，*Stigmosphaerostylus variospina*，*S. oumonhaoensis* 等；在孟连西部"曼信组"硅质岩中找到的 *Holoeciscus foremanae* 动物群还包含放射虫 *Stigmosphaerostylus oumonhaoensis*，*S. additiva*，*S.* sp. A，*Polyentactinia leptosphaera*，*Trilonche* cf. *echinata*，*Triaenosphaera* cf. *sicarius*，*T.* sp. A，*Haplentactinia* cf. *rhinophyusa* 等（Wang et al., 2000）。

在耿马县城至四排山公路 7.8 km 附近硅质岩中发现 *Albaillella indensis* 动物群，以 *A. indensis brauni* 为特征种，与其共存的放射虫包括 *Archocyrtium delicatum*，*A.* cf. *wonae*，*Deflandrellium* sp.，*Cyntisphaeractenium* sp. 等（Feng et al., 1997）。

新疆北部中天山温泉小区汗吉尕组硅质岩透镜体中 *Helenifore robustum* 动物群还包含放射虫 *Trilonche guangxiensis*，*T. minex*，*Triaenosphaera sicarius*，*Astroentactinia stellata*，*Stigmospharostylus variospina*，*Spongentactinella corynacantha* 等（Wang et al., 2013）；和丰牧场俄姆哈一带根那仁组硅质岩中的 *Holoeciscus foremanae* 动物群由放射虫 *H. foremanae*，*Palaeoscenidium cladophorum*，*Archocyrtium ormistoni*，*Entactinosphaera* sp. A，*Entactinia* sp. A，*E. oumonhaoensis* 等组成（Wang, 1997）。

Ishiga 和 Leitch（1988）把澳大利亚东部新英格兰造山带黑斯廷斯（Hastings）地块西部中、晚泥盆世放射虫分为 4 个带。弗拉期包括上、下 2 个带，即下带 *Helenifore laticlavium*（= *Helenifore robustum*）和上带 *Popofskyellum hastimgensis*；法门期晚期分成 2 带，即下带 *Holoeciscus foremanae* 带和上带 *Popofskyellum* sp. 带。*Helenifore laticlavium*（= *Helenifore robustum*）带还包括放射虫 *Haplentactinia hastingensis*，*Entacitnosphaera* sp. A，*E.* sp. B，*E.* sp. C 等。*Holoeciscus foremanae* 带还包括放射虫 *Archocyrtium* spp.，*Ceratoikiscum planistellare*，*Popofskyellum* sp. cf. *P. dumitricai*，*Archocyrtium* sp. cf. *A. delicatum* 等。*Helenifore robustum* 带同时也在加迈拉罗伊（Gamilaroi）地块（Stratford and Aitchison, 1997; Aitchison et al., 1999）和银沟（Silver Gully）山脉（Aitchison et al., 1999）发现。*Holoeciscus foremanae* 带也在阿奈湾（Anaiwan）地块找到，并与放射虫 *Ceratoikiscum* sp.，*Archocyrtium* spp.，*Cyrtentactinia* spp.，*Palaeoscenidium cladophorum* 等共存（Atichison and Flood, 1990; Aitchison, 1993）。在阿奈湾地块硅质岩中发现的 *Albaillella undulata - A. indensis* 动物群还含有放射虫 *Staurodruppa? prolata*，*Entactinosphaera palimbola* 等；*Albaillella indensis - Albaillella furcata* 动物群还包括放射虫内射虫类、古笼虫类和阿尔拜虫类（Aitchison, 1993）。

在美国西海岸内华达罗伯茨山脉（Roberts Moutains）肖拙尼岭（Shoshone Ridge）斯拉文（Slaven）硅质岩中发现的 *Durahelenifore robustum*（= *Helenifore robustum*）动物群还含有放射虫 *Ceratoikiscum* sp. cf. *C. planistellare*，*C.* sp.，*Palaeoscenidium* sp. cf. *P. tabernaculum*，*Polyentactinia* sp.，*Astroentactinia* sp.，*Spongentactinia* sp.，*Entactinia* sp. cf. *E. dissora*，*Entactinosphaera* sp.（Boundy-Sanders et al., 1999）；在阿拉斯加中南部丘利钠（Chulitna）地体"多曼（Do）"组合硅质岩中发现的 *Holoeciscus foremanae* 动物群还含有放射虫 *Popofskyellum? tetralongispina*，*P.* sp.，*Totollum?* sp.，

Cyrtentactinia? sp., *Ceratoikiscum mirum*, *C*. sp. cf. *C. sandbergi*, *Archocyrtium riedeli*, *A. verustum*, *Pylentonema* sp., *Entactinia variospina*, *Plenospongiosa*? *gourmelonae*, *Entactinosphaera palimbola*, *Palaeoscenidium planum*, *P.*? *chulitnaensis* 等（Won et al., 1999）；在阿拉斯加北布鲁克斯岭（Brooks Ridge）尼古断崖（Nigu Bluff）硅质岩中找到 *Albaillella indensis* 动物群，但没有报道共存的放射虫（Holdsworth and Murchey, 1988；Murchey, 1990）。另外，在马来西亚文冬-劳布（Bentong-Raub）缝合带 T3A-BR615、BR623 硅质岩中发现的 *Helenifore laticlavium*（= *Helenifore robustum*）动物群还含有放射虫 *Palaeoscenidium cladophorum*，在 KLK1-BR704、KLK1-BR705、KLK5-BR33、NS2-KP5 中发现的 *Holoeciscus foremanae* 动物群还包含放射虫 *Holoeciscus elongatus*, *Popofskyellum* sp., *P.* sp. cf. *P. bendricksi*, *Archocyrtium* sp. A。虽然在这个地区尚未发现 *Albaillella indensis* 动物群，但是位于此带之下的 *A. paradoxa* 带和之上的 *A. cartalla* 带都存在（Spiller, 2002）。

Sashida 等（1998）建立了泰国北部晚泥盆世至早石炭世 2 个放射虫带，弗拉期的 *Helenifore laticlavium*（= *Helenifore robustum*）带见于黎府（Loei）和帕伊（Pai）地区硅质岩中，并共存有放射虫 *Palaeoscenidium cladophorum*, *Ceratoikiscum* sp., *Trilonche palimbola*, *protalbaillella* sp., *Stigmosphaerostylus variospina* 等；法门期—杜内期的 *Stigmosphaerostylus variospina* 带还含有放射虫 *Archocyrtium coronaesimite*, *A. riedeli*, *Cyrtisphaeracterium* sp., *Palaeoscenidium cladophorum* 等；Feng 等（2004）在泰国西北部发现的 *Albaillella indensis* 动物群还含有放射虫 *Archocyrtium diductum*, *A. lagabriellei*, *Triaenosphaera sicarius*, *Astroentactinia multispinosa*, *Pylentonema antiqua*, *Entactinia vulgaris vulgaris*, *E.* sp. cf. *E. tortispina* 等。在泰国，尽管至今尚未发现 *Holoeciscus foremanae* 种，但这个动物群的许多共存放射虫已被发现（Saesaengseerung et al., 2007）。

在德国，目前尚未发现 *Helenifore robustum* 动物群，但在北巴伐利亚法兰肯塔尔（Frankenwald）地区找到了 *Holoeciscus foremanae* 带和 *Albaillella indensis* 带。前一个带还包含放射虫 *Huasha magnifica*, *Ceratoikiscum bujugum*, *C. extraordinarium*, *Holoeciscus brevis*, *H. elongatus*, *Archocyrtium* cf. *angulosum*, *A*. cf. *delicatum*, *A. diductum*, *A*. cf. *riedeli*, *A*. cf. *formosum*, *A. eupectum*, *A. ludicrum*, *A. effingi*, *Pylentonema* cf. *hindei*, *Popofskyellum defrendrei*, *Astroentactinia radiata*, *A. stellata*, *A. digitosa*, *A. multispinosa*, *Entactinia additiva*, *E. spongites*, *E. exilispina*, *E. tortispina*, *E. vulgaris*, *E. herculea*, *E. variospina*, *Entactinosphaera fredericki*, *E. riedeli*, *E. palimbola*, *Polyentactinia aranea*, *Tetrentactinia spinulosa*, *T. teuckestes*, *Triaenosphaera sicarius*, *T. hebes*, *Bisyllentactinia arrhinia* 等（Kiessling and Tragelehn, 1994），后一个带还含有放射虫 *Albaillella* sp. aff., *A. furcata*, *A. paradoxa*, *A. undulata*, *Ceratoikiscum formosa*, *C. umbraculum*, *Belowea variabilis*, *Astroentactinia digitosa*, *A.*? *mirousi*, *A. radiata*, *A. multispinosa*, *A. spatiosa*, *Callela conispinosa*, *C. hexactinia*, *C. stellaesimilia*, *Entactinia tortispina*, *E. variospina*, *E. vulgaris* 等（Braun and Schmidt-Effing, 1993）。

总之，羌塘双湖地区和青海可可西里地区晚泥盆世—早石炭世 3 个放射虫带完全可以同我国华南广西钦防地区板城一带、云南西部昌宁-孟连地区 3 个同名带以及新疆北部前 2 个化石带对比，这 3 个带也可以同美国西海岸、澳大利亚东部新英格兰造山带、马来西亚、泰国 3 个同名带以及德国后 2 个同名放射虫化石带对比（表 3.13）。

表 3.13 晚泥盆世—早石炭世放射虫动物群的国际对比表

Table 3.13 International correlation chart of the Late Devonian — Early Carboniferous radiolarian faunas

时代	化石带 地区	本书 羌塘北部	中国华南 Wang et al., 1994, 1998, 2007, 2012	澳大利亚新英格兰造山带 Aitchison, 1993	马来西亚 Spiller, 2002	泰国 Sashida et al., 1998; Feng et al., 2004; Caridroit et al., 1992	北美西海岸 Boundy-Sanders et al., 1999; Won et al., 1999; Holdsworth and Murchey, 1988	德匡 Kiessling and Tragelehn, 1994; Braun et al., 1993
早石炭世 维宪期			Latentifistula impella–L. turgida	Circulaforma omicron	A. nazarovi			A. nazarovi
					A. rockensis		Albaillella–3	A. rockensis
			A. cartalla	A. cartalla–A. thomasi	A. cartalla	A. cartalla	Pre–A. pennata–2	Latentifistula concentrica
			Eostylodictya rota			Eostylodictya rota		A. cartalla
								Eostylodictya rota
杜内期		Albaillella indensis	A. indensis	A. indensis–A. furcata	A. indensis	A. indensis	A. indensis	A. indensis
				A. undulata–A. indensis	A. deflandrei		(pre–A. pennata–1)	
			A. paradoxa– A. pseudoparadoxa	A. paradoxa	A. paradoxa		Albaillella–2	A. paradoxa
				Protoalbaillella anaiwaensis	Aibaillella 1 Ass.			
晚泥盆世 法门期		Holoeciscus foremanae	Holoeciscus foremanae	Ceratoikiscum umbraculum			Holoeciscus foremanae	Holoeciscus foremanae
				Paraholoeciscus bingaraensis	Holoeciscus 3 Ass.	Stigmosphaerostylus variospina		
				Holoeciscus foremanae	Holoeciscus 2 Ass.			
弗拉期		Helenifore robustum	Helenifore robustum	Protoholoeciscus hindea	Holoeciscus 1 Ass.		Durahelenifore robustum	
		Helenifore laticlavium	Helenifore laticlavium	Helenifore laticlavium– Ceratoikiscum planistellare	PreHoloeciscus Ass. (=Helenifore laticlavium)	Helenifore laticlavium		

西藏双湖和青海可可西里地区二叠纪的 6 个放射虫化石带，即 *Pseudoalbaillella lomentaria*-*Pseudoalbaillella sakmarensis* 带，*Pseudoalbaillella rhombothoracata* 带，*Albaillella xiadongensis* 带，*Pseudoalbaillella ishigai* 带，*Pseudoalbaillella globosa* 带和 *Neoalbaillella ornithoformis* 带，可以与我国华南地区、日本、马来西亚、泰国、俄罗斯远东地区、美国西海岸的同名放射虫带进行广泛的对比（表 3.14）。

Pseudoalbaillella lomentaria-*Pseudoalbaillella sakmarensis* 动物群发现于我国广西钦防地区板城石夹水库、田车、石梯水库、钦北区米拱等地板城组下部硅质岩中（Wang et al.，1994，1998，2007，2012；Shimakawa and Yao，2006）。除带的指引种外，此带还含有 *P. ornata*，*P. rhombothoracata*，*P. scalprata*，*P. postscalprata*，*Albaillella xiadongensis* 等。在云南东南部八布蛇绿混杂岩和云南西部昌宁-孟连地体硅质岩中也发现过这个动物群（Feng and Liu，2002；Wu and Li，1989）。

在日本西南部发现的 *P. lomentaria* 动物群含有放射虫 *P. sakmarensis*，*P. longicornis*，*P. scalprata*，*P. ornata* 等（Ishiga，1982，1990；Yao and Kuwahara，2004）。在美国俄勒冈地区，Blome 和 Reed（1992）发现的 *P. sakmarensis* 动物群还含有放射虫 *P. longicornis*。在泰国东北部，Sashida 等（1998）发现的 *P. lomentaria* 动物群包含放射虫 *P.* sp.，*Grandetortura* sp.。在马来西亚，Spiller（2002）发现的 *P. lomentaria* 动物群包括放射虫 *P. sakmarensis*，*P. ornata*，*P. scalprata m. scalprata*，*P. scalprata m. rhombothoracata*，*P.* sp. B，?*Copicyntra* sp.，*Stigmosphaerostylus* sp. cf. *S. itsukaichiensis*，*S.* sp. cf. *S. pycnoclada*，*Latentifistula* sp. aff. *L. crux*，*Latentibifistula triacanthophora*，*L.* sp. A，*L.* sp. B，*L.* sp. C，*L.* sp. D，*Polyfistula* sp.，*Quadriremis gliptoacus*，*Ruzhencevispongus uralicus*，*Pseudotormentus kamigoriensis* 等。在俄罗斯远东地区，Rudenko 和 Panasenko（1997a）发现 *P. sakmarensis* 动物群存有放射虫 *P.* sp. aff. *P. scalprata*，*P.* cf. *lomentaria* 等。

Pseudoalbaillella rhombothoracata 放射虫动物群常见于我国广西钦防地区板城石夹水库、田车、石梯水库、小董三泡岭、钦北区米拱等地板城组下部硅质岩中。这一动物群包括放射虫 *P. scalprata*，*P. longicornis*，*P. lomentaria*，*P. sakmarensis*，*P. postscalprata*，*Albaillella xiadongensis*，*P. ornata* 等（Wang et al.，1994，1998，2007，2012；Shimakawa and Yao，2006）。在日本，这个动物群还含有放射虫 *P.* sp. A，*P. elongata*，*P. scalprata*，*P. sakmarensis*，*Albaillella* sp. B 等（Ishiga et al.，1982b）；在马来西亚 *P. scalprata m. rhombothoracata* 动物群还含有放射虫 *P. scalprata m. scalprata*，*P. scalprata m. postscalprata*，*P. elongata*，*Albaillella asymmetrica* 等（Spiller，2002）；在泰国，与这个动物群相比较的是 *P. scalprata* 动物群，其中还含有放射虫 *Albaillella sinuata*，*P. scalprata postscalprata*，*P.* spp.（Sashida et al.，1998）；在俄罗斯远东地区，Rudenko 和 Panasenko（1997a）发现 *P. scalprata* 带与日本 *P. scalprata m. rhombothoracata* 带大部分和 *P. lomentaria* 带的上部相当，伴存放射虫包括 *P. sakmarensis*，*P. rhombothoracata*，*P. elongata* 等。

Albaillella xiadongensis 动物群常见于我国广西钦防地区板城石梯水库、小董三泡岭、钦北区米拱等地板城组下部硅质岩中（Wang et al.，1994，1998，2007，2012；Shimakawa and Yao，2006）。此动物群还包括放射虫 *P. scalprata*，*P. ishigai*，*P. rhombothoracata*，*P. longicornis*，*P. fusiformis*，*P. longtanesis*，*Albaillella sinuata*，*Latentifistula texana*。这次在青海可可西里地区发现的这个动物群，其中伴存的放射虫与上述各地十分相似，也包括了 *Albaillella sinuata*，*Latentifistula texana*，*L. patagilaterala*，*P. rhombothoracata*，*Pseudotormentus kamigoriensis*，*Hegleria mammilla* 等。

表 3.14 二叠纪放射虫动物群的国际对比表

Table 3.14 International correlation chart of the Permian radiolarian faunas

地区 化石带 时代	本书 羌塘北部	中国华南 Wang et al., 1994, 1998, 2007, 2012	马来西亚 Spiller, 2002	日本西南部 Ishiga, 1990; Kuwahara et al., 1998; Yao and Kuwahara, 2004	泰国 Sashida et al., 1998, 2000; Wonganan and Caridroit, 2007; Caridroit et al., 1992	俄罗斯远东地区 Rudenko and Panasenko, 1997	北美西海岸 Blome and Reed, 1992, 1995; Blome et al., 1986; Murchey, 1990	古特提斯区 Caridroit, 2001	Stage
晚二叠世 长兴期	N. ornithoformis	N. optima N. ornithoformis	N. ornithoformis	N. optima N. ornithoformis	N. optima N. ornithoformis	N. pseudogrypus Imotoella excelsa N. optima	N. ornithoformis N. optima	N. optima N. ornithoformis	Changhsingian
晚二叠世 吴家坪期		A. levis–A. excelsa A. protolevis Foremanhelena triangularis F. bipartitus–F. charveti–F. orthogonus	F. charveti	F. charveti–A. vamakitai	F. charveti–F. porrectus			Foremanhelena triangula–F. falx F. bipartitus–F. charveti	Wuchiapingian
中二叠世 茅口期	P. globosa	F. scholasticus–F. ventricosus P. monacantha P. globosa	F. porrectus F. monacanthus P. globosa	F. scholasticus–F. ventricosus F. monacanthus P. globosa	F. monacanthus	F. porrectus F.? monacanthus P. globosa	F. monacanthus P. globosa	F. scholasticus–F. ventricosus F.? monacanthus P. fusiformis	Maokouan
中二叠世 栖霞期	P. ishigai A. xiadongensis	P. ishigai A. sinuata A. xiadongensis	P. longtanensis A. sinuata	P. longtanensis A. sinuata		P. corniculata Spinodeflandrella acutata	P. sp. C A. sinuata	A. sinuata–Cauletella	Artinskian
早二叠世 隆林期	P. rhombothoracata	P. rhombothoracata	P. scalprata–m. rhombothoracata	P. scalprata m. rhombothoracata	P. scalprata	P. scalprata	P. scalprata m. rhombothoracata	P. scalprata	Sakmarian
早二叠世 马平期	P. lomentaria–P. sakmarensis	P. lomentaria–P. sakmarensis	P. lomentaria		P. lomentaria	P. sakmarensis	P. sakmarensis	P. sakmarensis	Asselian
早二叠世 马平期	P. u-forma–P. elegans P. bulbosa	P. u-forma–P. elegans P. bulbosa	P. u-forma m.II	P. u-forma m.II P. u-forma m.I P. bulbosa	P. elegans P. bulbosa	P. u-forma	P. bulbosa P. u-forma Paronaella? sp.	P. lomentaria P. anfractus–R. circumfusum	Gzhelian
晚石炭世	P. nodosa	P. nodosa						P. anfractus–Ruzhencevispongus visenda P. annulata	Kazimovia

Pseudoalbaillella ishigai 动物群发现于我国广西钦防地区板城石梯水库、小董三泡岭、钦北区米拱等地板城组中部硅质岩中（Wang et al.，1994，1998，2007，2012；Shimakawa and Yao，2006）。这一动物群还包括放射虫 *P. scalprata*，*P. rhombothoracata*，*Albaillella sinuata*，*Nazarovella gracilis*，*Stigmosphaerostylus* sp. B 等。在日本西南部，*Pseudoalbaillella ishigai* 种曾被日本学者 Ishiga 等（1982c）鉴定成 *P.* sp. C，在早期曾被建立为 *P.* sp. C 带，后来改成 *P. longtanensisi* 带，这个带还包含放射虫 *P. fusiformis*，*P. longicornis* 等（Ishiga，1990，Yao and Kuwahara，2004，Shimakawa and Yao，2006）。在俄罗斯远东地区，Rudenko 和 Panasenko（1997a）建立的 *P. corniculata* 带位于 *P. globosa* 带之下，这一地层位置与我国的 *P. ishigai* 带相当。这一化石带内还包含放射虫 *P.* sp. C，*P.* sp. D，*Albaillella asymmetrica* 等。*P. corniculata* 在本书中被视为 *P. longtanensis* Sheng et Wang 的同义名，因此，这个带也应改成 *P. longtanensis* 带。

Pseudoalbaillella globosa 动物群常见于我国广西钦防地区板城石屋（敬老院）、小董三泡岭等地板城组上部硅质岩中（Wang et al.，1994，1998，2007，2012），放射虫组成丰富，包括 *P. fusiformis*，*P. longicornis*，*P. yanaharensis*，*P. internata*，*P. longtanensis*，*Hegleria mammilla*，*Latentifistula texana*，*Stigmosphaerostylus itsukaichiensis*，*Copicyntra akikawaensis*，*Pseudotormentus monoporus* 等。这个带也发现于我国云南西部昌宁-孟连地区硅质岩中（Yao and Kuwahara，1999b），其中还包含放射虫 *Follicucullus ventricosus*，*F. monacanthus*，*Albaillella* sp.，*Latentifistula* sp.，*Nazarovella inflata* 等。在日本西南部，这个带还包含放射虫 *P. longicornis*，*P. fusiformis*，*P.* sp. C，*Albaillella asymmetrica*，*Follicucullus monacanthus* 等（Ishiga and Imoto，1982b，c；Ishiga et al.，1986；Ishiga，1990；Nishimura and Ishiga，1987；Sano，1988；Ishiga et al.，1992）。

在俄罗斯远东地区，Rudenko 和 Panasenko（1990，1997a）发现这个带还含有放射虫 *P. convexa* 及 *Albaillella* 的某些新种；在马来西亚 Bentong-Raub 缝合带，这个带还含有放射虫 *Albaillella asymmetrica*，*P. fusiformis*，*P.* sp. aff. *P. longicornis*，*Hegleria mammilla*，*Latentifistula* sp. aff. *L. crux*，*L. patagilaterala*，*Pseudotormentus* sp. 等（Spiller，2002）；在希腊克里特岛蓝片岩变质的千枚岩（phyllite）群中，Kozur 和 Krahl（1987）首次发现欧洲特提斯中二叠世 *Parafollicucullus fusiformis*-*P. globosa* 动物群还含有放射虫 *Pseudoalbaillella* cf. *eurasiatica*，*Phaenicosphaera mammilla*；在约旦安曼山脉 Hawasina 推覆体的硅质岩中，De Wever 等（1988）首次发现中西特提斯带二叠纪放射虫 *Follicucullus monacanthus*（= *Pseudoalbaillella fusiformis*）动物群还含有放射虫 *F. scholasticus* Ⅰ，*F. scholasticius* Ⅱ；根据 Murchey 和 Jones（1992）的资料，在北美西部二叠纪硅质岩沉积区多地发现 *P. fusiformis*-*P. globosa* 动物群：（1）在 Wrangellia 地体，*Parafollicucullus globosa* 还含有放射虫 *P. fusiformis*，*P. scalprata* m. *scalprata*，*P. scalprata* m. *postscalprata*，*P. longtanensis*，*Hegleria* sp. 等；（2）在加拿大西尔威斯特（Sylvester）外来体二叠纪硅质岩中发现 *Parafollicucullus globosa* 与放射虫 *P. fusiformis*，*Follicucullus* sp.，*Hegleria mammilla* 等伴存；（3）在加利福尼亚北内华达（Sierra）地体硅质泥岩中，*Parafollicucullus* sp. aff. *P. globosa* 与众多的放射虫共存，包括 *Pseudoalbaillella scalprata* m. *scalprata*，*P. sakmarensis*，*Pseudotormentus kamigoriensis*，*Ormistonella robusta*，*Quinqueremis robusta*，*Deflandrella* spp.，*Latentifistula patagilaterala*，*Polyfistula* sp.，？*Tormentum circumlusum* 等（Murchey，1989，1990；Harwood and Murchey，1990）。

Neoalbaillella ornithoformis 动物群在我国主要见于广西南丹龙王坡硅质岩中（Wang et al.，2006），其中还含有放射虫 *Albaillella triangularis*，*A. lauta*，*Ishigaum trifistis*，*Nazarovella inflata*，*Pseudotormentus kamigoriensis*，*Triplanospongos musashiensis* 等；在云南西部澜沧南畔硅质岩中，这个动物群还

含有放射虫 *Albaillella levis*，*A. triangularis*，*Copicyntra akikawaensis*，*C. shaiwaensis*，*Stigmosphaerostylus itsukaichiensis*，*Trilonche* aff. *crassispinosa*，*Follicucullus porrectus*，*Foremanhelena triangula*，*F. circula*，*Hegleria mammilla*，*Ishigaum trifustis*，*I. obesum*，*Nazarovella gracilis*，*Meschedea permica*，*Neoalbaillella gracilis*，*Pseudotormentus kamigoriensis*，*Spongosphaeradiscus shaiwaensis*，*Triplanosphaera minuta* 等（Wang et al.，2006）；在云南西部澜沧老厂，这个动物群还含有放射虫 *Copicyntra akikawaensis*，*C. shaiwaensis*，*Hegleria mammilla*，*Stigmosphaerostylus itsukaichiensis*，*Ishigaum trifustis*，*I.* sp. B，*Nazarocella inflata*，*N. scalae*，*Nabepecha leonardia*，*Neoalbaillella gracilis*，*Octatormentus floriferum*，*Quadricautis femoris*，*Triplanospongos musashiensis* 等（Wang et al.，2006）；在贵州紫云晒瓦组上部硅质岩中这个动物群还含有放射虫 *Albaillella triangularis*，*Copicyntra shaiwaensis*，*Deflandrella manica*，*Foremanhelena triangula*，*Quadriremis* sp.，*Spongosphaeradiscus shaiwaensis*，*Tetratormentum* sp. A，*Triplanospongos musashiensis*，*Ishigaum trifustis*，*I.* sp. B，*Nazarovella inflata*，*N. gracilis*，*N. phlogidea*，*N. scalae* 等伴存（Wang and Shang，2001；Wang et al.，2006）；在贵州罗甸沫阳南，Yao 和 Kuwahara（2000）发现 *N.* cf. *ornithoformis* 这个动物群还含有放射虫 *N.* cf. *optima*，*Copiellintra*? sp. A，*Hegleria* sp. CA，*H.* sp. CC，*Kashiwara*? sp.，*Gustefana* aff. *obliqueannulata*，*Octatormentum*? *floriferum* 等。

在国外，在泰国西北部清道（Chiang Dao）西北 10 km 滑塌堆积中的一块放射虫硅质岩（Th50）中发现的这个动物群还含有放射虫 *N. gracilis*，*Albaillella levis*，*Deflandrella manica*，*D. trifustis*，*Nazarovella gracilis*，*N. inflata*，*Ishigaum*? sp.，*Triplanospongos musashiensis* 等（Caridroit et al.，1992）；在泰国北部掸-泰（Shan-Thai）地块层状硅质岩中发现的这个动物群还含有放射虫 *Albaillella triangularis*，*A. levis*，*Entactinia itsukaichiensis*，*Entactinosphaera pseudocimelia*，*Hegleria mammilla*，*Copicyntra* sp.，*Latentibifistula apserspongiosa*，*Nazarovella gracilis*，*N. inflata*，*Ishigaum trifustis*，*Triplanospongos musashiensis*，*Pseudospongoprunum*? *chiangdaoensis*，*Grandetortura nipponica* 等（Sashida et al.，2000）；在马来西亚西北部塞芒加尔（Semanggal）组层状硅质岩中发现的 *N.* cf. *ornithoformis* 动物群还含有放射虫 *Albaillella excelsa*，*A. levis*，*Entactinosphaera pseudocimelia*，*E.* sp.，*Octatormentum*? sp.，*Nazarovella gracilis*，*N. inflata*，*Ishigaum*? sp.，*Triplanospongos musashiensis* 等（Sashida et al.，1995；Spiller，2002）；在菲律宾北巴拉望湾地块硅质岩中，Tumanda 等（1990）发现 *N. ornithoformis* 动物群还含有放射虫 *N. optima*，*Albaillella levis*，*A. triangularis*，*A. excelsa*，*Triplanospongos musashiensis*，*Kashiwara magna*，*Nazarovella scalae*，*Deflandrella manica*，*Tetratormentum acutum*，*Octatormentum* sp. A，*Copicyntra akikawaensis* 等；在俄罗斯远东地区，Rudenko 和 Panasenko（1997a）建立的 *N. ornithoformis* 动物群还包括放射虫 *Albaillella lauta*，*A. triangularis*，*Sphaeroidea* gen. et sp. indet.，*Hozmadia*? sp.，*Eptingium*? sp. 等；在日本西南部坦巴（Tamba）带，Takemura 和 Nakaseko（1981）首先报道 *Neoalbaillella ornithoformis* 还含有放射虫 *N. gracilis*，*N.* sp. A，*N.* sp. B，*Follicucullus ventricosus*，*F. scholasticus*，*Paronaella*? sp.，*Angulotracchia*? sp. 等；次年，Ishiga，Kito 和 Imoto（1982）在这个地区同样发现 *N. ornithoformis* 动物群还含有放射虫 *N. optima*，*Albaillella* spp.，*Follicucullus scholasticus*，*F. ventricosus* 等；Sashida 和 Tonishi（1985）发现日本中部关东山脉（Kanto Mountains）秩父（Chichibu）岩系中晚侏罗世黑色页岩具有许多不同大小的岩块，其中在硅质岩块中发现 *N. ornithoformis* 动物群还含有放射虫 *Follicucullus scholasticus*，*F. ventricosus*，*Albaillella levis*，*Neoalbaillella grypus*，*Entactinia itsukaichiensis*，*E. modesta*，*E. reticulata*，*Kashiwara magna*，*Entactinosphaera*? *orientalis*，

E. crassisphaera，*Tetragregnon japonicum*，*Meschedea permica*，*Helioentactinia nazarovi* 等。在北美内华达中北部黑岩地体奎恩河（Quinn River）组层状硅质岩中，Blome 和 Reed（1995）发现 *Neoalbaillella* aff. *ornithoformis* 与放射虫 *Triplanospongos musashiensis*，*Pseudotormentum* sp. 等伴存。

总之，西藏双湖地区、青海可可西里和四川理塘地区这 6 个放射虫带完全可以同中国华南、日本、北美西海岸、马来西亚、泰国和俄罗斯远东地区同名带相对比（表 3.14）。因此，西藏双湖地区、青海可可西里和四川理塘地区晚古生代放射虫动物群应该属于古特提斯放射虫动物群范畴。

第4章 晚古生代放射虫分类描述

放射虫亚门 Radiolaria Müller，1858
  多囊虫超目 Polycystina Ehrenberg，1838，emend. Riedel，1967
    阿尔拜虫目 Albaillellaria Deflandre，1953，emend. Holdsworth 1969
      古蓬虫超科 Palaeoscenidioidea Riedel，1967，emend. Nazarov et Rudenko，1981
        古蓬虫科 Palaeoscenidiidae Riedel，1967，emend. Nazarov et Rudenko，1981
          古蓬虫亚科 Palaeoscenidiinae Riedel，1967，emend. Nazarov et Rudenko，1981
            古蓬虫属 *Palaeoscenidium* Deflandre，1953

模式种：*Palaeoscenidium cladophorum* Deflandre，1953
鉴定要点：骨骼由 6—8 根射杆组成。射杆基部具帐篷状薄壳。基射杆较长，发育骨刺，顶射杆光滑。
时代与分布：中泥盆世艾菲尔期—早石炭世维宪期；世界各地。

具枝古蓬虫 *Palaeoscenidium cladophorum* Deflandre

(Pl. 2，fig. 24)

Palaeoscenidium cladophorum Deflandre，1953，Text-fig. 308，1960，pl. 1，fig. 21；Foreman，1963，pl. 8，fig. 10，pl. 9，fig. 6；Holdsworth，1973，pl. 1，fig. 19；Nazarov，1975，pl. 13，figs. 4，5，pl. 14，figs 5，6，1988，pl. 13，fig. 7，pl. 14，fig. 5；Nazarov et al.，1982，Fig. 5（D—F）；Nazarov and Ormiston，1983，pl. 2，figs. 6，7；Ishiga，1988，pl. 1，figs. 11，12；Ishiga et al.，1988，pl. 1，fig. 7；Schmidt-Effing，1988，pl. 1，figs. 11，12；Li and Wang，1991，pl. 2，figs. 18，19；Aitchison，1990，Figs. 2E，3M，I，J，1993，pl. 1，figs. 15—17，19，pl. 2，figs. 17，20；Wang，1991，pl. 2，fig. 15；Aitchison and Flood，1992，Fig. 3（7）；Kiessling and Tragelehn，1994，pl. 6，figs. 5—7，9，10；Stratford and Aitchison，1997，Fig. 2（O）；Wang，1997，pl. 1，fig. 6；Sashida et al.，1998，Fig. 8（11—14）；Wang et al.，2000，pl. 2，figs. 7—9，2003，pl. 3，figs. 14—17，2012，pl. 3，fig. 14；Aitchison et al.，1999，pl. 5，fig. O，pl. 6，figs. F，J；Afanasieva，2000，pl. 106，figs. 9—12.

描述：这个种的特点是基部具帐篷状构造，顶刺杆光滑，基射杆发育骨刺。
层位与产地：中泥盆统艾菲尔阶—下石炭统维宪阶；世界各地。

      角舍虫超科 Ceratoikiscoidea Holdsworth，1969
        石鱼虫科 Lapidopiscidae Deflandre，1958
          石鱼虫亚科 Lapidopiscinae Deflandre，1958
            全角虫属 *Holoeciscus* Foreman，1963

模式种：*Holoeciscus auceps* Foreman，1963
鉴定要点：a杆、b杆和交叉杆（i杆）形成三角形构架，其中a杆较短，腔肋侧向相连形成薄壁。
时代与分布：晚泥盆世法门期；世界各地。

长全角虫 *Holoeciscus elongatus* Kiessling et Tragelehn

(Pl. 2, fig. 1)

Holoeciscus elongatus Kiessling et Tragelehn, 1994, pl. 1, figs. 14—17; Spiller, 2002, pl. 2, figs. H, I, J; Wang et al., 2003, pl. 4, figs. 13, 14, 2012, pl. 3, figs. 4—6, 17, 18.

Holoeciscus longus Schwartzapfel et Holdsworth, 1996, pl. 12, figs. 3—6, 18—20; Spiller and Metcalfe, 1995a, Fig. 5 (b, c).

描述：只有1块标本，保存比较完整。at杆、bt杆和it杆被薄壳覆盖部分约为3/4，薄壳亚圆柱形。

比较：这个种和 *Holoeciscus foremanae* Cheng 的区别为后者薄壳侧视梯形，at杆、bt杆和it杆被覆盖部分约为2/3。

层位与产地：上泥盆统法门阶；德国，美国，马来西亚，中国华南和青海可可西里地区。

阿尔拜虫超科 Albaillellacea Cheng, 1986
阿尔拜虫科 Albaillellidae Cheng, 1986
阿尔拜虫亚科 Albaillellinae Cheng, 1986
阿尔拜虫属 *Albaillella* Deflandre, 1952, emend. Holdsworth, 1966

模式种：*Albaillella paradoxa* Deflandre, 1952

鉴定要点：壳体圆锥形，发育1个、2个或缺失翼，由横节或斜节组成，具H框架，腹小柱弱，背小柱较强壮，有时发育成对的侧刺。

时代与分布：石炭纪—二叠纪；世界各地。

英德阿尔拜虫 *Albaillella indensis* Won

(Pl. 17, figs. 1, 2)

Albaillella indensis Won, 1983, pl. 1, figs. 19, 20; Gourmelon, 1986, pl. 3, figs. 1, 2; 1987, pl. 12, figs. 6—10; Braun and Schmidt-Effing, 1988, figs. 10—14, 1993, pl. 2, fig. 4; Braun, 1989, pl. 1, figs. 1, 2, pl. 3, figs. 1, 2, 1990, pl. 1, fig. 2, 1993, pl. 2, fig. 4; Aitchison, 1993, pl. 1, fig. 10; Wang and Kuang, 1993, pl. 1, figs. 1—8; Wang et al., 1998, pl. 1, fig. 12, 2012, pl. 6, figs. 4—8, pl. 7, figs. 5, 13, 17—19, pl. 8, fig. 1, pl. 10, figs. 6, 12, 13; Holdsworth and Murchey, 1988, pl. 34. 1, figs. 16—20.

Albaillella indensis indensis Won; Braun, 1989, pl. 1, fig. 1, 1990, pl. 2, figs. 6—8, pl. 4, figs. 11, 12; Feng et al., 1997, pl. 1, fig. 5, pl. 3, figs. 7, 10, 2004, pl. 1, figs. 4, 5.

Albaillella furcata Won; Aitchison et al., 1992, Fig. 8 (B, F)

描述：这个种的特点是壳分节，一般有4—6节，形似烟囱。顶区发育背刺和腹刺。壳节斜向缠绕，各节间近乎平行，节间有浅沟。这个种已成为早石炭世杜内晚期—维宪早期一个标准放射虫化石带种。

层位与产地：下石炭统杜内阶上部—维宪阶下部；法国，德国，澳大利亚东部新英格兰造山带，美国阿拉斯加，中国华南和青海可可西里地区。

优美阿尔拜虫 *Albaillella lauta* Kuwahara

(Pl. 14, fig. 58)

Albaillella lauta Kuwahara (in Kuwahara et Sakamoto), 1992, pl. 1, figs. 4—6, pl. 2, figs. 8—12, 1999, pl. 2, figs. 13, 14; Wang et al., 2006, Fig. 13 (I, AA, CC), Fig. 14 (N); Kuwahara and Yao, 1998, pl. 1, fig. 3.

描述：这个种的特点是壳高锥形，壳表光滑无孔，具1个中空的腹翼，位于壳的中下部。

比较：这个种与 *Albaillella levis* Ishiga, Kito et Imoto 的区别为后者壳体短小，腹翼位于壳体较低

的位置。

层位与产地：上二叠统；日本，中国华南和西藏双湖地区。

光壳阿尔拜虫 *Albaillella levis* Ishiga, Kito et Imoto
(Pl. 14, figs. 50, 52—54, 59, 60)

Albaillella levis Ishiga, Kito et Imoto, 1982a, pl. 3, figs. 1—4; Kojima, 1982, pl. 3, figs. 5, 6; Nishizono and Murata, 1983, pl. 1, fig. 9; Sano, 1988, pl. 2, fig. 16; Wu and Li, 1989, pl. 1, fig. 10; Tumanda et al., 1990, pl. 2, fig. 24; Sashida and Tonishi, 1985, pl. 7, figs. 5, 6; Yoshida and Murata, 1985, pl. 2, figs. 11, 12; Ishiga, 1990, pl. 1, fig. 7; Rudenko and Panasenko, 1990, pl. 10, figs. 5—9; Noble and Renne, 1990, pl. 1, figs. 12—15; Kuwahara and Sakamoto, 1992, pl. 3, figs. 8, 9, 12; Feng and Liu, 1993, pl. 4, figs. 7, 8; Caridroit, 1993, pl. 1, figs. 12, 13; Sashida et al., 1993, pl. 1, figs. 20, 22—24, 1995, Fig. 10 (16, 17, 21), 2000a, Fig. 3 (9—12), 2000b, pl. 1, figs. 11, 12, 14, 15; Takemura and Yamakita, 1993, pl. 1, fig. 4; Yao et al., 1993, pl. 1, fig. 8; Wang et al., 1994, pl. 4, figs. 11—13, 2006, Fig. 13 (K), Fig. 14 (K—M); Blome and Reed, 1995, pl. 1, fig. 3; Spiller and Metcalfe, 1995, Fig. 5 (4); Feng and Ye, 1996a, pl. 11.1, fig. 5, 1996b, pl. 2.2, fig. 9; Yao and Kuwahara, 1999b, pl. 3, fig. 20; Kuwahara, Yao and An, 1997, pl. 1, fig. 19; Xia et al., 2004, pl. 2, figs. 8, 9, pl. 3, fig. 1; Sun and Xia, 2006, pl. 3, figs. 7—10; Jin et al., 2007, Fig. 6 (1—4); Wu and Feng, 2008, Fig. 3 (1).

Albaillella gen. et sp. indet.; Takumura and Nakaseko, 1981, pl. 34, fig. 10.

Albaillella aff. *levis* Ishiga, Kito et Imoto; Kuwahara and Sakamoto, 1992, pl. 3, figs. 10, 11; Rudenko et al., 1997b, pl. 16.1, figs. 14, 19, 24, Yao and Kuwahara, 1999b, pl. 3, fig. 19.

Albaillella sp. cf. *A. levis* Ishiga, Kito and Imoto; Cheng, 1989, pl. 1, figs. 5—7, pl. 2, figs. 1—4.

Albaillella triangularis Ishiga, Kito and Imoto; Rudenko et al., 1997b, pl. 16.1, figs. 3, 4.

Imotoella levis (Ishiga, Kito et Imoto), Rudenko and Panasenko, 1997a, pl. 3, fig. 12 (non 11).

描述：这个种的特点是圆锥形壳较短，顶部向腹面轻微弯曲，壳面光滑，壳的下部发育1根尖头状腹刺。

比较：这个种与*Albaillella triangularis* Ishiga, Kito et Imoto 的区别为后者壳表分节。

层位与产地：上二叠统；世界各地。

曲状阿尔拜虫 *Albaillella sinuata* Ishiga et Watase
(Pl. 7, figs. 15—18; pl. 10, figs. 9—12)

Albaillella sinuata Ishiga et Watase, 1986, pl. 1, figs. 1—8; Sano, 1988, pl. 2, fig. 5; Ishiga, 1990, pl. 1, fig. 15; Ujiie and Ota, 1991, pl. 2, figs. 7, 8; Blome and Reed, 1992, Fig. 9 (6—9); Sashida et al., 1993, Fig. 6 (7), 1998, Fig. 11 (19); Takemura and Yamakita 1993, pl. 1, fig. 7; Wang et al., 1994, pl. 2, figs. 13, 14, 1998, pl. 3, figs. 1, 2, 2012, pl. 14, figs. 1—3, 17, 24—26, 46, 47, pl. 15, figs. 22, 23, pl. 17, figs. 1, 2; Nagai and Ishikawa, 1995, pl. 2, figs. 1—4; Zhang et al., 1998, pl. 1, figs. 11; Spiller, 2002, pl. 1, fig. 2; Yao et al., 2004, pl. 1, figs. 22, 23; Shimakawa and Yao, 2006, pl. 2, fig. a; Xia and Zhang, 1998, pl. 1, figs. 1—4.

Albaillella sp. D; Ishiga et al., 1982, pl. 1, fig. 4.

Albaillella sp. cf. *A. sinuata* Ishiga et Watase; Ishiga et al., 1990, pl. 2, fig. 7; Ishida et al., 1992, pl. 2, fig. 7; Spiller and Metcalfe, 1995, Fig. 5 (5), Spiller, 1996, pl. 4, fig. 1, Saesaengseerung et al., 2009, Fig. 7 (18, 19).

Albaillella aff. *sinuata* Ishiga et Watase; Tomooka et al., 1998, pl. 1, figs. 8, 9.

Spinodeflandrella sinuata (Ishiga et Watase); Rudenko and Panasenko, 1990, pl. 19, fig. 5.

描述：这个种的特点是锥形壳由8节或更多节组成，顶锥针形，壳波状，从腹边至背边，各节微斜，光滑。对称的杆形背刺和腹刺水平伸展。背杆和腹杆垂直向下延伸。

比较：这个种与 *Albaillella asymmetrica* Ishiga et Imoto 的区别为前者具有对称的杆形翼刺。

层位与产地：中二叠统伦纳德阶；世界各地。

三角形阿尔拜虫 *Albaillella triangularis* Ishiga, Kito et Imoto
(Pl. 14, figs. 42—49, 51, 55—57, 61—65)

Albaillella triangularis Ishiga, Kito et Imoto, 1982a, pl. 2, figs. 8—11, 1982b, pl. 2, figs. 17, 18; Wakita, 1983, pl. 6, fig. 8; Caridroit et al., 1985, pl. 1, fig. 1; Ishiga, 1985, pl. 2, figs. 13—19; Yoshida and Murata, 1985, pl. 2, figs. 1—5; Ishiga and Miyamoto, 1986, pl. 64, fig. 13; Sano, 1988, pl. 2, fig. 15; Wu and Li, 1989, pl. 1, fig. 14; Tumande et al., 1990, pl. 2, fig. 18; Kuwahara et al., 1991, Fig. 4 (1, 2); Kuwahara and Sakamoto, 1992, pl. 3, figs. 5, 6; Yao et al., 1993, pl. 1, fig. 3; Wang et al., 1994, pl. 4, figs. 9, 10, 2006, Fig. 13 (J, DD—FF), Fig. 14 (O—Q); Wang and Li, 1994, pl. 1, fig. 17; Caridroit and De Wever, 1986, pl. 1, figs. 1—5; Wang and Shang, 2001, pl. 1, figs. 17—23; Rudenko and Panasenko, 1990, pl. 10, figs. 9—11, Sashida et al., 1993, pl. 1, figs. 15—17, 19, 1995, Fig. 10 (18—20), 2000a, pl. 1, figs. 6—10, 13, 2000b, Fig. 7 (13—16); Yu Jie, 1996, pl. 1, fig. 10; Kuwahara et al., 1997, pl. 1, fig. 6; Yao and Kuwahara, 1996b, pl. 3, fig. 22, 2000, pl. 1, figs. 1—3; Xia et al., 2004, pl. 3, figs. 2, 3, 6—9; Kuwahara, 1999, pl. 3, figs. 6—9; Jin et al., 2007, Fig. 4 (1—6); Kuwahara, Yao and An, 1997, pl. 1, fig. 16.

Albaillella sp. C Ishiga et Imoto, 1980, pl. 5, figs. 11—16; Nishizono and Nakaseko, 1981, pl. 34, fig. 9.

Albaillella sp. cf. *A. triangularis* Ishiga, Kito et Imoto; Cheng, 1989, pl. 5, figs. 6—9, 11, 12; Blome and Reed, 1992, pl. 9, fig. 10, 1995, pl. 1, fig. 6; Yao and Kuwahara, 1981, pl. 34, fig. 9.

Albaillellidae gen. et sp. indet.; Takemura and Nakaseko, 1981, pl. 34, fig. 9.

Imotoella triangularis (Ishiga, Kito et Imoto); Kozur, 1999, pl. 2, fig. 13; Rudenko and Panasenko, 1997a, pl. 3, fig. 8.

Imotoella levis (Ishiga, Kito et Imoto); Rudenko and Panasenko, 1997a, pl. 3, fig. 11.

描述：这个种的特点是壳体三角形，短小，具有1个脊状的H框架，壳上横节清楚。

比较：这个种与 *Albaillella levis* Ishiga, Kito et Imoto 的区别为后者壳体光滑，无横节。

层位与产地：上二叠统；世界各地。

波状阿尔拜虫 *Albaillella undulata* Deflandre
(Pl. 17, fig. 3)

Albaillella undulata Deflandre, 1952, Figs. 8, 9, 1953b, pl. 7, fig. 13 (8, 9), 1960, pl. 1, fig. 24; Gourmelon, 1985, pl. 2, fig. 21; Braun, 1990, pl. 4, figs. 9, 10; Aitchison, 1993, pl. 1, fig. 9; Aitchison and Flood, 1990, Fig. 5 (A—G); Holdsworth, 1973, pl. 1, fig. 13; Aitchison et al., 1992, Fig. 8 (A).

Albaillella sp. aff. *A. undulata* Deflandre; Noble et al., 2008, Fig. 4 (3, 4)

Albaillella sp. cf. *A. undulata* Deflandre; Stewart et al., 1986, Fig. 1 (H, I, J).

Albaillella cf. *undelata* Deflandre; Feng et al., 1997, pl. 1, figs. 13, 14.

描述：这个种的特点是壳角锥形，分头节和后头节，壳边缘呈波状起伏。各节未见穿孔。H框架未见。

比较：这个种与 *Albaillella crenulata* Won 的区别为后者壳边缘呈锯齿状，每个节上发育1列圆孔。

层位与产地：下石炭统杜内阶；美国，法国，德国，澳大利亚，土耳其，中国云南和青海可可西里地区。

小董阿尔拜虫 *Albaillella xiaodongensis* Wang

(Pl. 7, figs. 3, 11—14)

Albaillella xiaodongensis Wang, 1994, pl. 2, figs. 11, 12; Wang et al., 1998, pl. 3, figs. 14, 15, 2012, pl. 15, figs. 8—11, 24—28, 32, 39, pl. 16, figs. 1, 5—10; Shimakawa and Yao, 2006, pl. 2, figs. 2, 3.

Albaillella asymmetrica m. Ⅱ Ishiga et Imoto; Xia and Zhang, 1998, pl. 1, figs. 12, 13.

描述：这个种的特点是壳呈宝塔状，小，顶锥直或向腹边微曲。壳体分节，主体由7—9个水平环状节和凹带组成，每个环状节具有1对规则的梯状侧翼，凹带和侧翼间发育大的多边形或椭圆形孔。

比较：这个种与 *Albaillella sinuata* Ishiga et Watase 的区别为前者壳小，具有规则的梯状侧翼。这个种将成为中二叠世栖霞早期一个标准放射虫带种。

层位与产地：中二叠统栖霞阶下部；中国广西和青海可可西里地区。

丑巾虫超科 Follicucullacea Cheng, 1986

假阿尔拜虫科 Pseudoalbaillellidae Cheng, 1986

假阿尔拜虫属 *Pseudoalbaillella* Holdsworth et Jones, 1980

模式种：*Pseudoalbaillella scalprata* Holdsworth et Jones, 1980

鉴定要点：壳体两侧对称，不穿孔，由三部分组成：顶锥、假胸节和假腹节。顶锥和假腹节有时进一步分节。假胸节膨大，有两翼、一翼或无翼。

时代与分布：早中二叠世；世界各地。

长形假阿尔拜虫 *Pseudoalbaillella elongata* Ishiga et Imoto

(Pl. 8, figs. 1—5; pl. 16, figs. 1, 2)

Pseudoalbaillella elongata Ishiga et Imoto, 1980, pl. 4, figs. 1—4; Ishiga et al., 1982b, pl. 1, figs. 15, 16; Spiller and Metcalfe, 1995b, Fig. 5 (10); Spiller, 1996, pl. 4, figs. 3, 4, 2002, pl. 3, fig. L; Ujiie and Oba, 1991, pl. 2, figs, 5, 6; Li and Bai, 1993, pl. 3, fig. 13.

Pseudoalbaillella cf. *elongata* Ishiga et Imoto; Li and Bai, 1993, pl. 3, fig. 12.

Pseudoalbaillella (*Kitoconus*) *elongata* Ishiga et Imoto; Catalano et al., 1991, pl. 2, figs. 6, 11, 17.

描述：这个种的特点是假腹节长，圆柱形，未分节。假胸节小，微凸，具2个锥形翼。顶锥未分节。

比较：这个种与 *Pseudoalbaillella elegans* Ishiga et Imoto 的区别为后者假腹节细长，微弯曲。

层位与产地：下二叠统狼营阶上部；日本，马来西亚，意大利西西里岛，中国青海可可西里地区。

纺锤形假阿尔拜虫 *Pseudoalbaillella fusiformis* (Holdsworth et Jones)

(Pl. 4, figs. 11—16)

Pseudoalbaillella fusiformis (Holdsworth et Jones); Ishiga et al., 1982b, pl. 2, figs. 1, 2, 1982c, pl. 4, figs. 10, 11, 1986, pl. 1, figs. 16—18; Tazawa et al., 1984, Fig. 2 (1, 5, 8); Yoshida and Murata, 1985, pl. 1, fig. 14; Nishimura and Ishiga, 1987, pl. 3, figs. 1—7; Blome et al., 1986, pl. 8, 1, figs. 3—6, 11; Murchey, 1990, pl. 1, fig. 4; Ishiga, 1990, pl. 1, fig. 12; Tumanda et al., 1990, pl. 1, figs. 2, 3; Wang, 1991, pl. 3, fig. 3; Blome and Reed, 1992, Fig. 9 (21—23), Fig. 10 (1—4); Wang et al., 1994, pl. 2, figs. 5, 6, 1998, pl. 3, figs. 16, 17, 2012, pl. 14, fig. 31, pl. 16, figs. 3, 4, 13, 14, pl. 17, figs. 6, 7, 9, 10; Wu et al., 1994, pl. 2, fig. 14; Wang and Qi, 1995, pl. 1, figs. 6—8, 12, pl. 4, figs. 12, 13; Nagai and Zhu, 1992, pl. 2, figs. 10—12; Feng et al., 1996, pl. 11.1, fig. 2; Spiller, 1996, pl. 4, figs. 9, 10, 2002, pl. 3, figs. N, O; Nagai et al., 1998, pl. 1, fig. 7; Xia and Zhang, 1998, pl. 3, figs. 1—5; Yao and Kuwahara, 1999a, pl. 1, figs. 1, 2, 1999c, pl. 1, figs. 1, 5—11,

13; Takemura et al., 1999, Fig. 4 (I); Kawahara and Yao, 1998, pl. 1, fig. 34, 2001 pl. 1, fig. 3; Kuwahara and Yamakita, 2001, pl. 1, fig. 1; Kametaka et al., 2009, Fig. 6 (6, 7); Wang and Yang, 2003, pl. 1, figs. 20—23, 26—30; Yao et al., 2004, pl. 2, fig. 12; Ito et al., 2013, Fig. 4 (1—3).

Pseudoalbaillella nanjingensis Sheng et Wang, 1985, pl. 1, figs. 1—5, 7—10; Yao and Kuwahara, 1999c, pl. 1, fig. 12.

Pseudoalbaillella cf. *fusiformis* (Holdsworth et Jones); Nishimura and Ishiga, 1987, pl. 1, figs. 13—18; Kozur et al., 1987, Fig. 2; Ishida et al., 1992, pl. 2, fig. 4; Wang R J, 1993b, pl. 1, figs. 14, 15; Yao et al., 2004, pl. 1, fig. 15; Kuwahara et al., 2008, Fig. 8 (13, 14).

Pseudoalbaillella aff. *fusiformis* (Holdsworth et Jones); Blome et al., 1986, pl. 8, fig. 12; Sano, 1988, pl. 2, fig. 8.

Pseudoalbaillella sakmarensis (Kozur); Xia and Zhang, 1998, pl. 2, figs. 15—18.

Follicucullus monacanthus Ishiga et Imoto; De Wever et al., 1988, Fig. 2 (D).

Parafollicucullus fusiformis Holdsworth et Jones, 1980, Fig. 1 (D, E); Harms and Murchey, 1992, pl. 1, fig. E.

Pseudoalbaillella convexus Rudenko et Panasenko; Xia and Zhang, 1998, pl. 3, figs. 13—15; Wang R J, 1995, pl. 1, fig. 3.

Pseudoalbaillella sp. nov. Rudenko et al., 1997b, pl. 16, fig. 2 (non 1).

Pseudoalbaillella delawarensis Maldonado et Noble, 2010, pl. 3, figs. 3—9; Nestell and Nestell, 2010, pl. 2, figs. 23—25.

描述：这个种的特点是壳体分节。顶锥弯曲，未分节。假胸节小，近球形，具2个侧翼。假腹节3节，首节短小，瓶颈状，中节凸，桶状，第3节短，裙边状。

比较：这个种与 *Pseudoalbaillella monacantha* (Ishiga et Imoto) 的区别为后者只发育1个背翼。这个种已成为中二叠世瓜德鲁普期早期一个标准化石带种。

讨论：最近，Maldonado 和 Noble（2010）建立的新种 *Pseudoalbaillella delawarensis* 的特点是具有1个光滑的、无节的、桶形假腹节。但从发表的图影观察，其中 pl. 3, figs. 3, 9 的假腹节明显地由短的首节、桶形中节和裙边状末节3节组成，这些特征与 *Pseudoalbaillella fusiformis* (Holdsworth et Jones) 的假腹节特征一致，因此，我们把这个新种视作 *Pseudoalbaillella fusiformis* 的同义名。

层位与产地：中二叠统瓜德鲁普阶；世界各地。

球形假阿尔拜虫 *Pseudoalbaillella globosa* Ishiga et Imoto

(Pl. 3, figs. 42—44, 47—49; pl. 4, figs. 17—30)

Pseudoalbaillella globosa Ishiga et Imoto, 1982b, pl. 2, figs. 3, 4, 1982c, pl. 1, figs. 1—8; Ishiga et al., 1986, pl. 2, figs. 1—3; Nishimura and Ishiga, 1987, pl. 4, figs. 1—5; Sano, 1988, pl. 2, fig. 10; Ishiga, 1990, pl. 1, fig. 14; Rudenko and Panasenko, 1990, pl. 9, fig. 3, 1997, pl. 2, figs. 5, 6; Ishida et al., 1992, pl. 2, fig. 6; Wang et al., 1994, pl. 2, figs. 9, 10, 1998, pl. 3, figs. 10, 11, 2012, pl. 16, figs. 11, 12, 33, 34, pl. 17, fig. 8; Spiller, 2002, pl. 3, fig. 9; Kuwahara and Yao, 1998, pl. 1, fig. 35; Xia et al., 2004, pl. 2, fig. 22; Xie et al., 2011, Fig. 2 (F, K, L); Xia and Zhang, 1998, pl. 3, figs. 6—12; Yao et al., 2004, pl. 2, figs. 9, 11; Kametaka et al., 2009, Fig. 6 (10); Sun and Xie, 2006, pl. 2, fig. 3; Wang and Yang, 2003, pl. 1, figs. 24, 25.

Parafollicucullus globosa (Ishiga et Imoto); Cornell and Simpson, 1985, pl. 1, figs. 4, 8; Kozur and Krahl, 1987, Fig. 3; Harms and Murchey, 1992, pl. 1, fig. 1.

Pseudoalbaillella sp. B Ishiga et Imoto; Yao and Kuwahara, 1999a, pl. 1, figs. 3, 4, 1999b, pl. 3, fig. 1, 1999c, pl. 1, figs. 14, 15; Yao et al., 2004, pl. 1, figs. 18, 19; Kuwahara et al., 2007a, pl. 1, figs. 4—6, 2007b, Fig. 7 (8, 9).

Pseudoalbaillella sp. aff. *P. globosa* Ishiga et Imoto; Blome and Reed, 1992, Fig. 10 (5, 6); Kametaka et al., 2009, Fig. 6 (11, 12).

描述：这个种的特点是壳体分节。顶锥强壮，微曲，分节或未分节。假胸节大，球形，具2根鸟嘴状侧翼，向下延伸。假腹节2节，一般较短。背杆和腹杆短小，向下伸展。

比较：这个种与 *Pseudoalbaillella scalprata* (Holdsworth et Jones) 外形十分相似，区别在于后者假腹节只有 1 节。这个种已成为中二叠世瓜德鲁普期一个标准化石带种。

层位与产地：中二叠统瓜德鲁普阶；世界各地。

石贺裕明假阿尔拜虫 *Pseudoalbaillella ishigai* Wang
(Pl. 10, figs. 15, 16, 18, 20—22, 24, 27, 30, 31, 41)

Pseudoalbaillella sp. C Ishiga, Kito and Imoto, 1982b, pl. 1, fig. 19, 1982c, pl. 4, figs. 8, 9; Ishiga et al., 1986, pl. 1, fig. 19; Sano, 1988, pl. 2, fig. 7; Wang, 1991, pl. 3, fig. 1; Rudenko and Panasenko, 1997, pl. 2, fig. 1.

Pseudoalbaillella ishigai Wang, 1994, pl. 2, figs. 1, 2; Wang and Qi, 1995, pl. 1, fig. 1; Wang et al., 1998, pl. 3, figs. 5, 6, 2012, pl. 14, figs. 4—7, 13, 15, 16, 19—23, 37—44, pl. 15, figs. 12—14.

Pseudoalbaillella banchengensis Xia et Zhang, 1998, pl. 2, figs. 1—6.

Pseudoalbaillella longtanensis Sheng et Wang; Wang et al., 1998, pl. 3, figs. 8, 9; Xie et al., 2011, Fig. 2 (Q, R); Sun and Xie, 2006, pl. 2, figs. 1, 2.

描述：这个种的特点是壳体分节。顶锥未分节。假胸节小，近球形，具 2 个平的侧翼。假腹节分 5 节，首节比其他各节小，其余各节圆柱形，长度相似，末节裙边状。背杆向下向腹边倾斜，腹杆向下伸展。

比较：这个种与 *Pseudoalbaillella longtanensis* Sheng et Wang 的区别为后者假腹节只由 4 节组成。这个种已成为中二叠世伦纳德期晚期一个标准化石带种。

讨论：根据 Xie 等 (2011)、Sun 和 Xie (2006) 所鉴定的 *Pseudoalbaillella longtanensis* Sheng et Wang 图影，假腹节具有 5 节，因此，本书把它归入 *Pseudoalbaillella ishigai* Wang 之中。

时代与分布：中二叠统伦纳德阶；日本，俄罗斯远东地区，中国华南、青海和西藏。

理塘假阿尔拜虫（新种） *Pseudoalbaillella litangensis* Wang sp. nov.
(Pl. 3, figs. 31—40)

词源：litang，标本产地，四川省理塘县。

描述：壳锥形，分节，由三部分组成：顶锥较强壮，不分节；假胸节小，微凸，具 2 根短翼；假腹节 3 节，首节较短，中节桶形，末节圆柱形。首节和中节间、中节和末节间收缩明显。背杆和腹杆未保存。

比较：这一新种形状和构造与 *Pseudoalbaillella fusiformis* (Holdsworth et Jones) 十分相似，区别在于后者假腹节的末节裙边状。

层位与产地：中二叠统茅口阶下部；中国四川理塘地区。

豆荚状假阿尔拜虫 *Pseudoalbaillella lomentaria* Ishiga et Imoto
(Pl. 11, figs. 7—18; pl. 16, figs. 3—9)

Pseudoalbaillella lomentaria Ishiga et Imoto, 1980, pl. 2, figs. 9—15; Ishiga et al., 1982b, pl. 1, figs. 4—6; Sano, 1988, pl. 2, fig. 1; Ujiie and Ota, 1991, pl. 1, figs. 2, 3; Kuwahara, 1992, pl. 2, fig. 14; Sashida et al., 1993, Fig. 6 (5, 6), 1998, Fig. 11 (16—18), 2002, Fig. 3 (4); Wu et al., 1994, pl. 2, fig. 8; Wang et al., 1994, pl. 1, figs. 12, 13, 2012, pl. 11, figs. 19, 20, pl. 12, figs. 21—24, 45—47, pl. 13, figs. 1—3, 16—18, 41, pl. 15, figs. 1, 2, 4; Sashida, 1995, Fig. 1 (18, 19); Feng and Liu, 2002, pl. 1, figs. 7, 8; Shimakawa and Yao, 2006, pl. 1, figs. 12, 13.

Pseudoalbaillella sp. cf. *P. lomentaria* Ishiga et Imoto; Blome and Reed, 1992, pl. 10, figs. 7, 8; Rudenko and Panasenko, 1997, pl. 1, fig. 2; Spiller, 2002, pl. 3, fig. Q.

Parafollicucullus lomentaria (Ishiga et Imoto); Saesaengseerung et al., 2009, Fig. 7 (14).

描述：这个种的特点是顶锥分节或未分节。假胸节球形，具 2 个平缓的侧翼。假腹节圆柱形，有 3 节。

比较：这个种与 *Pseudoalbaillella nodosa* Ishiga 的区别为后者假腹节的各节向背边斜交。

层位与产地：下二叠统狼营阶或萨克马尔阶；美国，日本，泰国，马来西亚，俄罗斯远东地区，中国华南和西藏双湖地区。

龙潭假阿尔拜虫 *Pseudoalbaillella longtanensis* Sheng et Wang

(Pl. 4, figs. 1—10)

Pseudoalbaillella longtanensis Sheng et Wang, 1985, pl. 2, figs. 3, 4; Ishiga, 1990, pl. 1, fig. 18; Wang, 1991, pl. 3, fig. 2; Wang R J, 1993a, pl. 1, figs. 3—5, 1993b, pl. 1, fig. 2, 1995, pl. 1, fig. 2; Wang et al., 1994, pl. 2, figs. 3, 4, 1998, pl. 3, fig. 7 (non 8, 9), 2012, pl. 15, figs. 29—31, 38, pl. 16, figs. 35—37; Wang and Qi, 1995, pl. 1, figs. 2—5; Spiller and Metcalfe, 1995b, Fig. 5 (14); Spiller, 1996, pl. 4, figs. 11, 12, 2002, pl. 3, fig. T; Maldonado and Noble, 2010, pl. 3, figs. 10—14; Kametaka et al., 2009, Fig. 6 (3—5); Sun and Xie, 2006, pl. 2, figs. 1, 2; Xia and Zhang, 1998, pl. 1, figs. 14—17, pl. 3, figs. 16—20.

Pseudoalbaillella corniculata Rudenko et Panasenko, 1997, pl. 2, figs. 3, 4.

Pseudoalbaillella sp. aff. *P. longtanensis* Sheng et Wang; Nishimura and Ishiga, 1987, pl. 3, figs. 8—12; Yao and Kuwahara, 1999b, pl. 3, fig. 3; Xia and Zhang, 1998, pl. 1, figs. 18, 19.

Pseudoalbaillella sp. cf. *P. longtanensis* Sheng et Wang; Ishiga et al., 1992, pl. 2, fig. 5; Kuwahara et al., 2007a, pl. 1, fig. 1, pl. 2, figs. 7—9, 2007b, Fig. 7 (15); Wang and Fan, 1997, pl. 1, figs. 5, 6.

描述：这个种的特点是壳体分节。顶锥分节或未分节。假胸节小，近球形，具 2 个侧翼。假腹节由 4 节组成，首节的宽度和长度都比其他节小。这个种已成为中二叠世瓜德鲁普期早期一个标准化石带种。

层位与产地：中二叠统瓜德鲁普阶下部；日本，马来西亚，俄罗斯远东地区，中国华南和青海可可西里地区。

单翼假阿尔拜虫（新种） *Pseudoalbaillella monopteryla* Wang sp. nov.

(Pl. 10, figs. 17, 19, 23, 25, 26, 33—36, 38, 40, 42)

词源：mono, 拉丁词，单；pteryl, 拉丁词，翼。

描述：壳体分节。顶锥未分节。假胸节近球形，具有 1 根侧翼。假腹节有 5 节，首节小，其他 4 节较宽，各节高度相近，宽度渐宽。背杆和腹杆未保存。

比较：这一新种与 *Pseudoalbaillella ishigai* Wang 在形状和构造上十分相似，区别在于后者有 2 个侧翼；这个新种具有单翼与 *Pseudoalbaillella monacantha* 相同，区别在于后者的假腹节只有 3 节。

层位与产地：中二叠统伦纳德阶；中国西藏双湖地区。

无翼假阿尔拜虫（新种） *Pseudoalbaillella nonpteryla* Wang sp. nov.

(Pl. 10, figs. 28, 29, 32, 37, 39, 43—48)

词源：non, 拉丁词，无；pteryl, 拉丁词，翼。

描述：壳体分节。顶锥未分节。假胸节椭圆形，无明显侧翼，但有轻微的突起。假腹节有 5 节，首节小，其他各节宽度逐渐增大，高度不等。背杆和腹杆未保存。

比较：新种以其假胸节上无明显双翼且有轻微突起区别于这个属的其他种。

层位与产地：中二叠统伦纳德阶；中国西藏双湖地区。

后锐边假阿尔拜虫 *Pseudoalbaillella postscalprata* Ishiga

(Pl. 15, figs. 13—17, 19, 24)

Pseudoalbaillella scalprata m. *postscalprata* Ishiga, 1983, pl. 2, figs. 1—16; Hori, 2004, Fig. 1 (5, 6); Yao et al., 2004, pl. 1, fig. 14; Xie et al., 2011, Fig. 2 (B, C).

Pseudoalbaillella scalprata m. *scalprata* Ishiga, 1983, pl. 1, figs. 5—7, 13.

Pseudoalbaillella sp. F Ishiga et al., 1982, pl. 1, fig. 13.

Pseudoalbaillella rhombothoraeata Ishiga et Imoto, 1980, pl. 3, fig. 12; Yamanaka, 2001, pl. 1, figs. 11—13.

描述：这个种的特点是顶锥未分节，向腹边轻微弯曲。假胸节微凸，近菱形，具背翼和腹翼。假腹节较长。顶锥和假胸节间、假胸节和假腹节间具弱缢。

比较：这个种与 *Pseudoalbaillella scalprata* Holdsworth et Jones 的区别为后者假胸节三角形，假腹节较短。

层位与产地：下二叠统；日本，中国广西和四川。

菱形假阿尔拜虫 *Pseudoalbaillella rhombothoracata* Ishiga et Imoto

(Pl. 7, figs. 4—10; pl. 8, figs. 6—17, 35)

Pseudoalbaillella rhombothoracata Ishiga et Imoto, 1980, pl. 3, figs. 9—11 (non 12); Ishiga, Kito and Imoto, 1982c, pl. 1, fig. 14; Nishizono et al., 1982, pl. 2, fig. 1; Yoshida and Murata, 1985, pl. 1, fig. 10; Rudenko and Panasenko, 1997, pl. 1, fig. 5; Liu et al., 1991, pl. 1, figs. 11—13; Wang et al., 2012, pl. 14, figs. 9—12, 18, 45, pl. 15, figs. 15—21.

Pseudoalbaillella scalprata m. *rhombothoracata* Ishiga, 1983, pl. 3, figs. 1—12, 1990, pl. 2, fig. 4; Wu et al., 1994, pl. 2, fig. 10; Wang et al., 1994, pl. 1, figs. 14, 15, 1998, pl. 3, figs. 3, 4; Kurihara and Sashida, 1998, pl. 1, fig. 19; Feng and Ye, 1996b, pl. 2, 3, figs. 3, 5; Spiller, 1996, pl. 4, fig. 2, 2002, pl. 4, figs. F, G; Xie et al., 2011, Fig. 2 (D, E).

Pseudoalbaillella sp. cf. *P. rhombothoracata* Ishiga et Imoto; Sano, 1988, pl. 2, fig. 4; Wang R J, 1993a, pl. 1, figs. 11, 12.

描述：这个种的特点是壳体分节。顶锥未分节。假胸节菱形，具2个侧翼。假腹节长，未分节，与假胸节间无缢。

比较：这个种与 *Pseudoalbaillella scalprata* Holdsworth et Jones 的区别为后者假胸节三角形或近球形，假腹节较短。

层位与产地：下二叠统狼营阶上部；日本，马来西亚，俄罗斯远东地区，中国华南和青海可可西里地区。

萨克马尔假阿尔拜虫 *Pseudoalbaillella sakmarensis* (Kozur)

(Pl. 8, figs. 18—32; pl. 11, figs. 5, 6, 19—39, 41; pl. 15, figs. 1—9)

Parafollicucullus sakmarensis Kozur, 1981, pl. 1, figs. 1, 3.

Pseudoalbaillella sakmarensis (Kozur); Ishiga, Kito and Imoto, 1982b, pl. 1, fig. 8; Kojima, 1982, Fig. 4 (3, 5, 6); Nishizono and Murata, 1983, pl. 1, fig. 2; Ishiga, 1985, pl. 1, figs. 2, 3; Yoshida and Murata, 1985, pl. 1, figs. 4, 5; Ishiga, 1990, pl. 2, fig. 8; Ujiee and Oba, 1991, pl. 1, fig. 1; Kuwahara, 1992, pl. 2, fig. 13; Wang et al., 1994, pl. 1, figs. 9—11, 2012, pl. 11, figs. 3, 4, 36—38, pl. 12, figs. 11—13, pl. 13, figs. 4—8, 32—34, pl. 15, figs. 3, 40—43; Wang, 1995, pl. 15, figs. 1—8, pl. 16, figs. 7—9; Spiller and Metcalfe, 1995, Fig. 5 (18); Feng et al., 1996, pls. 2, 3, fig. 2; Spiller, 1996, pl. 3, fig. 3, 2002, pl. 4, figs. D, E; Feng and Liu, 2002, pl. 1, figs. 3, 4, 6; Rudenko and Panasenko, 1997a, pl. 1, figs. 3, 4; Yao et al., 2004, pl. 2, figs. 2, 6, 7; Shimakawa and Yao, 2006, pl. 1, figs. 14, 15.

Pseudoalbaillella sp. cf. *P. sakmarensis* (Kozur); Samo, 1988, pl. 2, fig. 3.

Pseudoalbaillella cf. *sakmarensis* (Kozur); Wu and Li, 1989, pl. 1, fig. 17; Isozaki and Tamura, 1989, pl. 1, fig. 10; Sano, 1988, pl. 2, fig. 3.

Pseudoalbaillella aff. *sakmarensis* (Kozur); Kurihara and Sashida, 1998, figs. 15, 16.

Pseudoalbaillella sp. A Ishiga, Kito et Imoto, 1980, pl. 2, figs. 16—19, pl. 3, figs. 1, 2; Nishizono et al., 1982, Fig. 2 (2).

描述：这个种的特点是顶锥分节或未分节。假胸节近球形，具2个不对称的侧翼。假腹节长，有3节，最后一节强烈弯曲。

比较：这个种与*Pseudoalbaillella chilensis* Ling et Forsythe 的区别为后者假腹节不分节。

层位与产地：下二叠统狼营阶或萨克马尔阶；日本，马来西亚，俄罗斯南乌拉尔地区和远东地区，中国华南和青海可可西里地区。

锐边假阿尔拜虫 *Pseudoalbaillella scalprata* Holdsworth et Jones

(Pl. 3, figs. 41, 45, 46, 50; pl. 4, figs. 31—37; pl. 15. figs. 10—12, 18, 20—23)

Pseudoalbaillella scalprata Holdsworth et Jones, 1980, Figs. A, B; Ishiga et al., 1982c, pl. 1, figs. 11, 12; Yoshida and Murata, 1985, pl. 1, figs. 8, 9; Cornell and Simpson, 1985, pl. 1, fig. 5; Sheng and Wang, 1985, pl. 2, figs. 9—12; Steward et al., 1988, Fig. 1 (D); Rudenko and Panasenko, 1990, pl. 19, fig. 4, 1997, pl. 1, fig. 6; Wang, 1991, pl. 3, fig. 10; Liu et al., 1991, pl. 1, figs. 7, 8; Wang R J, 1993a, pl. 1, figs. 12, 13, 1993b, pl. 1, figs. 1, 2, 1995, pl. 1, fig. 1; Ishida et al., 1992, pl. 2, fig. 8; Nazarov and Ormiston, 1993, pl. 7, fig. 10; Wang et al., 1994, pl. 1, figs. 20—22, 2012, pl. 12, figs. 19, 20, pl. 14, figs. 8, 14, 32—36, pl. 15, figs. 7, 44, 45; Sashida, 1995, Fig. 5 (16, 17); Sashida et al., 1998, Fig. 11 (11, 22, 23), 2002, Fig. 3 (6, 7); Tomoka et al., 1998, pl. 1, fig. 5; Yamanaka, 2001, pl. 1, figs. 4, 5; Shimakawa and Yao, 2006, pl. 1, figs. 16—18.

Pseudoalbaillella scalprata m. *scalprata* Holdsworth et Jones; Ishiga, 1983, pl. 1, figs. 1—4, 8—10, 14—18; Blome and Reed, 1992, Fig. 10 (13—17); Wu et al., 1994, pl. 2, fig. 11; Spiller, 1996, pl. 3, figs. 6, 7, 2002, pl. 4, figs. H, I; Feng and Ye, 1996a, pl. 11.1, fig. 1, 1996b, pl. 2.3, fig. 1; Spiller and Metcalfe, 1995, Fig. 5 (19, 20); Yao et al., 2004, pl. 1, figs. 12, 13, pl. 2, figs. 4, 5, 8; Wang et al., 1994, pl. 1, fig. 22; Miyamoto et al., 1997, pl. 1, figs. 1—3; Hori, 2004, Figs. 1 (4), 2 (1); Xie et al., 2011, Fig. A (non S); Feng and Liu, 2002, pl. 1, figs. 1, 2; Saesaengseerung et al., 2009, Fig. 7 (28, 29).

Pseudoalbaillella sp. cf. *P. scalprata* Holdsworth et Jones; Ishiga and Imoto, 1988, pl. 2, figs. 4—8.

Pseudoalbaillella sp. aff. *P. scalprata* Holdsworth et Jones; Ishiga et al., 1982b, pl. 2, fig. 7, 1984, pl. 1, figs. 23—25; Sano, 1988, pl. 2, fig. 2; Miyamoto and Tanimoto, 1992, pl. 2, figs. 13—15; Miyamoto et al., 1997, pl. 1, figs. 1—3.

Pseudoalbaillella cf. *scalprata* m. *postscalprata* Ishiga; Sashida et al., 1993, Fig. 6 (10, 11, 14).

Pseudoalbaillella globosa Ishiga et Imoto; Wang et al., 2003, pl. 1, figs. 24, 25.

描述：这个种的特点是壳体分节。顶锥小，未分节，明显向腹边倾斜。假胸节近球形，具2个平缓的侧翼，与假腹节间无缢。假腹节短，未分节。背杆和腹杆向下延伸。

比较：这一种与*Pseudoalbaillella postscalprata* Ishiga 的区别为后者假腹节较长，假胸节和假腹节间有缢；与*Pseudoalbaillella globosa* Ishiga et Imoto 的区别为后者假胸节为凸球形，假腹节有2节。

层位与产地：中二叠统伦纳德阶—瓜德鲁普阶；世界各地。

简单假阿尔拜虫 *Pseudoalbaillella simplex* Ishiga et Imoto

(Pl. 8, figs. 33, 34, 36—40)

Pseudoalbaillella simplex Ishiga et Imoto, 1980, pl. 1, figs. 13—18; Hattori and Yoshimura, 1982, pl. 1, fig. 2; Ishiga et al., 1984, pl. 1, figs. 17—22; Ling et al., 1985, Fig. 3 (L, M); Ling and Forsythe, 1987, pl. 1, figs. 10, 11; Yoshida and Murata, 1985, pl. 1, fig. 1; Ishiga, 1990, pl. 2, fig. 3; Kuwahara, 1992, pl. 2, figs. 10, 11; Wang et al., 1994, pl. 1, figs. 3, 4, 1998, pl. 2, figs. 16, 17, 2012, pl. 11, figs. 28, 29; Sashida, 1995, Fig. 5 (1—6); Tomooka et al., 1998, pl. 1, fig. 4; Sashida et al., 2002, Fig. 3 (3); Spiller, 2002, pl. 4, fig. J; Yao et al., 2004, pl. 1, fig. 7; Xie et al., 2011, Fig. 2 (I, J).

G. et sp. indet.; De Wever et al., 1984, pl. 1, figs. 14, 15, 26, 28.

描述：这个种的特点是壳体小，简单。顶锥未分节，长，约占壳长的1/2。假胸节微凸，具2个小的侧翼。假腹节1节，短小，与假胸节间有1个缢。

比较：这个种与 *Pseudoalbaillella scalprata* Holdsworth et Jones 的区别为前者壳体较小，假胸节和假腹节间有1个缢。

层位与产地：下二叠统狼营阶或萨克马尔阶；世界各地。

新阿尔拜虫科 Neoalbaillellidae Cheng, 1986
新阿尔拜虫属 *Neoalbaillella* Takemura et Nakaseko, 1981

模式种：*Neoalbaillella ornithoformis* Takemura et Nakaseko, 1981

鉴定要点：壳锥形或亚圆柱形，近乎两侧对称。顶锥弯曲。壳的上部具无孔的两翼，下部具横向排列的窗孔。壳的下方发育背杆和腹杆，杆上有2—3根（或更多）侧刺。

时代与分布：晚二叠世；日本，美国西海岸，菲律宾，马来西亚，泰国，俄罗斯远东地区，中国华南、云南和西藏。

纤细新阿尔拜虫 *Neoalbaillella gracilis* Takemura et Nakaseko

(Pl. 13, figs. 15—19)

Neoalbaillella gracilis Takemura et Nakaseko, 1981, pl. 33, figs. 7—10, pl. 34, fig. 1; Wang et al., 1994, pl. 4, figs. 4, 5, 2006, Fig. 14 (C, D), Fig. 15 (E—I); Feng and Liu, 1993, pl. 4, figs. 5, 6; Kuwahara and Yao, 1998, pl. 1, fig. 27.

Neoalbaillella sp. cf. *N. gracilis* Takemura et Nakaseko; Ishiga, Kito and Imoto, 1982a, pl. 2, figs. 2, 3, 1982b, pl. 2, fig. 15.

描述：这个种的特点是壳体细弱，顶锥未穿孔，直或向腹边轻微弯曲。双翼细弱。具横向排列的窗孔，窗孔长方形。

比较：这个种与 *Neoalbaillella ornithoformis* Takemura et Nakaseko 的区别为后者的两翼特别强壮，壳体和窗孔较大。

层位与产地：上二叠统；日本，中国华南和西藏双湖地区。

最优新阿尔拜虫 *Neoalbaillella optima* Ishiga, Kito et Imoto

(Pl. 13, figs. 10, 24)

Neoalbaillella optima Ishiga, Kito et Imoto, 1982a, pl. 1, figs. 1—5, 1982b, pl. 2, figs. 12, 13; Nishizono and Murata, 1983, pl. 1, fig. 12; Ishiga, 1990, pl. 1, fig. 6; Tumanda et al., 1990, pl. 2, fig. 17; Wang et al., 1994, pl. 4, figs. 1—4, 2006, pl. 13, fig. B, pl. 14, figs. A, B; Yao and Kuwahara, 1996b, pl. 3, fig. 24; Sashida et al., 2000a,

Fig. 7 (1—5), 2000b, pl. 1, figs. 1—4; Wakita, 1983, pl. 5, figs. 1—3; Kuwahara et al., 1991, Fig. 4 (4); Yao, Yu and An, 1993, pl. 2, fig. 9; Rudenko and Panasenko, 1990, pl. 11, fig. 4; Kuwahara and Yao, 1998, pl. 1, fig. 24, 2001, pl. 4, fig. 3; Yu Jie, 1996, pl. 1, fig. 12; Feng and Liu, 1993, pl. 4, figs. 1, 2; Jin et al., 2007, Fig. 6 (7—10); Wu and Feng, 2008, Fig. 3 (4).

Neoalbaillella sp. cf. *N. optima* Ishiga, Kito et Imoto, Nishizono et al., 1982, pl. 2, fig. 8; Sano, 1988, pl. 2, fig. 17; Cheng, 1989, pl. 4, figs. 13—16, pl. 5, figs. 1—3; Yao and Kuwahara, 2000, pl. 1, fig. 4; He et al., 2011, Fig. 5 (M, N).

Neoalbaillella sp. B Takemura et Nakaseko, 1981, pl. 34, figs. 4, 5.

Neoalbaillella sp. Cheng, 1989, pl. 5, fig. 4.

Neoalbaillella sp. cf. *N. ornithoformis* Takemura et Nakaseko; Ishida, Yamashita and Ishiga, 1992, pl. 1, fig. 3.

Neoalbaillella cribrosa Rudenko et Panasenko, 1990, pl. 11, fig. 5, 1997a, pl. 3, fig. 15.

Neoalbaillella aff. *optima* Ishiga, Kito et Imoto; Kuwahara and Yao, 1998, pl. 1, fig. 25.

Neoalbaillella ornithofomis Takemura et Nakaseko; Feng and Liu, 1993, pl. 4, figs. 3, 4.

描述：这个种的特点是顶锥未穿孔，微向腹边弯曲。壳体具有梯形两翼和8—10列横向排列的窗孔，每横列有8—12个窗孔。当前标本由于保存欠佳，梯状两翼不完整。

层位与产地：上二叠统；日本，菲律宾，泰国，俄罗斯远东地区，中国华南和西藏双湖地区。

鸟形新阿尔拜虫　*Neoalbaillella ornithoformis* Takemura et Nakaseko
(Pl. 12, figs. 31—34; pl. 13, figs. 1—5, 7, 8, 11—13, 20—22, 25—27)

Neoalbaillella ornithoformis Takemura et Nakaseko, 1981, pl. 33, figs. 1—6; Nishizono et al., 1982, pl. 2, fig. 7; Ishiga, Kito and Imoto, 1982a, pl. 1, figs. 6—8, pl. 2, fig. 1, 1982b, pl. 2, figs. 14, 16; Nishizono and Murata, 1983, pl. 1, fig. 11; Sashida and Tonishi, 1985, pl. 7, figs. 8, 9; Tumanda et al., 1990, pl. 2, fig. 21; Ishiga et al., 1992, pl. 1, fig. 2; Ishiga, 1990, pl. 1, fig. 5; Ishida et al., 1992, pl. 1, fig. 2; Wang et al., 1994, pl. 4, figs. 6—8, 2006, Figs. 13 (A), 14 (E, F), 15 (A—D); Feng and Liu, 1993, pl. 4, figs. 3, 4; Yu, 1996, pl. 1, fig. 11; Rudenko et al., 1997b, pl. 16, figs. 5, 7 (non 6); Sashida et al., 2000, Fig. 7 (6, 7); Feng and Ye, 1996a, pl. 11, fig. 14, 1996b, pl. 2.2, fig. 3; Xia et al., 2004, pl. 2, fig. 1; Kuwahara and Yao, 1998, pl. 1, fig. 26, 2001, pl. 3, fig. 2; Spiller, 2002, pl. 3, fig. I; Wang and Shang, 2001, pl. 2, fig. 18.

Neoalbaillella sp. Kojima, 1982, pl. 2, figs. 8, 9, pl. 3, fig. 1.

Neoalbaillella cf. *ornithoformis* Takemura et Nakaseko; Sashida et al., 1995, Fig. 10 (1—4); Yao and Kuwahara, 2000, pl. 1, fig. 5.

Neoalbaillella sp. cf. *N. ornithoformis* Takemura et Nakaseko; Kojima, 1982, pl. 2, fig. 10; Sano, 1988, pl. 2, fig. 18.

描述：这个种的特点是顶锥未穿孔，向腹边轻微弯曲。壳体近圆柱形，两翼强壮，三角形，通常只发育1个翼，翼的形状像鸟嘴一样。具4—5列横向排列的窗孔，每列有8—10个窗孔。

层位与产地：上二叠统；日本，菲律宾，马来西亚，泰国，俄罗斯远东地区，中国华南和西藏双湖地区。

假鹰钩新阿尔拜虫　*Neoalbaillella pseudogrypa* Sashida et Tonishi
(Pl. 13, fig. 14)

Neoalbaillella pseudogrypa Sashida et Tonishi, 1988, Fig. 9 (1—6); Blome and Reed, 1992, Fig. 10 (1); Kuwahara et al., 1997, pl. 2, figs. 1, 2; Wang et al., 2006, Fig. 14 (G, H), 15 (J—L); Rudenko et al., 1997a, pl. 3, fig. 14, 1997b, pl. 16.2, fig. 8; Yao and Kawahara, 1996, pl. 3, fig. 23.

Neoalbaillella grypus Ishiga, Kito et Imoto; Sashida and Tonishi, 1985, pl. 7, fig. 2.

Neoalbaillella ornithoformis Takemura et Nakaseko; Wang et al., 1994, pl. 4, fig. 6.

描述：这个种的特点是顶锥向腹边强烈弯曲。壳体圆柱形，具有 3—4 列横向排列的窗孔，窗孔较大，方形至长方形。

比较：这个种与 *Neoalbaillella grupus* Ishiga, Kito et Imoto 都具有向腹边强烈弯曲的顶锥，区别在于前者壳体较长，有较多长方形或卵形的窗孔，而后者壳体短，窗孔小，圆形至方形。

层位与产地：上二叠统；日本，美国俄勒冈，俄罗斯远东地区，中国华南和西藏双湖地区。

隐管虫目 Latentifistularia Caridroit, De Wever et Dumitrica, 1999
隐管虫超科 Latentifistuloidea Nazarov et Ormiston, 1999
隐管虫科 Latentifistulidae Nazarov et Ormiston, 1983
隐管虫亚科 Latentifistulinae Nazarov et Ormiston, 1983
隐管虫属 *Latentifistula* Nazarov et Ormiston, 1983

模式种：*Latentifistula crux* Nazarov et Ormiston, 1983

鉴定要点：内骨架为中空无孔球体，3 个臂从球体中伸出，各臂间约有 120°夹角。壳表海绵状，近三角形。臂刺有或无。

时代与分布：石炭纪—二叠纪；世界各地。

圆锥形隐管虫（新种） *Latentifistula conica* Wang sp. nov.
(Pl. 11, figs. 1—4; pl. 16, figs. 14, 18)

词源：conic, 希腊词，圆锥形的。

描述：三射形外壳小，由 3 个大小和形状相似的海绵臂组成，臂细弱，较短，两臂夹角 120°，每个臂的基部最宽，向远端逐渐变细，至顶端变窄呈锥形。壳孔较密，大小不均匀。

比较：新种与 *L. patagilaterala* Nazarov et Ormiston 的区别为臂的形状不同，后者的臂基部较窄，向远端逐渐变宽，至末端又变窄呈矛状。这个新种与早石炭世的 *L. turgida* (Ormiston et Lane) 的区别为后者的臂较肿大，孔构多边形，不甚规则。

层位与产地：下二叠统；西藏双湖地区。

侧翼隐管虫 *Latentifistula patagilaterala* Nazarov et Ormiston
(Pl. 7, figs. 19, 20, 22, 25—28; pl. 10, figs. 2, 3)

Latentifistula patagilaterala Nazarov et Ormiston, 1985, pl. 4, fig. 1; Blome and Reed, 1992, Fig. 13 (8); Ujiie and Oba, 1991, pl. 3, fig. 8; Wang R J, 1993, pl. 3, figs. 4—6; Wang and Qi, 1995, pl. 4, figs. 4—6; Spiller, 2002, pl. 7, fig. O; Wang et al., 2012, pl. 14, figs. 27, 28, pl. 16, fig. 40, pl. 17, fig. 11; Jasin and Ali, 1997, pl. 1, fig. 1; Saesaengseerung et al., 2009, Fig. 8 (4).

Latentifistula crux Nazarov et Ormiston; Blome and Jones, 1986, Fig. 8 (19).

描述：这个种的特点是 3 个海绵臂位于同一平面，臂较粗壮，臂自基部向远端逐渐变粗，至末端又变窄，呈矛尖状。

比较：这个种与 *Latentifistula texana* Nazarov et Ormiston 的区别为后者臂的末端膨大呈棒形。

层位与产地：中二叠统；日本，美国西海岸，马来西亚，泰国，中国华南、青海可可西里地区和西藏双湖地区。

隐管虫（未定种 A） *Latentifistula* sp. A

(Pl. 7, fig. 21)

描述：只保存 2 个海绵臂，臂孔小，每个臂的近端部分较窄，至远端宽度逐渐加大，至末端又变窄，顶端圆钝。

比较：这个未定种 A 与 *Latentifitula crux* Nazarov et Ormiston 的区别为后者臂短，末端臂更宽。

层位与产地：中二叠统下部；中国青海可可西里地区。

隐管虫（未定种 B） *Latentifistula* sp. B

(Pl. 10, fig. 8)

描述：壳体由 3 个海绵臂组成，臂短，较宽，从基部至远端近乎相等，至顶端变窄。臂间夹角不等。

比较：这个未定种 B 与 *Latentifistula patagilaterala* Nazarov et Ormiston 的区别为后者臂较窄，臂间夹角近乎相等，约 120°。

层位与产地：中二叠统下部；中国青海可可西里地区。

德克萨斯隐管虫 *Latentifistula texana* Nazarov et Ormiston

(Pl. 7, figs. 23, 24; pl. 10, figs. 1, 4, 5; pl. 16, fig. 13)

Latentifistula texana Nazarov et Ormiston, 1985, pl. 4, fig. 2; Wang, 1991, pl. 3, fig. 8; Blome and Reed, 1992, Fig. 13 (6); Feng and Liu, 1993, pl. 6, fig. 1; Wang R J, 1993a, pl. 3, fig. 8, 1993b, pl. 1, figs. 17—21; Wang and Li, 1994, pl. 1, figs. 15, 16; Wang et al., 1994, pl. 3, fig. 26, 1998, pl. 3, fig. 20, 2012, pl. 15, figs. 6, 36, 37, pl. 16, figs. 2, 19—21, 38, 39, pl. 17, figs. 25, 26; Wang and Qi, 1995, pl. 4, figs. 1—3; Wang and Yang, 2003, pl. 1, figs. 1—5; Saesaengseerung et al., 2009, Fig. 8 (3); Nestella and Nestell, 2010, pl. 12, fig. 6; Maldonado and Noble, 2010, pl. 6, fig. 10; Kametaka et al., 2009, Fig. 8 (1).

Latentifistula sp. cf. *L. texana* Nazarov et Ormiston; Cheng, 1989, pl. 3, fig. 5.

Latentifistula sp. cf. *L. texana* Nazarov et Ormiston; Wu et al., 1994, pl. 3, fig. 1.

Paronaella? sp. Sano et al., 1982, pl. 1, fig. 1.

描述：这个种的特点是 3 个海绵臂棒槌状，每个臂的远端膨大。3 个臂的大小、形状和长度大体相似，臂间夹角 120°。

比较：这个种与 *Latentifistula crux* Nazarov et Ormiston 的区别为后者海绵臂粗短，末端膨大。

层位与产地：中二叠统；日本，美国，菲律宾，中国华南、青海可可西里地区和西藏双湖地区。

双隐管虫亚科 Latentibifistulinae Afanasieva, 2000

双隐管虫属 *Latentibifistula* Nazarov et Ormiston, 1983

模式种：*Latentibifistula triacanthophora* Nazarov et Ormiston, 1983

鉴定要点：壳体中央有一个未穿孔球体，从球上放射状产生 3 个中空海绵臂，臂间夹角 120°，并发育海绵层。

时代与分布：二叠纪；世界各地。

粗海绵双隐管虫 *Latentibifistula asperspongiosa* Sashida et Tonishi

(Pl. 8, figs. 41, 42, 45)

Latentibifistula asperspongiosa Sashida et Tonishi, 1986, pl. 1, figs. 4—6, 8, 11; Tumanda et al., 1990, pl. 1, fig. 11; Kuwahara and Yao, 1998, pl. 3, fig. 97; Sashida et al., 2000, Fig. 8 (4, 5); Kuwahara et al., 2004, pl. 1, fig. 12;

Yao et al., 2005, pl. 1, fig. 5; Saesaengseerung et al., 2009, Fig. 8 (5, 6); Wang et al., 2012, pl. 19, fig. 5, pl. 20, figs. 4, 23, 24.

Latentibifistula sp. cf. *L. asperspongiosa* Sashida et Tonishi; Blome and Reed, 1992, Fig. 13 (1); Wu et al., 1994, pl. 3, fig. 3; Rudenko and Panasenko, 1997, pl. 16, fig. 17.

Latentifistula sp. cf. *L. crux* Nazarov et Ormiston; Cheng, 1989, pl. 3, fig. 6.

Latentifistula texana Nazarov et Ormiston; Wang and Li, 1994, pl. 1, figs. 15, 16.

Latentifistula sp. aff. *L. asperspongiosa* Sashida et Tonishi; Blome and Reed, 1992, Fig. 13 (1).

Latentifistula sp. Sashida et Tonishi, 1986, pl. 1, figs. 2, 3, 7, 9, 10.

描述：这个种的特点是壳体由2层海绵组成，内海绵层粗糙，外海绵层细小。内层具有3个形状和大小相似的中空海绵臂，从中央部分以120°夹角伸出，外海绵连接各臂。

比较：这个种与 *Latentifistula texana* Nazarov et Ormiston 的区别为前者具有外海绵层连接3个内海绵臂。

层位与产地：二叠系；日本，菲律宾，泰国，美国西海岸，俄罗斯远东地区，中国华南和青海可可西里地区。

三刺双隐管虫 *Latentibifistula triacanthophora* Nazarov et Ormiston
(Pl. 4, figs. 38—45)

Latentibifistula triacanthophora Nazarov et Ormiston, 1983, pl. 1, figs. 4, 5, 1985, pl. 3, figs. 12—14.

描述：这个种的特点是壳体具有2层，内海绵层由3个形状和大小相似的海绵长臂组成，臂上孔很小，排列尚规则。外海绵层发育在内海绵臂之上，往往保存不全。

比较：这个种与 *Latentifistula patagilaterala* Nazarov et Ormiston 形状十分相似，区别在于后者的壳体只由一层海绵层组成。

层位与产地：下中二叠统萨马尔阶—瓜德鲁普阶；俄罗斯南乌拉尔地区，中国华南和青海可可西里地区。

考勒特虫科 Cauletellidae Caridroit, De Wever et Dumitrica, 1999
考勒特虫属 *Cauletella* Caridroit, De Wever et Dumitrica, 1999

模式种：*Deflandrella manica* De Wever et Caridroit, 1984

鉴定要点：具有3个喇叭状张开的臂。顶刺对称排列。壳体中部光滑，有时具少量孔。某些隔壁 (lamellae) 同臂和壳壁相连，并将臂分成若干小壳室。

时代与分布：中二叠世；美国西海岸，东亚地区。

袖状考勒特虫 *Cauletella manica* (De Wever et Caridroit)
(Pl. 13, fig. 37)

Deflandrella manica De Wever et Caridroit, 1984, pl. 1, figs. 1—7; Caridroit et al., 1985, pl. 1, fig. 9; Caridroit and De Wever, 1986, pl. 2, figs. 20—25, pl. 3, figs. 1, 2; Tumanda et al., 1990, pl. 2, fig. 12; Blome and Reed, 1992, Fig. 12 (3, 4); Caridroit, 1993, pl. 2, figs. 6, 7; Kuwahara and Yao, 2001, pl. 1, fig. 16; Wang and Shang, 2001, pl. 2, figs. 15—17; Wang et al., 2006, Fig. 11 (pp), Fig. 13 (p), 2012, pl. 18, figs. 42, 43, 52, pl. 21, fig. 44; Maldonado and Noble, 2010, pl. 6, figs. 1—3.

Deflandrella sp. A; Kuwahara et al., 1997, pl. 2, fig. 12; Yao and Kuwahara, 2000, pl. 2, fig. 1.

Deflandrella sp. B; Ishiga et al., 1986, pl. 3, fig. 5.

Deflandrella sp. cf. *D. manica* De Wever et Caridroit；Cheng，1989，pl. 4，figs. 6—8.

Deflandrella sp. Naka et Ishiga，1985，pl. 1，figs. 20，21；Sashida and Tonishi，1986，pl. 3，figs. 7—9.

Ishigaum sp. Spiller，2002，pl. 7，fig. H（non I）.

Cauletella wangi Caridroit et Shang；Shang et al.，2001，pl. 2，figs. 9，10.

Cauletella manica（De Wever et Caridroit）；Caridroit et al.，1999，Figs. 1，2；Shang et al.，2001，pl. 3，fig. 21；De Wever et al.，2001，Fig. 56（3）；Feng et al.，2006b，Fig. 7（9，11）；Wang et al.，2006，Fig. 11（pp），Fig. 13（p）；Maldonado and Noble，2010，pl. 6，figs. 1—3.

描述：这个种的特点是3个同面臂喇叭状张开，较短，大小和形状相似，臂间夹角约120°。臂壳薄，较光滑。臂的远端较宽，具顶刺。

比较：这个种与 *Ishigaum trifistis* De Wever et Caridroit 的区别为后者的臂较长，臂呈棒槌状。

层位与产地：中上二叠统；日本，美国西海岸，菲律宾，马来西亚，泰国，中国华南和西藏双湖地区。

三面海绵虫属 *Triplanospongos* Sashida et Tonishi，1988

模式种：*Triplanospongos musashiensis* Sashida et Tonishi，1988

鉴定要点：壳三角形，被海绵组织覆盖。发育3个双脊臂，臂的末端变成单脊，具臂刺。

时代与分布：晚二叠世；世界各地。

武藏三面海绵虫 *Triplanospongos musashiensis* Sashida et Tonishi

(Pl. 12，figs. 35，36，39；pl. 13，figs. 35，45—48；pl. 14，figs. 1—3，5，6)

Triplanospongos musashiensis Sashida et Tonishi，1988，pl. 9，figs. 7—12；Tumanda et al.，1990，pl. 2，fig. 14；Wang，1991，pl. 3，fig. 9；Feng and Liu，1993，pl. 6，figs. 4—8；Wang and Li，1994，pl. 3，figs. 1—4；Wang et al.，1994，pl. 4，figs. 16，17，2006，Figs. 11（MM—OO），13（N，O，KK），14（GG，HH），15（P—R），2012，pl. 18，figs. 47—49，51，pl. 21，figs. 11，32，39—43；Sashida et al.，1995，Fig. 11（12，13），2000a，Fig. 8（6，7），2000b，pl. 2，figs. 6，7；Rudenko and Panasenko，1997，pl. 16.1，figs. 12，23，pl. 16.2，fig. 12；Kuwahara et al.，1997，pl. 3，figs. 7—9；Feng et al.，1998，Fig. 5（A，B）；Kuwahara and Yao，1998，pl. 4，fig. 131；Yao and Kuwahara，2000，pl. 2，fig. 24，pl. 3，fig. 27；Wang and Shang，2001，pl. 1，figs. 14—16；Shang et al.，2001，pl. 1，figs. 1—5；Sashida and Salyapongse，2002，Fig. 3（34）；Yao et al.，2005，pl. 1，figs. 8，9；Spiller，2002，pl. 9，figs. a，b；Saesaengseerung et al.，2009，Fig. 8（13，14）.

? *Trifidospongus dekkasensis* Noble et Renne，1990，pl. 1，figs. 1—3；Blome and Reed，1992，Fig. 13（13，14）.

? *Trifidospongus angustus* Noble et Renne，1990，pl. 1，figs. 4—6.

Triplanospongos dekkasensis（Noble et Renne）；Bolme and Reed，1992，Fig. 13（13，14）；Yao and Kuwahara，2000，pl. 2，fig. 23；Feng and Gu，2002，Fig. 7（16，17）.

Triplanospongos angustus（Noble et Renne），Kuwahara and Yamakita，2001，pl. 1，fig. 22.

Triplanospongos sp. cf. *T. musashiensis* Sashida et Tonishi；Cheng，1989，pl. 3，figs. 13—16；Wu et al.，1994，pl. 3，fig. 14.

Triplanospongos aff. *dekkasensis*（Noble et Renne）；Kuwahara et al.，2004，pl. 1，fig. 15.

Angulobracchia（?）sp. Takemura et Nakaseko，1981，pl. 34，fig. 11；Yoshida and Murata，1985，pl. 2，fig. 18.

Triplanospongos aff. *musashiensis* Sashida et Tonishi；Kuwahara and Yao，1998，pl. 4，fig. 132.

? *Paronaella* sp. A Wakita，1983，pl. 7，fig. 7.

Latentifistula sp. cf. *L. similicutis* Caridroit et De Wever；Cheng，1989，pl. 1，figs. 13，14，16.

Foremanhelena musahiensis（Sashida et Tonishi）；De Wever et al.，2001，Fig. 56（2，7）.

描述：这个种的特点是三角形壳面由 3 个双脊臂和覆盖其上的海绵组织构成。每个臂从中心以 120°角伸出，远端具膨大的海绵组织，顶端发育臂刺。由于双脊臂外的海绵组织容易脱落，因此，有些作者将海绵组织保存完整的标本鉴定成 *Triplanospongos musashiensis*，而将海绵组织脱落的标木鉴定成另一些种，如 *Trifidospongus dekkasensis*，*T. angustus* 等。我们认为这个种的特点是具有 3 个双脊形臂，末端变成单脊，臂端被海绵组织覆盖形成三角形，具臂刺。海绵组织保存多少不能作为鉴定属种的标准。

层位与产地：上二叠统；世界各地。

石贺裕明虫属 *Ishigaum* De Wever et Caridroit, 1984

模式种：*Ishigaum trifistis* De Wever et Caridroit, 1984

鉴定要点：壳体由 3 个同面臂组成，每个臂中空。近端部分未穿孔，远端部分具一些长孔或圆孔，末端海绵状，具顶刺。臂间夹角 120°。

时代与分布：中晚二叠世；世界各地。

格状石贺裕明虫 *Ishigaum craticula* Shang, Caridroit et Wang

(Pl. 14, figs. 12, 16, 21, 22, 27—29, 34, 36, 38)

Ishigaum craticula Shang, Caridroit et Wang, 2001, pl. 1, figs. 12—15.

描述：这个种的特点是 3 个同面臂以 120°角相间。每个臂细长管状，近端具规则的长孔，远端圆柱形。通常具 2—3 层海绵层。中央部分光滑，有 4 个或多个小孔。

比较：这个种与 *Ishigaum trifistis* De Wever et Caridroit 的区别为前者具细长管状臂，臂的近端具规则的长孔，远端发育较长的海绵构造。

层位与产地：上二叠统；中国广西和西藏双湖地区。

肥胖石贺裕明虫 *Ishigaum obesum* De Wever et Caridroit

(Pl. 13, fig. 36; pl. 14, figs. 15, 18, 23, 41)

Ishigaum obesum De Wever et Caridroit, 1984, pl. 1, figs. 8, 9; Caridroit and De Wever, 1986, pl. 3, figs. 6, 7; Wang, 1991, pl. 3, fig. 13; Wang and Li, 1994, pl. 3, figs. 12, 18, 19; Wang et al., 1994, pl. 3, fig. 23, 1998, pl. 4, fig. 8, 2006, Figs. 13（LL）, 14（T）, 2012, pl. 8, fig. 41; Wu et al., 1994, pl. 3, fig. 5; Kuwahara and Yao, 1998, pl. 4, fig. 94, 2001, pl. 4, fig. 10; Yao and Kuwahara, 1999a, pl. 1, fig. 10, 1999b, pl. 4, fig. 2, 1999c, pl. 2, fig. 10; Shang et al., 2001, pl. 3, fig. 11.

Ishigaum sp. cf. *I. obesum* De Wever et Caridroit; Blome and Reed, 1995, pl. 1, fig. 21.

Ishigaum trifistis De Wever et Caridroit; Blome and Reed, 1992, Fig. 12（12）; Feng and Liu, 1993, pl. 6, fig. 16（non 15, 17, 18）; Kuwahara et al., 1997, pl. 2, fig. 10; Shang et al., 2001, pl. 3, fig. 12（non 9, 10）.

Ishigaum sp.; Wang R J, 1993, pl. 4, fig. 21; Rudenko and Panasenko, 1997, pl. 16.2, fig. 11（non 5, 9, 14）.

Ishigaum sp. A Sashida et Tonishi, 1986, pl. 3, figs. 1—4.

Ishigaum sp. nov. 2 Cheng, 1989, pl. 3, figs. 4, 7, 8, 12.

? *Tormentum* sp. Stewart et al., 1986, Fig. 1（P）.

描述：这个种的特点是 3 个同面臂处于不规则位置，臂间夹角变化。中央部分少孔。臂的近端部分无孔，中部具长孔，远端部分发育宽海绵层。保存好的标本具顶刺。

比较：这个种与 *Ishigaum trifistis* De Wever et Caridroit 的区别为后者的臂棒槌状。

层位与产地：上二叠统；世界各地。

三棍石贺裕明虫 *Ishigaum trifistis* De Wever et Caridroit

(Pl. 12, figs. 40, 42; pl. 13, fig. 34; pl. 14, figs. 4, 7—11, 19, 26, 35)

Ishigaum trifistis De Wever et Caridroit, 1984, pl. 1, figs. 10—13, 16; Caridroit et al., 1985, pl. 1, fig. 8; Caridroit and De Wever, 1986, pl. 3, figs. 3—5; Ishiga, 1985, pl. 2, fig. 24; Miyamoto and Tanimoto, 1986, pl. 1, figs. 6, 7; Wang, 1991, pl. 3, fig. 17; Blome and Reed, 1992, Fig. 12 (12); Feng and Liu, 1993, pl. 6, figs. 15, 17, 18; Wang et al., 1994, pl. 3, fig. 18, pl. 4, figs. 21—23, 1998, pl. 4, figs. 1, 2, 10, 11, 2006, Figs. 11 (Y-AA, QQ, ZG), 13 (M, II), 14 (II), 15 (W—Z), 2012, pl. 18, figs. 38, 39, 50, pl. 19, figs. 6—8, 35, 36, pl. 20, figs. 20—22, 27, 28, pl. 21, fig. 38; Wang and Li, 1994, pl. 3, figs. 7, 15; Wu et al., 1994, pl. 3, fig. 6; Wang and Fan, 1997, pl. 1, figs. 10—13; Kuwahara and Yao, 1998, pl. 3, fig. 93, 2001, pl. 4, fig. 11; Yao and Kuwahara, 1999a, pl. 1, fig. 9, 1999b, pl. 4, fig. 3, 1999c, pl. 2, fig. 11, 2000, pl. 2, fig. 7, pl. 3, fig. 23; Sashida and Salyapongse, 2002, Fig. 3 (26); Feng et al., 1998, Fig. 4 (O—Q); Sashida et al., 2000, Fig. 8 (9); Wang and Shang, 2001, pl. 2, figs. 26—32; Shang et al., 2001, pl. 3, figs. 9, 10, 12; Kuwahara et al., 2003, pl. 2, fig. 4, 2004, pl. 1, fig. 13, 2005, pl. 1, figs. 5, 6; Maldonado and Noble, 2010, pl. 6, figs. 6—9.

Ishigaum cf. *trifistis* De Wever et Caridroit; Miyamoto and Tanimoto, 1985, Fig. 3 (non 10—12); Sashida and Tonishi, 1986, pl. 2, figs. 4—6; Cheng, 1989, pl. 1, figs. 4, 15, pl. 3, fig. 3, pl. 4, figs. 1, 2, 5; Wang R J, 1993, pl. 4, figs. 18—20.

Ishigaum sp. Ishiga et al., 1986, pl. 3, figs. 6, 7; Noble and Renne, 1990, pl. 1, fig. 11; Tumanda et al., 1990, pl. 2, fig. 20.

Paronaella (?) sp. Takemura et Nakaseko, 1981, pl. 34, fig. 12.

Ishigaum? sp. Sashida et al., 1985, Fig. 11 (7, 8, 10, 11, non 9).

描述：这个种的特点是壳体由3个同面棒槌状臂组成，每个臂的近端部分无孔，中部具长孔，末端海绵状，具顶刺。中央部分光滑无孔，或有时发育2—4个小孔。

层位与产地：上二叠统；世界各地。

石贺裕明虫（未定种A） *Ishigaum* sp. A Wang et Li

(Pl. 14, fig. 30)

Ishigaum sp. A Wang et Li, 1994, pl. 2, figs. 17—19.

描述：只保存1个臂，蜡烛状，穿孔，臂的下半部圆柱形，上半部宽度逐渐减少，至顶部变窄，有一个较长的顶刺。

层位与产地：上二叠统；中国广西和西藏双湖地区。

四桨虫科 Quadriremidae Afanasieva, 2000
四桨虫亚科 Quadrireminae Afanasieva, 2000
四桨虫属 *Quadriremis* Nazarov et Ormiston, 1985

模式种：*Quadriremis gliptoacus* Nazarov et Ormiston, 1985

鉴定要点：内骨架由1个无孔球和4个中空射组成，其中3射从球体伸出，相互间以120°角相交，第4射与之垂直。外壳格状，很少海绵状，4个臂与内射相连。

时代与分布：二叠纪；俄罗斯南乌拉尔地区，美国，中国青海。

微小四桨虫 *Quadriremis minima* Nazarov et Ormiston

(Pl. 7, fig. 46)

Quadriremis minima Nazarov et Ormiston, 1985, pl. 4, fig. 13.

Quadriremis sp. cf. *Q. minima* Nazarov et Ormiston; Nazarov and Ormiston, 1993, pl. 7, fig. 7.

描述：这个种的特点是壳的中心微突，3根细长的臂以120°角相间，第4臂与之垂直。臂的近端部分具一突槽，远端部分圆柱形。

层位与产地：中二叠统伦纳德阶；美国德克萨斯，中国青海可可西里地区。

鲁仁采夫海绵虫超科 Ruzhencevispongoidea Nazarov et Ormiston, 1983
石片虫科 Tormentidae Nazarov et Ormiston, 1983
假石片虫属 *Pseudotormentus* De Wever et Caridroit, 1984

模式种：*Pseudotormentus kamigoriensis* De Wever et Caridroit, 1984

鉴定要点：具有3个同面臂，每个臂近端部分中空管状，远端具6—8个纵向窄脊和较宽的浅槽。每个浅槽中发育1根纵梁和许多短的横杆隔成许多双孔。

时代与分布：中晚二叠世；世界各地。

上郡假石片虫 *Pseudotormentus kamigoriensis* De Wever et Caridroit

(Pl. 5, figs. 44, 45, 53, 54; pl. 7, figs. 39—43; pl. 8, figs. 43, 44; pl. 10, figs. 13, 14)

Pseudotormentus kamigoriensis De Wever et Caridroit, 1984, pl. 2, figs. 1—7; Caridroit et al., 1985, pl. 1, figs. 10—12; Ishiga, 1985, pl. 2, figs. 20, 21; Caridroit and De Wever, 1986, pl. 5, figs. 7—11; Blome and Reed, 1992, Fig. 12 (13—18, 21); Wu et al., 1994, pl. 3, fig. 12; Wang and Li, 1994, pl. 2, figs. 13—16; Wang et al., 1994, pl. 3, fig. 22, 1998, pl. 4, fig. 12, 2012, pl. 16, fig. 23, pl. 17, fig. 22, pl. 18, fig. 26; Feng and Liu, 1993, pl. 6, fig. 9; Wang and Fan, 1997, pl. 1, figs. 18—20; Kuwahara et al., 1997, pl. 3, figs. 5, 6, 2004, pl. 1, fig. 18; Kuwahara and Yao, 2001, pl. 1, fig. 22; Spiller, 2002, pl. 8, figs. N, O.

Pseudotormentus sp. cf. *P. kamigoriensis* De wever et Caridroit; Sashida and Tonishi, 1986, pl. 4, figs. 8, 9; Tumanda et al., 1990, pl. 1, fig. 5; Spiller, 2002, pl. 8, figs. P, Q; Blome and Reed, 1995, pl. 1, figs. 13, 14.

Pseudotormentus sp. Ishiga et al., 1986, pl. 3, figs. 8, 9.

Nazarovispongus (?) sp. A Ishiga et Suzuki, 1984, pl. 1, fig. 21.

描述：这个种的特点是壳体由1个小的球形中央部分和3个同面长臂组成。每个臂近端部分管状，未穿孔，远端有6—8个纵向窄脊和较宽的浅槽组成的网格。每个浅槽由1根纵梁和许多短的横杆形成双孔。

比较：这个种与*Pseudotormentus monoporus* Wang的区别为在后者臂的远端部分几个纵向窄脊和浅槽中的横杆形成单孔。

层位与产地：中上二叠统；日本，美国西海岸，菲律宾，马来西亚，中国华南、内蒙古、青海可可西里地区和西藏双湖地区。

奥米斯顿虫科 Ormistonellidae De Wever et Caridroit, 1984, emend. De Wever et al., 2001
奥米斯顿虫亚科 Ormistonellinae De Wever et Caridroit, 1984, emend. De Wever et al., 2001
奥米斯顿虫属 *Ormistonella* De Wever et Caridroit, 1984

模式种：*Ormistonella robusta* De Wever et Caridroit, 1984

鉴定要点：4个未穿孔臂呈四面体形。中央部分无孔，为一球形四面体。臂长，管状，开放，近端切面圆，远端呈 U 字形。

时代与分布：中晚二叠世；日本，菲律宾，中国。

强壮奥米斯顿虫 *Ormistonella robusta* De Wever et Caridroit

(Pl. 14, figs. 17, 24, 25, 37, 40)

Ormistonella robusta De Wever et Caridroit, 1984, pl. 2, figs. 8, 9; Caridroit and De Wever, 1986, pl. 4, figs. 7, 8; Wang, 1991, pl. 3, fig. 13; Wang and Li, 1994, pl. 3, figs. 12, 18, 19; Wang et al., 1994, pl. 3, fig. 17, pl. 4, fig. 19; Wang and Fan, 1997, pl. 1, figs. 7—9.

Ormistonella sp. cf. *O. robusta* De Wever et Caridroit; Cheng, 1989, pl. 1, fig. 1; Wang and Li, 1994, pl. 1, figs. 19—21.

描述：这个种的特点是具有 4 个呈四面体状排列的短的强壮臂。臂管状，无孔。每个臂远端轻微膨大。

层位与产地：二叠系；日本，菲律宾，中国华南、内蒙古、青海可可西里地区和西藏双湖地区。

卡里特罗伊特虫属 *Raciditor* Sugiyama, 2000

模式种：*Nazarovella gracilis* De Wever et Caridroit, 1984

鉴定要点：4 个发育不一的管状臂从中央一个无孔小球中伸出，其中 3 个臂形状和大小相似，位于同一平面，彼此以 120°角相间，第 4 臂短小，与其他臂垂直，呈四面体状。每个臂的近端光滑或具浅沟，远端发育一个海绵体，形状多样，如球形、椭圆形、火焰状、扁球形，具顶刺。

时代与分布：中晚二叠世；世界各地。

纤细卡里特罗伊特虫 *Raciditor gracilis* (De Wever et Caridroit)

(Pl. 6, figs. 1, 3, 5—10; pl. 10, fig. 7; pl. 12, fig. 41; pl. 14, figs. 31—33)

Nazarovella gracilis De Wever et Caridroit, 1984, pl. 1, figs. 14, 15, 17; Ishiga, 1985, pl. 2, figs. 22, 23; Naka and Ishiga, 1985, pl. 1, figs. 14, 15; Caridroit and De Wever, 1986, pl. 4, figs. 9—15; Yamakita, 1986, pl. 1, figs. 12, 13; Ishiga and Miyamoto, 1986, pl. 64, fig. 16; Sashida and Tonishi, 1986, pl. 3, figs. 10—12, pl. 4, fig. 7; Tumanda et al., 1990, pl. 1, fig. 27; Blome and Reed, 1992, Fig. 13 (9, 10); Wang R J, 1993a, pl. 4, figs. 15, 16, 1993b, pl. 2, figs. 5, 6; Feng and Liu, 1993, pl. 6, fig. 14; Sashida et al., 1995, pl. 11, figs. 17, 19, 1997, Fig. 5 (20—24), 2000, pl. 18, fig. 12; Kuwahara et al., 1997, pl. 3, figs. 1, 2; Kuwahara and Yao, 1998, pl. 4, fig. 118; Feng et al., 1998, Fig. 5 (E—H); Yao and Kuwahara, 1999a, pl. 2, fig. 9, 1999b, pl. 1, fig. 12, pl. 2, figs. 19—21, 2000, pl. 2, figs. 16, 17; Shang et al., 2001, pl. 4, figs. 5, 6; Sashida and Salyapongse, 2002, Fig. 3 (25); Wang and Yang, 2003, pl. 1, fig. 9; Wang et al., 2006, Figs. 14 (V, W), 15 (BB), 2012, pl. 18, figs. 44, 45, pl. 19, figs. 9, 10, 21, pl. 21, figs. 12, 22.

Nazarovella cf. *gracilis* De Wever et Caridroit; Cheng, 1989, pl. 4, fig. 4; Wang and Li, 1994, pl. 2, figs. 4, 5, 7, 8; Wu et al., 1994, pl. 3, fig. 9; Blome and Reed, 1995, pl. 1, fig. 12.

Nazarovella spp. Cheng, 1989, pl. 2, fig. 12.

Nazarovella sp. Ishiga et al., 1986, pl. 3, figs. 11, 12 (non 13)

Nazarovispongus? sp. B Ishiga et Suzuki, 1984, pl. 1, figs. 17—20.

Raciditor gracilis (De Wever et Caridroit); Nestell and Nestell, 2010, p. 36; Saesaengseerung et al., 2009, Fig. 8 (23).

Raciditor cf. *gracilis* (De Wever et Caridroit); Kuwahara et al., 2003, pl. 2, fig. 11; Yao et al., 2007, Fig. 14 (20, 21).

描述：这个种的特点是壳体纤细，4 个未穿孔的长臂从中央部分伸出，臂呈四面体状排列，其中 3 个臂位于同一平面，管状，末端发育一个球状海绵体，第 4 臂短小，与其他 3 个臂垂直。

比较：这个种与 *Raciditor inflata*（Sashida et Tonishi）的区别为后者臂上发育 1—2 列圆形至卵圆形孔，臂末端海绵体呈透镜形。

层位与产地：中上二叠统；世界各地。

膨胀卡里特罗伊特虫 *Raciditor inflata*（Sashida et Tonishi）

(Pl. 10, fig. 6; pl. 14, figs. 20, 39)

Nazarovella inflata Sashida et Tonishi, 1986, pl. 4, figs. 1—6, 10—12; Tumanda et al., 1990, pl. 1, fig. 20; Wang, 1991, pl. 3, figs. 14, 15, pl. 4, fig. 2; Wang R J, 1993, pl. 4, figs. 10—14; Wang and Li, 1994, pl. 2, figs. 1—3, 6; Wang et al., 1994, pl. 3, figs. 15, 16, 1998, pl. 4, figs. 4—7, 2006, Figs. 11 (KK, LL), 13 (U), 15 (FF, GG), 2012, pl. 17, fig. 36, pl. 18, fig. 53, pl. 20, fig. 36, pl. 21, fig. 45; Wang and Qi, 1995, pl. 3, figs. 11—13; Sashida et al., 1995, Fig. 11 (15, 16), 2000, Fig. 8 (10, 11); Kuwahara et al., 1997, pl. 3, fig. 3; Kuwahara and Yao, 1998, pl. 4, fig. 119, 2001, pl. 1, fig. 21; Yao and Kuwahara, 1999, pl. 4, fig. 11, 2000, pl. 2, figs. 18, 19, pl. 3, fig. 26; Feng and Gu, 2002, Fig. 7 (14); Wang and Yang, 2003, pl. 1, fig. 10.

Ormistonella sp. Ishiga et al., 1986, pl. 3, fig. 10.

Nazarovella cf. *inflata* Sashida et Tonishi; Wang R J, 1993, pl. 2, figs. 1, 2.

Unidentified latentifistulid Blome et Reed, 1992, Fig. 13 (21).

Nazarovella sp. Feng et Liu, 1993, pl. 6, figs. 11—13.

Nazarovella spp. Cheng, 1989, pl. 2, fig. 10 (non 12).

Quadricaulis femoris Caridroit et De Wever; Shang et al., 2001, pl. 4, figs. 3, 4.

Raciditor inflata (Sashida et Tonishi); Kuwahara et al., 2003, pl. 2, fig. 12, 2004, pl. 1, fig. 21.

Raciditor aff. *inflata* (Sashida et Tonishi); Kuwahara et al., 2004, pl. 1, fig. 22.

描述：这个种的特点是壳体由 4 个管状臂组成，其中 3 个臂位于同一平面，第 4 臂短，与之垂直。每个臂上发育 1—2 列圆形至卵圆形孔，臂的末端发育 1 个透镜状海绵体。

比较：这个种与 *Raciditor scalae*（Cari-droit et De Wever）的区别为后者臂的远端呈梯形。

层位与产地：中上二叠统；世界各地。

扁球形卡里特罗伊特虫（新种） *Raciditor oblatum* Wang sp. nov.

(Pl. 5, fig. 43; pl. 6, figs. 11—19; pl. 7, fig. 44; pl. 16, figs. 11, 15—17, 20)

Latentifistulidae gen. et sp. indet. Spiller, 2002, pl. 8, figs. I, J (non K); Saesaengseerung et al., 2009, Fig. 8 (21, 22).

gen. et sp. indet. Kada, 1990, pl. 1, fig. 17.

incertae sedis E Kuwahara et Yao, 1998, pl. 4, fig. 139.

Nazarovella sp. Feng et Liu, 2002, pl. 2, figs. 6, 7.

Nazarovella phlogidea Wang; Wang et al., 2012, pl. 20, fig. 31 (non 32).

Katroma（?）sp. Xia et Zhang, 1998, pl. 4, figs. 11—14.

Ishigaum? sp. Xia et Zhang, 1998, pl. 4, figs. 15, 16; Feng and Liu, 2002, pl. 2, fig. 8 (non 9, 10).

Quadreiremis sp. Li et Bian, 1993, pl. 2, fig. 9.

词源：oblatus，拉丁词，扁球形。

描述：4 个发育不同的管状臂，其中 3 个臂形状和大小相似，位于同一平面，相互间以 120°角相间，第 4 臂短小，与其他臂垂直。臂的上部发育一个扁球形的海绵体，顶端具一个较发达的顶刺。

比较：这一新种与 *Raciditor gracilis*（De Wever et Caridroit）的区别为后者臂上部海绵体较小，近球形，顶刺不发育；新种与 *R. phlogidea*（Wang）的区别为后者臂上部海绵体较长，火焰状，顶刺较短。

层位与产地：二叠系；日本，马来西亚，泰国，中国华南和青海可可西里地区。

火焰状卡里特罗伊特虫 *Raciditor phlogidea* (Wang)

(Pl. 5, figs. 32—42; pl. 6, figs. 2, 4)

Nazarovella phlogidea Wang, 1994 (in Wang et Li), pl. 2, figs. 9—11, 20—24; Wang et al., 1994, pl. 4, fig. 24, 2012, pl. 20, fig. 32, (non 31); Wang and Shang, 2001, pl. 2, figs. 1—7.

Nazarovella? sp. Sashida et al., 2000, Fig. 8 (8); Shang et al., 2001, pl. 3, figs. 16, 20.

Entactinosphaera? sp. Feng et Liu, 1993, pl. 5, figs. 16, 17.

Ishigaum sp. A Kuwahara et al., 1997, pl. 2, fig. 11.

Nazarovella sp. cf. *N. phlogidea* Wang; Feng et al., 1998, Fig. 5 (m—o).

Latentifistulidae gen. et sp. indet. Kuwahara et al., 1997, pl. 3, fig. 12.

Quadriremis? sp. 1 Nestell et Nestell, 2010, pl. 12, figs. 7, 8.

描述：这个种的特点是长度不等的4个臂从无孔的中央伸出，其中3个臂大小和形状相似，有3条纵脊和3条纵沟相间，在纵沟中具孔。臂的末端发育1个火焰状的海绵构造，具顶刺。第4臂短。

比较：这个种与 *Raciditor inflata* (Sashida et Tonishi) 的区别为后者的臂为圆柱形的长臂，臂上发育1—2列圆形至卵圆形孔。

层位与产地：中上二叠统；日本，泰国，中国华南和青海可可西里地区。

卡里特罗伊特虫（未定种A） *Raciditor* sp. A

(Pl. 16, fig. 10)

描述：中央球小，无孔。4根管状臂呈四面体状排列。3根长臂位于同一平面，相互间以120°角相间，第4臂短小，与其他臂垂直。由于化石保存较差，臂的上部构造未显示。很可能这个未定种A属于 *R. scalae* (Caridroit et De Wever)。

层位与产地：下二叠统；中国青海可可西里地区。

四茎虫属 *Quadricaulis* Caridroit et De Wever, 1986

模式种：*Quadricaulis femoris* Caridroit et De Wever, 1986

鉴定要点：4个等长的长臂四面体状排列。内骨架为1个有4个射杆的中心球体被封闭在一个海绵壳中。

时代与分布：晚二叠世；东亚地区。

股状四茎虫 *Quadricaulis femoris* Caridroit et De Wever

(Pl. 5, figs. 46—52; pl. 16, fig. 12)

Quadricaulis femoris Caridroit et De Wever, 1986, pl. 4, figs. 16—19; Wang et al., 2012, pl. 9, figs. 19, 20, 43.

描述：这个种的特点是壳体具有4个长臂呈四面体状排列。长臂的近端部分孔略大，至远端部分孔越来越密，呈海绵状，具臂顶刺。

层位与产地：上二叠统；日本，中国华南和青海可可西里地区。

五浆虫属 *Quinqueremis* Nazarov et Ormiston, 1983

模式种：*Quinqueremis arundinea* Nazarov et Ormiston, 1983

鉴定要点：内骨架由 1 个无孔球体组成，从球中放射状伸出 5 个中空臂，其中 4 个臂位于同一平面中，相互间以 90°角相间，第 5 臂亚圆柱形，与其余 4 个臂垂直。

时代与分布：晚石炭世—三叠纪；美国西海岸，中国华南和青海可可西里地区。

芦苇状五浆虫　*Quinqueremis arundinea* Nazarov et Ormiston
(Pl. 8, figs. 46—50)

Quinqueremis arundinea Nazarov et Ormiston, 1983, pl. 1, figs. 6, 7, 1985, pl. 4, figs. 5—7; Nazarov, 1988, pl. 28, fig. 10; Amon et al., 1990, pl. 1, figs. 1, 2.

描述：这个种的特点是壳体由 5 个海绵臂组成。每个臂在基部较窄，至中部臂逐渐膨大，到远端宽度最大，顶部又变窄，膨大部分呈棒槌状。

层位与产地：中上二叠统；俄罗斯南乌拉尔地区，中国青海可可西里地区。

强壮五浆虫　*Quinqueremis robusta* Nazarov et Ormiston
(Pl. 6, figs. 20, 21; pl. 16, fig. 19)

Quinqueremis robusta Nazarov et Ormiston, 1985, pl. 4, fig. 11; Wang and Yang, 2003, pl. 1, figs. 11, 12; Wang et al., 2012, pl. 17, figs. 17, 21, pl. 18, fig. 36.

Quinqueremis flata Wang R J, 1993, pl. 2, figs. 3, 4.

Quinqueremis sp. aff. *Q. robusta* Nazarov et Ormiston; Blome and Reed, 1992, Figs. 11 (17, 18), 12 (1, 2).

Quinqueremis sp. cf. *Q. robusta* Nazarov et Ormiston; Wang, 1995, pl. 16, fig. 19.

Quinqueremis (?) sp. De Wever et Yielynek, 1984, pl. 1, fig. 6; Blome et al., 1986, pl. 8, fig. 18.

描述：这个种的特点是壳体具 5 个臂，其中 4 个臂位于同一平面中，彼此夹角 90°，第 5 臂与之垂直。每个臂细长，平直，在基部无孔，其他地方具小孔。

比较：这个种与 *Quinqueremis arundinea* Nazarov et Ormiston 的区别为两者臂的形状不同，后者的臂海绵状，从基部向远端宽度逐渐增大，至末端又变窄，具顶刺。

层位与产地：下中二叠统；美国西海岸，俄罗斯南乌拉尔地区，中国华南和青海可可西里地区。

五浆虫（未定种 A）　*Quinqueremis* sp. A
(Pl. 19, fig. 20)

描述：壳体具有 5 个臂，4 个臂位于同一平面，相互间呈 90°角相间，第 5 臂与之垂直。臂较平直。

层位与产地：中三叠统拉丁阶上部；中国青海可可西里地区。

内射虫目　Entactinaria Kozur et Mostler, 1982
内射虫超科　Entactinioidea Riedel, 1967
内射虫科　Entactiniidae Riedel, 1967, emend. Nazarov, 1975
内射虫亚科　Entactiniinae Riedel, 1967, emend. Nazarov, 1975

斑点球桩虫属　*Stigmosphaerostylus* **Rüst, 1892, emend. Foreman, 1963, Aitchison et Stratford, 1997**

同义名：*Entactinia* Foreman, 1963

模式种：*Stigmosphaerostylus notabilis* Rüst, 1892

鉴定要点：一个格状球形外壳。6 根三棱形主刺。内骨架由中梁上产生的 6 射纤细小骨针构成。

时代与分布：晚古生代；世界各地。

十字形斑点球桩虫（新种） *Stigmosphaerostylus cruciformis* Wang sp. nov.

(Pl. 5, figs. 12, 20—22; pl. 7, figs. 30—34)

词源：cruciform，拉丁词，十字形的。

描述：格状单壳小，近球形。具有6根三棱形主刺，脊窄沟宽，主刺粗壮，彼此间以90°角相间。刺的长度为壳径的2倍以上，孔中等大小，圆形，卵圆形，孔的接点发育小结节。辅刺不发育。

比较：这一新种与 *Stigmosphaerostylus modesta* (Sashida et Tonishi) 的区别为后者的壳体较大，主刺较短，辅刺发育。

层位与产地：中二叠统；中国青海可可西里地区。

分异斑点球桩虫 *Stigmosphaerostylus diversitus*（Nazarov）

(Pl. 1, figs. 3, 4, 9, 29—31, 33—35)

Entactinia diversita Nazarov, 1973, pl. 1, fig. 1, 1975, pl. 1, figs. 1—3, pl. 2, figs. 1—3; Wang, 1997, pl. 3, figs. 11—13.

描述：这个种的特点是单个格状壳球形，具有6根三棱形主刺，其中1根特别粗壮，其余5根较弱。长刺长度是壳径的2—3倍，辅刺细小。

比较：这个种与 *Stigmosphaerostylus proceraspina* (Aitchison) 的区别为后者具有7根主刺，长度更长，辅刺更发育。

层位与产地：上泥盆统弗拉阶—法门阶；俄罗斯南乌拉尔地区，中国新疆和青海可可西里地区。

细弱斑点球桩虫（新种） *Stigmosphaerostylus gracilentus* Wang sp. nov.

(Pl. 5, figs. 23, 24, 29, 30)

词源：gracilentus，拉丁词，纤细的。

描述：格状单壳小，球形。具有6根长的主刺，主刺平直，具窄脊，纤细，长度大于壳径，一般为壳径的2倍以上。壳孔大，圆形至多边形。

比较：这个新种与 *Stigmosphaerostylus* sp. B, Umeda et al., 2004 (= *Borisella* cf. *maksimovae* Afanasieva, 2000; Obut et al., 2008) 在外形上十分相似，区别在于后者的主刺是杆形的。

层位与产地：上泥盆统法门阶；中国青海可可西里地区。

市川浩一郎斑点球桩虫 *Stigmosphaerostylus ichikawai*（Caridroit et De Wever）

(Pl. 6, fig. 22)

Haplentactinia? *ichikawai* Caridroit et De Wever, 1984, pl. 1, figs. 30—35, 1986, pl. 3, figs. 8—11; Ishiga, 1990, pl. 1, fig. 2; Blome and Reed, 1992, Fig. 11 (14—16); Ishiga and Miyamoto, 1986, pl. 64, figs. 14, 15; Kuwahara et al., 2004, pl. 2, fig. 12; Maldonado and Noble, 2010, pl. 5, figs. 1—4; Noble and Jin, 2010, pl. 2, figs. 1—3.

描述：这个种的特点是单个格状壳小，具有6根很长的、强壮的三棱形主刺，在主刺的中部发育冠状骨刺，辅刺不发育。

比较：这个种与 *Stigmosphaerostylus itsukaichiensis* (Sashida et Tonishi) 在外形上十分相似，区别在于后者主刺上不发育骨刺。

层位与产地：中上二叠统；日本，美国西海岸，中国华南和青海可可西里地区。

五日市斑点球桩虫 *Stigmosphaerostylus itsukaichiensis* (Sashida et Tonishi)
(Pl. 5, figs. 2—11, 14, 17)

Entactinia itsukaichiensis Sashida et Tonishi, 1985, pl. 1, figs. 1—10; Ishiga, 1990, pl. 1, fig. 1; Tumanda et al., 1990, pl. 1, figs. 16; Blome and Reed, 1992, Fig. 11 (2—5), 1995, pl. 1, figs. 25, 26; Feng and Liu, 1993, pl. 5, figs. 10, 11; Feng et al., 1998, Fig. 3 (C, D), 2001, pl. 1, figs. 1, 2; Kuwahara and Yao, 1998, pl. 2, fig. 59, 2001, pl. 3, fig. 7; Sashida and Salyapongse, 2002, Fig. 3 (31); Kuwahara et al., 1997, pl. 1, fig. 17, pl. 2, fig. 4, 2004, pl. 2, fig. 6; Nestell and Nestell, 2010, pl. 4, figs. 1—4.

Haplentactinia ichikawai Caridroit et De Wever; Ujiie and Oba, 1991, pl. 4, fig. 2; Blome and Reed, 1992, Fig. 11 (14—16).

Haplentactinia? aff. *ichikawai* Caridroit et De Wever; Sashida et al., 1997, Fig. 6 (16, 18—21).

Unnamed entactiniid Nishimura et Ishiga, 1987, pl. 4, figs. 12, 13.

Stigmosphaerostylus itsukaichiensis (Sashida et Tonishi); Kuwahara et al., 1997, pl. 2, fig. 5.

Stauracontium sp. Zhang et al., 1992, pl. 3, figs. 6, 9.

Stigmosphaerostylus sp. cf. *S. itsukaichiensis* (Sashida et Tonishi); Spiller, 2002, pl. 6, figs. B, C, D.

Stigmosphaerostylus itsukaichiensis (Sashida et Tonishi); Kozur, 1993, pl. 3, fig. 13; Wang et al., 2006, Figs. 11 (ZC), 13 (C), 14 (AA, BB), 15 (S, T, U, V), 2012, pl. 16, figs. 18, 27, 28, pl. 17, fig. 23.

描述：这个种的特点是球形格状壳小，具有6根三棱形主刺，主刺长，强壮，窄棱和宽沟相间，长度大约是壳径的2倍。辅刺小，针状。

比较：这个种与 *Stigmosphaerostylus modesta* (Sashida et Tonishi) 的区别在于后者壳体较大，主刺较短，辅刺不发育。

层位与产地：中上二叠统；世界各地。

碎屑状斑点球桩虫 *Stigmosphaerostylus micula* (Foreman)
(Pl. 2, figs. 27—31, 43, 45; pl. 3, fig. 27)

Entactinia micula Foreman, 1963, pl. 2, fig. 8; Gourmelon, 1985, pl. 2, fig. 2, 1987, pl. 2, figs. 11, 12.

Cubosphaera significans Deflandre, 1958, Fig. 1, 1960, pl. 1, fig. 1.

Entactinia sp. 2, cf. *E. micula* Foreman; Nazarov and Ormiston, 1983, pl. 1, fig. 13.

描述：这个种的特点是格状单壳近球形。6根三棱形主刺大小和形状相似。刺短，长度与壳径相近。壳孔大小不一，圆形至多边形。辅刺不发育。

比较：这个种与 *Stigmosphaerostylus additiva* (Foreman) 的区别为后者主刺有7根，刺的长度不等，其中有1根长度大于壳径。

层位与产地：上泥盆统弗拉阶—下石炭统杜内阶；法国，澳大利亚西部，美国俄亥俄，中国青海可可西里地区。

中等斑点球桩虫 *Stigmosphaerostylus modestus* (Sashida et Tonishi)
(Pl. 5, figs. 1, 13, 15, 16, 18, 19)

Entactinia modesta Sashida et Tomishi, 1985, pl. 1, figs. 11, 12, pl. 2, figs. 1—6; Kuwahara et al., 2004, pl. 2, fig. 7; Feng et al., 2007, pl. 1, figs. 5, 6; Nestell and Nestell, 2010, pl. 5, figs. 1, 2.

描述：这个种的特点是具有1个中等大小、球形格状外壳和6根三棱形主刺，主刺长度略大于壳径，辅刺不甚发育。

层位与产地：中上二叠统；日本，美国，中国华南和青海可可西里地区。

大无畏斑点球桩虫　*Stigmosphaerostylus pantotolma* (Braun)

(Pl. 3, fig. 24)

Entactinia (?) *pantotolma* Braun, 1989, pl. 2, figs. 6, 7; Kiessling and Tragelehn, 1994, pl. 4, fig. 14.
Stigmosphaerostylus pantotolma (Braun); Shu et al., 2007, pl. 1, figs. 4, 5.

描述：这个种的特点是单个格状壳近球形，具 6 根规则排列的角锥形刺，刺短，较粗壮，长度约为壳径的一半。壳孔较大，圆形至卵圆形，每半圈有 10—12 个孔。辅刺不发育。

比较：这个种与 *Stigmosphaerostylus variospina* (Won) 的区别为后者的主刺较长，长度大于壳径的一半。

层位与产地：上泥盆统弗拉阶—下石炭统杜内阶；德国，中国新疆和青海可可西里地区。

长刺斑点球桩虫　*Stigmosphaerostylus proceraspina* (Aitchison)

(Pl. 1, fig. 32)

Entactinia proceraspina Aitchison, 1993, pl. 6, figs. 1, 5, pl. 7, fig. 1.

描述：这个种的特点是单个格状壳球形，具 7 根三棱形主刺，其中 1 根特别长，长度为壳径的 2—3 倍，其余主刺较短，大小和形状相似，长度一般小于或等于壳径，辅刺不发育。

层位与产地：上泥盆统弗拉阶—下石炭统杜内阶；澳大利亚西部卡宁盆地，中国青海可可西里地区。

华丽斑点球桩虫（新种）　*Stigmosphaerostylus spiciosus* Wang sp. nov.

(Pl. 1, figs. 39—41)

词源：spicios，拉丁词，华丽，美丽。

描述：格状单壳近球形，壳孔较大，圆形至多边形。6 根三棱形主刺长度不一，其中 1 根略长，长度略大于或等于壳径，其余 5 根形状和大小相似，长度都小于壳径。辅刺发育，细长。

比较：这个新种与 *Stigmosphaerostylus cometes* (Foreman) 的区别为后者的主长刺特别发育，长度大于壳径，辅刺也特别发育。

层位与产地：上泥盆统法门阶；中国青海可可西里地区。

斑点球桩虫（未定种 A）　*Stigmosphaerostylus* sp. A

(Pl. 6, figs. 25, 32; pl. 7, figs. 35—38)

描述：格状单壳中等大小，近球形或椭圆形，具有 2 根以上三棱形主刺，形状和大小相似，刺短，刺长与壳径相近。刺在基部较宽，向远端逐渐变细，3 根纵脊和 3 条纵沟相间。辅刺不发育。壳孔大，圆形至多边形。每半圈 12—14 个孔。

比较：这个未定种 A 与 *Stigmosphaerostylus itsukaichiensis* (Sashida et Tonishi) 的区别为后者的主刺较长，长度约为壳径的 2 倍，辅刺也很发育。

层位与产地：中二叠统；中国青海可可西里地区。

斑点球桩虫（未定种 B）　*Stigmosphaerostylus* sp. B

(Pl. 6, figs. 23, 24, 26, 27)

描述：格状单壳小，近球形。具有 2—6 根三棱形主刺，刺细长，平直，刺的棱窄，沟较宽。刺的长度大于壳的长径。辅刺不发育。

比较：这个未定种 B 与 *Stigmosphaerostylus modesta* (Sashida et Tonishi) 的区别为后者的主刺较

短，刺的基部较宽，向远端逐渐变窄，辅刺发育。

层位与产地：中二叠统；中国青海可可西里地区。

斑点球桩虫（未定种 C） *Stigmosphaerostylus* sp. C
(Pl. 6, figs. 34, 35, 46, 47)

描述：格状单壳小，近球形。具有 3—6 根纤细的杆形主刺，长度等于或略大于壳径。辅刺不发育。壳孔大小不匀，圆形，卵圆形至多边形。每半圈有 8—10 个孔。

层位与产地：中二叠统；中国青海可可西里地区。

斑点球桩虫（未定种 D） *Stigmosphaerostylus* sp. D
(Pl. 16, fig. 21)

描述：格状单壳小，近球形。具有 6 根较短的三棱形主刺，每个刺的基部较宽，向远端逐渐变窄。主刺长度略小于壳径。壳孔较大，近圆形至多边形。每半圈有 10—12 个孔。

层位与产地：下二叠统；中国青海可可西里地区。

变刺斑点球桩虫 *Stigmosphaerostylus variospina* (Won)
(Pl. 1, figs. 1, 10; pl. 3, figs. 5, 13—15, 19, 22; pl. 9, figs. 26—28; pl. 17, fig. 4)

Palaeoxyphostylus variospina Won, 1983, pl. 3, figs. 11, 22, pl. 8, fig. 30, pl. 9, fig. 12, pl. 10, figs. 29—30.

Entactinia variospina (Won); Gourmelon, 1986, pl. 4, fig. 1, 1987, pl. 3, figs. 6—11; Braun, 1989, pl. 2, figs. 3, 4, pl. 4, fig. 5, 1990a, pl. 1, fig. 4, 1990b, pl. 7, figs. 4—6; Wang and Kuang, 1993, pl. 3, figs. 6—9; Sashida et al., 1993, Fig. 4 (1, 13, 14); Kiessling and Tragelehn, 1994, pl. 4, figs. 23, 24; Feng et al., 1997a, pl. 2, fig. 5, 1997b, pl. 3, figs. 12, 13; Spiller, 1996, pl. 2, figs. 7, 8; Wang et al., 1998, pl. 2, figs. 6, 7; Sashida et al., 1998, Fig. 9 (1—10), 2000, Fig. 6 (5—12).

Entactinia cf. *variospina* (Won); Li and Wang, 1991, pl. 1, figs. 16, 17.

Stigmosphaerostylus variospina (Won); Wang et al., 2000, pl. 2, figs. 10—13, pl. 3, figs. 19, 20, 2003, pl. 4, figs. 1—7, 2012, pl. 1, fig. 27, pl. 3, figs. 11, 22, pl. 8, fig. 30, pl. 9, fig. 12, pl. 10, figs. 29, 30; Sashida et al., 2002, pl. 1, figs. 18, 22, 23, pl. 3, figs. 12, 15; Saesaengseerung et al., 2007, Fig. 8 (7, 17).

描述：这个种的特点是格状单壳孔较大，具有 2—5 根三棱形主刺，其中有 2 根强壮主极刺的标本最普通。

层位与产地：上泥盆统—下石炭统；世界各地。

古老斑点球桩虫（新种） *Stigmosphaerostylus vetulus* Wang sp. nov.
(Pl. 5, figs. 25—28, 31; pl. 6, figs. 36—38)

词源：vetulus, 拉丁词，古老的。

描述：格状壳中等大小，近球形。具有 6 根平直的三棱形主刺，刺细长，长度略大于壳径。3 条纵脊与 3 条纵沟相间。壳孔较小。辅刺稀少，短针状。

比较：这个新种与 *S. itsukaichiensis* (Sashida et Tonishi) 的区别为后者主刺较长，长度约为壳径的 2 倍。辅刺十分发育。

层位与产地：中二叠统茅口阶下部；中国青海可可西里地区。

普通斑点球桩虫 *Stigmosphaerostylus vulgaris*（Won）

(Pl. 17, figs. 8—15, 17)

Entactinia vulgaris Won, 1983, pl. 4, figs. 1—3; Gourmelon, 1986, pl. 2, fig. 4, 1987, pl. 4, figs. 1—6; Giese and Schmidt-Effing, 1989, pl. 2, figs. 4, 9; Braun and Schmidt-Effing, 1993, pl. 1, fig. 8; Wang et al., 2012, pl. 8, fig. 31, pl. 9, fig. 7, pl. 10, fig. 18.

Entactinia vulgaris vulgaris Won; Braun, 1989, pl. 2, figs. 1—3, 1990, pl. 7, figs. 7—10; Wang and Kuang, 1993, pl. 1, figs. 16, 17, pl. 3, fig. 5; Feng et al., 1997, pl. 2, fig. 7, 2004, pl. 2, fig. 11.

描述：这个种的特点是壳近球形，具有 6 根三棱形主刺，主刺长度与壳径相近。壳孔圆形至多边形，半圈壳孔数目 10 个以上。

层位与产地：上泥盆统—下石炭统；德国，法国，中国华南和青海可可西里地区。

三矛虫属 *Trilonche* Hinde, 1899, emend. Foreman, 1963, Aitchison et Stratford, 1997

模式种：*Trilonche vetusta* Hinde, 1899

鉴定要点：具有 2 个格状壳和 6 根三棱形主刺，内骨架由内射双针上发育的 6 根小射刺组成。

时代与分布：晚古生代；世界各地。

达维特三矛虫 *Trilonche davidi*（Hinde）

(Pl. 9, figs. 6, 7, 9, 24, 29, 32, 33)

Staurodonche davidi Hinde, 1899, pl. 8, fig. 13.
Staurodonche tenella Hinde, 1899, pl. 8, fig. 14.
Staurodonche laterna Hinde, 1899, pl. 8, fig. 15.
Staurodruppa nucula Hinde, 1899, pl. 9, fig. 7.
Staurolonchidium grandis Nazarov, 1975, pl. 5, figs. 11, 12, pl. 7, figs. 1—7; Umeda et al., 2004, pl. 1, figs. 1, 2, pl. 2, figs. 6, 7, 11.
Trilonche davidi（Hinde）; Aitchison and Stratford, 1997, Figs. 2 (8), 3 (4); Wang et al., 2000, pl. 2, figs. 16, 17, 2003, pl. 1, figs. 14, 15, pl. 3, fig. 21, 2013, pl. 1, figs. 43—46; Luo et al., 2002, pl. 2, figs. 7, 8; Aitchison et al., 1999, pl. 3, figs. K, L, pl. 4, figs. C, D, F, G, K, L, pl. 5, figs. M, N, pl. 6, figs. C, S; Obut et al., 2008, Fig. 4 (E—G); Wang and Luo, 2009, pl. 1, fig. 11.
Entactinosphaera obliquum Hinde, 1899, pl. 8, fig. 17.

描述：这个种的特点是具有 2 个格状壳和 6 根三棱形主刺，其中有 1 根比较强壮，其余 5 根大小和形状相似，次强壮。辅刺不发育。

比较：这个种与 *Trilonche echinata*（Hinde）的区别为后者的其余 5 根主刺较弱。

层位与产地：中上泥盆统；澳大利亚，俄罗斯南乌拉尔地区和鲁德内阿尔泰地区，中国云南、贵州、广西、新疆和西藏双湖地区。

刺三矛虫 *Trilonche echinatum*（Hinde）

(Pl. 9, figs. 8, 10—16)

Heliosome echinatum Hinde, 1899, pl. 9, figs. 1, 2.
Heliosome paronae Hinde, 1899, pl. 9, figs. 3.
Entactinosphaera palimbola Foreman, 1963, pl. 2, figs. 7a—e, pl. 3, figs. 3a—d; Schmidt-Effing, 1988, pl. 1, figs. 5, 6; Braun, 1990, pl. 2, figs. 5, 6; Kiessling and Tragelehn, 1994, pl. 5, figs. 2, 4, 5.

Entactinosphaera echinata? (Hinde); Foreman, 1963, pl. 3, fig. 10, pl. 4, figs. 12a, b.

Entactinosphaera echinata (Hinde); Nazarov, 1975, pl. 3, figs. 1—3, pl. 4, figs. 1—4; Nazarov and Ormiston, 1983, pl. 1, figs. 6, 7; Aitchison, 1993, pl. 5, figs. 6, 11, 14; Wang, 1997, pl. 3, figs. 4—10, pl. 4, figs. 4, 7.

Entactinosphaera assidera Nazarov, 1975, pl. 5, figs. 6, 7, pl. 6, figs. 6—8.

Entactinosphaera cf. *assidera* Nazarov; Li and Wang, 1991, pl. 1, fig. 17; Aitchison, 1993, pl. 6, figs. 2, 6.

Trilonche echinata (Hinde); Aitchison and Stratford, 1997, pl. 1, figs. 5, 6, 10; Wang et al., 2000, pl. 1, fig. 24, pl. 3, figs. 9—11, 26, 2003, pl. 1, figs. 16—19, pl. 5, figs. 26—30, 2012, pl. 3, fig. 20, 2013, pl. 1, figs. 35, 36; Luo et al., 2002, pl. 1, figs. 13, 14, pl. 2, figs. 4, 5; Obut et al., 2008, Fig. 4 (L); Wang and Luo, 2009, pl. 1, figs. 19, 20; Wonganan and Caridroit, 2005, pl. 2, figs. 17, 18; Hara et al., 2010, Fig. 6 (10, 16).

描述：这个种的特点是具有 2 个格状壳，1 根三棱形主刺长且特别强壮，其他主刺弱小。辅刺发育，细长。

层位与产地：中上泥盆统；世界各地。

华美三矛虫 *Trilonche elegans* Hinde
(Pl. 9, figs. 17, 35)

Trilonche elegans Hinde, 1899, pl. 8, fig. 22; Aitchison and Stratford, 1997, Figs. 2 (7), 3 (6); Aitchison et al., 1999, pl. 1, figs. C, D, F, P, pl. 4, figs. E, I, O; Wang et al., 2003, pl. 1, figs. 22, 23, 2013, pl. 1, figs. 37—42; Umeda et al., 2004, pl. 1, fig. 10; Obut et al., 2008, pl. 2, fig. 3.

Staurodruppa nana Hinde, 1899, pl. 9, fig. 8.

Entactinosphaera symphypora Foreman, 1963, pl. 2, figs. 5a—c, pl. 3, fig. 4; Kiessling and Tragelehn, 1994, pl. 5, fig. 7.

描述：这个种的特点是具有 2 个球形格状壳和 6 根三棱形主刺，主刺形状和大小相似，中等强壮。辅刺不发育。

比较：这个种与 *Trilonche davidi* (Hinde) 的区别为后者的主刺中有 1 根比较粗壮。

层位与产地：中上泥盆统；澳大利亚东南部，美国俄亥俄，德国，俄罗斯南乌拉尔地区和鲁德内阿尔泰地区，中国云南、新疆和西藏双湖地区。

小三矛虫 *Trilonche minax* (Hinde)
(Pl. 9, fig. 5)

Xiphosphaera minax Hinde, 1899, pl. 8, fig. 8.

Stylosphaera obtusa Hinde, 1899, pl. 8, fig. 9.

Staurodruppa praelonga Hinde, 1899, pl. 9, fig. 6.

Staurodruppa prolata Foreman, 1963, pl. 8, figs. 1a, b, pl. 9, fig. 12.

Entactinosphaera egindyensis Nazarov, 1975, pl. 5, figs. 2, 4, pl. 6, figs. 1, 5; Li and Wang, 1991, pl. 1, figs. 3, 4; Yao and Kuwahara, 1999, pl. 1, fig. 12.

Trilonche minax (Hinde); Aitchison and Stratford, 1997, Fig. 2 (3, 4), Fig. 3 (3, 5); Aitchison et al., 1999, pl. 2, figs. H, I, pl. 5, figs. J, K; Luo et al., 2002, pl. 1, figs. 10—12; Wang et al., 2000, pl. 2, fig. 24, 2003, pl. 2, figs. 6, 7, 10 (non 8, 9), pl. 5, figs. 16, 17, 19 (non 15, 18); Umeda et al., 2004, pl. 1, figs. 7, 8, pl. 2, fig. 14; Obut et al., 2008, fig. 4 (M); Wonganan and Caridroit, 2005, pl. 3, figs. 13, 14, 23.

Trilonche sp. B Wang et al., 2003, pl. 3, fig. 22.

描述：这个种的特点是具有 2 个格状壳和 2 根特别强壮的三棱形主刺，位于相对位置，呈一条直线状。

比较：这个种与 *T. pittmani* Hinde 的区别为后者的 2 根强壮的三棱形主刺相向而不在同一直线上，两刺间的交角小于 180°。

层位与产地：中上泥盆统；澳大利亚东南部，俄罗斯南乌拉尔地区和鲁德内阿尔泰地区，美国俄亥俄，泰国，哈萨克斯坦，中国云南、广西、新疆和西藏双湖地区。

皮特曼氏三矛虫　*Trilonche pittmani* Hinde
(Pl. 9, figs. 2—4)

Trilonche pittmani Hinde, 1899, pl. 8, fig. 20；Wang et al., 2012, pl. 1, fig. 18, 2013, pl. 1, figs. 18, 27—30.

Trilonche vetusta Hinde；Aitchison and Stratford, 1997, Fig. 2 (1, 2)；Aitchison et al., 1999, pl. 2, figs. 12, 13；Wonganan and Caridroit, 2005, pl. 3, figs. 6—9, 19；Wang et al., 2000, pl. 2, figs. 25, 26, 2003, pl. 1, fig. 30；Luo et al., 2002, pl. 1, figs. 18, 19；Obut et al., 2008, Fig. 4 (O).

Entactinosphaera aitpaiensis Nazarov, 1973, pl. 1, figs. 4, 5, 1975, pl. 5, figs. 3, 5, pl. 6, figs. 2—4；Li and Wang, 1991, pl. 1, figs. 8—10.

Trilonche minax (Hinde)；Wang et al., 2003, pl. 2, figs. 8, 9 (non 6, 7, 10), pl. 5, figs. 15, 18 (non 16, 17, 19).

Bientactinosphaera pittmani (Hinde)；Afanasieva, 2000, pl. 44, figs. 4—11, pl. 126, fig. 7.

Entactiniidae gen. et sp. indet. sp. F Stratford et Aitchison, 1996, pl. 7, fig. E.

Entactinosphaera sp. A Wang, 1997, pl. 4, figs. 14, 15.

描述：这个种的特点是有 2 个格状壳，外壳上发育 2 根特别强壮的三棱形主刺，长度近乎相等，相对而生，但不在同一条直线上，两刺之间有一个较大的交角，小于 180°，大于 120°。辅刺不发育。

比较：这个种与 *T. minax* (Hinde) 的区别为后者的 2 根主刺位于同一直线上。

层位与产地：中上泥盆统；澳大利亚，俄罗斯南乌拉尔地区和鲁德内阿尔泰地区，泰国，哈萨克斯坦，中国云南、广西、新疆和西藏双湖地区。

假美丽三矛虫　*Trilonche pseudocimelia* (Sashida et Tonishi)
(Pl. 12, fig. 37；pl. 14, figs. 13, 14)

Entactinosphaera pseudocimelia Sashida et Tonishi, 1988, Fig. 7 (1—3, 6—8)；Sashida et al., 1995, Fig. 11 (1—3), 2000, Fig. 7 (17—19)；Kuwahara et al., 1997, pl. 2, figs. 7, 8；Kuwahara and Yao, 1998, pl. 3, fig. 77, 2001, pl. 2, fig. 9；Shang et al., 2001, pl. 4, figs. 1, 2；Suzuki and Kuwahara, 2003, pl. 1, fig. 9.

Entactinosphaera cimelia Nazarov et Ormiston；Wang and Li, 1994, pl. 1, fig. 18.

Trilonche pseudocimelia (Sashida et Tonishi)；Wang et al., 2012, pl. 19, fig. 13, pl. 21, figs. 23—25.

描述：这个种的特点是外壳球形，具有 2 根三棱形主刺，相对而生。外壳表面发育多边形房室。当前标本内部构造未显示，但外形与正模十分相似。

层位与产地：上二叠统；日本，泰国，中国华南、青海可可西里地区和西藏双湖地区。

三矛虫（未定种 B）　*Trilonche* sp. B
(Pl. 7, fig. 45)

描述：格状外壳近球形，壳小，壳孔较大，大小不匀，近圆形、卵形或多边形。具有 2 根强壮的三棱形主刺，长度不相等，长刺长度与壳径相近，短刺的长度小于壳径。

比较：这个未定种 B 与 *Trilonche pseudocimelia* (Sashida et Tonishi) 的区别为后者壳表发育多边形房室，2 根主刺长度相等。

层位与产地：中二叠统；青海可可西里地区。

三射三矛虫 *Trilonche tretactinia* (Foreman)

(Pl. 9, figs. 31, 34, 36—40, 43)

Entactinosphaera tretactinia Foreman, 1963, pl. 2, figs. 6a, b, pl. 6, figs. 2a, b; Li and Wang, 1991, pl. 2, figs. 6—8.
Trilonche tretactinia (Foreman); Wang et al., 2003, pl. 5, fig. 24, 2012, pl. 4, figs. 27—30.

描述：这个种的特点是具有2个格状壳，内外壳都较小。6根三棱形主刺较粗壮但较短，长度一般小于外壳直径。

层位与产地：上泥盆统法门阶；美国俄亥俄，中国华南和西藏双湖地区。

星内射虫科 Astroentactiniidae Nazarov et Ormiston, 1985
星内射虫亚科 Astroentactiniinae Nazarov et Ormiston, 1985
星内射虫属 *Astroentactinia* Nazarov, 1975

模式种：*Astroentactinia stellata* Nazarov, 1975

鉴定要点：外壳球形，单壳。内骨架由从中梁上产生的纤细6—7射小骨针组成，通常射上发育骨刺。外部主刺数目多，一般为8—24根，杆形，锥形或三棱形。

时代与分布：晚泥盆世—早石炭世；世界各地。

双尖星内射虫 *Astroentactinia biaciculata* Nazarov

(Pl. 1, figs. 11, 15, 16, 20, 22, 24, 25; pl. 2, figs. 37, 39—42, 44; pl. 17, figs. 18—20)

Astroentactinia biaciculata Nazarov, 1975, pl. 8, fig. 8, pl. 10, figs. 6, 7; Schmidt-Effing, 1988, pl. 1, fig. 8, pl. 2, fig. 3; Wang, 1995, pl. 17, fig. 18; Kiessling and Tragelehn, 1994, pl. 4, fig. 6; Nazarov, 1988, pl. 13, fig. 3; Spiller, 2002, pl. 5, figs. I, J, L, M; Braun and Schmidt-Effing, 1988, Fig. 18, 1993, pl. 1, fig. 2; Braun, 1989, pl. 2, fig. 8; Gourmelon, 1987, pl. 8, figs. 1—5.

描述：这个种的特点是单个格状壳球形，具有12—16根杆形主刺，刺短，长度小于壳径，壳孔较大。

比较：这个种与*Astroentactinia paronae* (Hinde)的区别为后者主刺中有1根比其他主刺长。

层位与产地：中上泥盆统—下石炭统；澳大利亚，俄罗斯乌拉尔地区，马来西亚，德国，法国，中国青海可可西里地区。

多刺星内射虫 *Astroentactinia multispinosa* (Won)

(Pl. 3, figs. 26, 28—30)

Entactinia? multispinosa Won, 1983, pl. 2, figs. 15, 16.
Entactinia? brilonensis Won, 1983, pl. 6, figs. 13—16, 1998, pl. 7, figs. 4, 5; Braun, 1989, pl. 2, fig. 14.
Astroentacinia multispinosa (Won); Gourmelon, 1986, pl. 4, fig. 2, 1987, pl. 8, fig. 10; Giese and Schmidt-Effing, 1989, pl. 2, fig. 2; Braun, 1990, pl. 9, figs. 1—3, pl. 12, fig. 1; Li and Bian, 1993, pl. 3, figs. 9, 10; Feng et al., 1997, pl. 4, figs. 1, 4, 2004, pl. 2, figs. 9, 10; Wang et al., 2012, pl. 6, figs. 29—31, pl. 9, fig. 9; Saesaengseerung et al., 2007, Fig. 8 (15, 16).
Entactinosphaera? trendalli Won; Braun, 1989, pl. 2, fig. 13.

描述：这个种的特点是格状单壳球形，孔密，近圆形。主刺锥形，数目多，在同一平面中有20根以上。主刺基部较宽但较短。

比较：这个种与*Astroentactinia radiata* Braun十分相似，区别在于后者的主刺为杆形。

层位与产地：上泥盆统—下石炭统；德国，法国，泰国，中国华南和青海可可西里地区。

帕罗纳氏星内射虫 *Astroentactinia paronae* (Hinde)

(Pl. 1, figs. 21, 23, 26—28, 36—38; pl. 2, figs. 11, 13—15, 25)

Heliosome paronae Hinde, 1899, pl. 9, fig. 3.

Astroentactinia paronae (Hinde); Nazarov, 1975, pl. 8, fig. 7, pl. 10, figs. 4, 5, 13, 1988, pl. 14, fig. 5; Nazarov and Ormiston, 1983, pl. 1, fig. 3; Umeda et al., 2004, pl. 1, fig. 13.

Astroentactinia stellata Nazarov; Nazarov and Ormiston, 1983, pl. 1, figs. 8, 9; Aitchison, 1993, Fig. 6 (4).

Astroentactinia sp. aff. *A. paronae* (Hinde); Kiessling and Tragelehn, 1994, pl. 4, fig. 5.

Entactinia sexradiata Won, 1997, pl. 1, figs. 5—7, 13, 14, 17.

描述：这个种的特点是格状单壳小，球形。发育许多杆形主刺，其中有 1 根主刺比其他刺强壮，刺细长。

层位与产地：上泥盆统；澳大利亚，俄罗斯南乌拉尔地区和鲁德内阿尔泰地区，德国，中国青海可可西里地区。

星状星内射虫 *Astroentactinia stellata* Nazarov

(Pl. 1, figs. 12—14, 17—19; pl. 2, fig. 16)

Astroentactinia stellata Nazarov, 1975, pl. 8, fig. 6, pl. 10, figs. 1—3; Nazarov and Ormiston, 1983, pl. 1, figs. 8, 9; Aitchison, 1993, pl. 6, fig. 4; Kiessling and Tragelehn, 1994, pl. 4, fig. 7; Wang et al., 2003, pl. 3, figs. 27, 28, 2012, pl. 3, fig. 21, pl. 4, fig. 26; Wonganan and Caridroit, 2005, pl. 3, figs. 26, 27; Nazarov, 1988, pl. 14, fig. 9; Saesaengseerung et al., 2007, Fig. 8 (13, 14); Obut et al., 2008, Fig. 4 (A, B); Umeda et al., 2004, pl. 1, fig. 12, pl. 2, figs. 9, 10, pl. 3, fig. 1.

Astroentactinia biaciculata Nazarov; Braun, 1990, pl. 2, fig. 8.

描述：这个种的特点是格状单壳小，球形。主刺数目多，较短，角锥形，大小和形状相似，基部具角卵形孔。内部小骨针一般位于偏心位置。

比较：这个种与 *Astroentactinia paronae* (Hinde) 的区别在于后者的主刺中有 1 根比其他主刺长。

层位与产地：上泥盆统弗拉阶—法门阶；俄罗斯南乌拉尔地区和鲁德内阿尔泰地区、澳大利亚、德国、泰国、中国华南和青海可可西里地区。

球虫目 Sphaerellaria Haeckel, 1887
多内射虫超科 Polyentactinioidea Nazarov, 1975
多内射虫科 Polyentactiniidae Nazarov, 1975
多内射虫亚科 Polyentactiniinae Nazarov, 1975
多内射虫属 *Polyentactinia* Foreman, 1963

模式种：*Polyentactinia craticulata* Foreman, 1963

鉴定要点：具有一个粗网状的球形外壳。主刺 6 根或更多，杆形或三棱形。内骨架由中梁上产生的 6—8 射小骨针组成。

时代与分布：中奥陶世—早二叠世；世界各地。

蜘蛛多内射虫 *Polyentactinia aranea* Gourmelon

(Pl. 2, fig. 17)

Polyentactinia aranea Gourmelon, 1986, pl. 4, fig. 5, 1987, pl. 10, figs. 7—9; Braun, 1990b, pl. 14, figs. 4, 5, 10, pl. 15, figs. 13—15; Kiessling and Tragelehn, 1994, pl. 5, fig. 10; Wang et al., 2003, pl. 5, figs. 8—12, 2012, pl. 8, figs. 11—13.

Polyentactinia sp. A Gourmelon, 1986, pl. 4, fig. 5.

Polyentactinia sp. aff. *P. aranea* Gourmelon; Braun, 1989, pl. 1, fig. 10.

  描述：这个种的特点是格状单壳近球形。壳孔大，多边形，每半圈有 7—9 个孔。主刺杆形，细长，长度大于壳径。辅刺短针状。

  比较：这个种与 *Polyentactinia polygonia* Foreman 的区别为后者的主刺短，长度一般小于壳径，壳孔大，数量少。

  层位与产地：上泥盆统法门阶—下石炭统杜内阶；法国，德国，中国华南和青海可可西里地区。

单内射虫科   Haplentactiniidae Nazarov, 1980
单内射虫亚科   Haplentactiniinae Nazarov, 1980
三叉球虫属 *Triaenosphaera* Deflandre, 1963

  模式种：*Triaenosphaera sicarius* Deflandre, 1963

  鉴定要点：格状单壳球形。4 根主刺三棱形或杆形，呈四面体状排列。内骨架由一点产生的 4 射内骨针组成。

  时代与分布：晚泥盆世—早石炭世；世界各地。

年轻三叉球虫   *Triaenosphaera hebes* Won
(Pl. 2, fig. 21; pl. 3, figs. 17, 21)

Triaenosphaera hebes Won, 1983, pl. 3, figs. 12, 13; Gourmelon, 1986, pl. 2, fig. 3, 1987, pl. 7, figs. 1—5; Braun, 1990b, pl. 11, fig. 10, pl. 12, fig. 12.

  描述：这个种的特点是格状单壳球形。4 根三棱形主刺较强壮，呈四面体状排列，长度小于壳径。

  层位与产地：上泥盆统—下石炭统；德国，法国，中国青海可可西里地区。

粗刺三叉球虫（新种）   *Triaenosphaera robustispina* Wang sp. nov.
(Pl. 9, figs. 22, 23)

  词源：robustispinus，拉丁词，粗刺的。

  描述：格状壳中等大小，近球形。具有 4 根粗壮的三棱形主刺，呈四面体状排列，形状和大小相似。主刺基部最宽，向远端迅速变窄，至顶部呈锥状。壳孔较大，每半圈有 7—8 个孔。主刺长度大于壳径。

  比较：这个新种与 *T. sicarius* Deflandre 的区别为后者主刺较短，基部宽度较小，壳孔较多。

  层位与产地：上泥盆统弗拉阶；中国西藏双湖地区。

剑三叉球虫   *Triaenosphaera sicarius* Deflandre
(Pl. 2, figs. 19, 20, 22; pl. 9, figs. 19—21, 25, 30, 41, 42)

Triaenosphaera sicarius Deflandre, 1973, pl. 2, figs. 3, 4; Holdsworth et al., 1978, Fig. 2a—c; Gourmelon, 1986, pl. 2, fig. 1, 1987, pl. 6, figs. 1—4; Schmidt-Effing, 1988, pl. 3, fig. 5; Braun, 1990b, pl. 11, figs. 8, 9; Kiessling and Tragelehn, 1994, pl. 5, figs. 21, 22; Wang et al., 2003, pl. 5, fig. 7, 2012, pl. 1, figs. 29, 31, 32.

  描述：这个种的特点是格状单壳球形。4 根三棱形主刺呈四面体状排列。主刺特别强壮，长度略大于壳径。壳孔圆形至多边形。无辅刺。

  比较：这个种与 *Triaenosphaera hebes* Won 的区别为后者的主刺欠强壮，较短，长度一般小于壳径。

层位与产地：上泥盆统—下石炭统；法国，德国，中国华南和青海可可西里地区。

海绵内射虫超科 Spongentactinioidea Nazarov，1975
海绵内射虫科 Spongentactiniidae Nazarov，1975
海绵内射虫亚科 Spongentactiniinae Nazarov，1975
四内射虫属 *Tetrentactinia* Foreman，1963

模式种：*Tetrentactinia barysphaera* Foreman，1963

鉴定要点：外壳海绵状，单壳，4根三棱形主刺呈四面体状排列。内骨架由从中心产生的纤细4射小骨针组成。

时代与分布：志留纪—早石炭世；俄罗斯，哈萨克斯坦，白俄罗斯，美国，中国华南和青海可可西里地区。

大壳四内射虫（新种） *Tetrentactinia gigantia* Wang sp. nov.
(Pl. 3，fig. 20)

词源：gigant，希腊词，巨大的。

描述：海绵外壳极大，近球形。具有4根三棱形主刺，呈四面体状排列。主刺形状和大小相似，长度小于壳径。辅刺不发育。内部构造未显示。壳径450 μm，主刺长300 μm，基部宽100 μm。

比较：这个种以其特别巨大的壳径和主刺长度区别于这个属的其他种。

层位与产地：上泥盆统法门阶；中国青海可可西里地区。

海绵四内射虫 *Tetrentactinia spongacea* Foreman
(Pl. 2，figs. 12，18，23；pl. 9，fig. 18)

Tetrentactinia spongacea Foreman，1963，pl. 7，fig. 5；Wang et al.，2012，pl. 2，figs. 23—25，27，28，pl. 4，figs. 7，14—16.

描述：这个种的特点是单个海绵壳近球形。4根相等的三棱形主刺呈四面体状排列。主刺长度与壳径相近。海绵层较厚，壳孔较小。当海绵层破损时，可见较大的壳孔。辅刺不发育。

比较：这个种与*Spongentactinia spongites*（Foreman）的区别为后者的主刺有6根，相互间相互垂直。

层位与产地：上泥盆统法门阶；美国俄亥俄，中国青海可可西里地区。

海绵内射虫属 *Spongentactinia* Nazarov，1975

模式种：*Spongentactinia fungosa* Nazarov，1975

鉴定要点：外壳具海绵层，内壳格状。具有6根主刺与4—6射小骨针相接。内壳和外壳邻近。辅刺有或无。

时代与分布：晚泥盆世；美国，澳大利亚，俄罗斯南乌拉尔地区，中国青海可可西里地区。

弱刺海绵内射虫 *Spongentactinia exilispina*（Foreman）
(Pl. 2，figs. 26，32—36，38)

Entactinia exilispina Foreman，1963，pl. 1，fig. 8；Kiessling and Tragelehn，1994，pl. 4，fig. 16.

描述：这个种的特点是海绵外壳近球形。6根形状和大小相近的三棱形主刺细长，长度大于壳径。辅刺细小。

比较：这个种与 *Spongentactinia spongites* (Foreman) 的区别为后者的壳体较小，主刺较短。

层位与产地：上泥盆统法门阶；美国俄亥俄，德国，中国青海可可西里地区。

欠发育海绵内射虫 *Spongentactinia indisserta* Nazarov
(Pl. 3, figs. 16, 25)

Spongentactinia indisserta Nazarov, 1975, pl. 11, figs. 7, 8.

描述：这个种的特点是单个海绵壳较大，近球形。具有6根三棱形主刺，其中2根相对的极主刺较长，长度略大于壳径。与正模相比，我们的标本辅刺不发育。

层位与产地：上泥盆统；俄罗斯南乌拉尔地区，中国青海可可西里地区。

海绵状海绵内射虫 *Spongentactinia spongites* (Foreman)
(Pl. 3, figs. 18, 23)

Entactinia spongites Foreman, 1963, pl. 1, fig. 7; Kiessling and Tragelehn, 1994, pl. 4, fig. 15.
? *Spongentactinia spongites* (Foreman); Nazarov, 1975, p. 75; Nazarov et al., 1982, Fig. 4 (G).

描述：这个种的特点是海绵单壳近球形。6根三棱形主刺较短，较粗壮，长度小于壳径。主刺基部较宽，向远端迅速变窄。辅刺不发育。

层位与产地：上泥盆统法门阶；美国俄亥俄，澳大利亚，德国，中国青海可可西里地区。

海绵内射虫（未定种A） *Spongentactinia* sp. A
(Pl. 3, figs. 6—11)

描述：单个海绵壳较大，近球形。2根三棱形主刺相对而生，位于同一直线或有轻微的交角。刺短小，长度小于壳径。辅刺不发育。

层位与产地：上泥盆统法门阶；中国青海可可西里地区。

海绵内射虫（未定种B） *Spongentactinia* sp. B
(Pl. 1, figs. 2, 5, 6)

描述：海绵外壳椭圆形。具有2根长度不等的三棱形主刺，主刺长度一般大于壳的长径。辅刺不发育。

层位与产地：上泥盆统法门阶；中国青海可可西里地区。

海绵内射虫（未定种C） *Spongentactinia* sp. C
(Pl. 1, figs. 7, 8; pl. 3, fig. 12)

描述：壳体由2层组成，外壳海绵状，近球形。内层格状，孔较大。两壳相邻。具有2根三棱形主刺，相对而生，主刺短小，长度小于壳径。辅刺细小，不甚发育。

层位与产地：上泥盆统法门阶；中国青海可可西里地区。

泡沫虫目 Spumellaria Ehrenberg, 1875, emend. De Wever et al., 2001
海绵尾虫超科 Sponguracea Haeckel, 1862, emend. De Wever et al., 2001
奥特尔海绵虫科 Oertlispongidae Kozur et Mostler, 1980

多内圆虫亚科 Copicyntrinae Kozur et Mostler, 1989
多内圆虫属 *Copicyntra* Nazarov et Ormiston, 1985

模式种：*Copicyntra cuspidata* Nazarov et Ormiston, 1985

鉴定要点：海绵外壳和格状内球间发育 8—13 个同心壳，很少有更多的薄海绵同心壳。所有各壳由放射状横梁相连。壳表发育许多三棱形主刺。

时代与分布：晚石炭世—二叠纪；俄罗斯南乌拉尔地区，日本，泰国，菲律宾，美国德克萨斯，中国华南和青海可可西里地区。

秋川多内圆虫 *Copicyntra akikawaensis* Sashida et Tonishi
(Pl. 6, fig. 33; pl. 13, figs. 38—43)

Copicyntra akikawaensis Sashida et Tonishi, 1988, Fig. 7 (14—17); Tumanda et al., 1990, pl. 2, fig. 11; Feng and Liu, 1993, pl. 5, figs. 1, 2; Kuwahara and Yao, 1998, pl. 2, fig. 43, 2001, pl. 1, fig. 4; Shang et al., 2001, pl. 4, fig. 10; Sashida and Salyapongse, 2002, Fig. 3 (27); Wang et al., 2006, Fig. 11 (EE, ZB), 12 (P, Q), 15 (kk), 2012, pl. 17, fig. 3, pl. 19, fig. 41, pl. 20, fig. 25.

Copicyntra cf. *akikawaensis* Sashida et Tonishi; Kuwahara et al., 2004, pl. 2, fig. 1.

Copicyntra sp. Rudenko et Panasenko, 1997, pl. 16, fig. 16; Sashida et al., 2000, Fig. 7 (29, 30).

描述：这个种的特点是球形海绵外壳具有许多锥形刺，半圈中有 20 个以上，其长度小于壳径的一半。内部构造未显现。

层位与产地：上二叠统；日本，菲律宾，泰国，俄罗斯远东地区，中国华南和西藏双湖地区。

长刺多内圆虫 *Copicyntra cuspidata* Nazarov et Ormiston
(Pl. 6, figs. 40, 45—47)

Copicyntra cuspidata Nazarov et Ormiston, 1985, pl. 2, fig. 14; Maldonado and Noble, 2010, pl. 8, fig. 13; Nestell and Nestell, 2010, pl. 10, fig. 4.

描述：这个种的特点是海绵外壳近球形。具有很多细长的三棱形主刺，刺的末端变尖。由于缺乏破损标本，外壳和内球间的壳圈数目未显现，但当前标本外部主要特征与这个种的正模十分相似。

层位与产地：中二叠统伦纳德阶；美国德克萨斯，俄罗斯南乌拉尔地区，中国青海可可西里地区。

多内椭圆虫属 *Copiellintra* Nazarov et Ormiston, 1985

模式种：*Copiellintra diploacantha* Nazarov et Ormiston, 1985

鉴定要点：海绵外壳椭圆形，内球多孔。内外壳间发育 10 个以上椭圆形海绵薄壳，2 根三棱形主刺平直或扭曲。

时代与分布：晚石炭世—二叠纪；俄罗斯南乌拉尔地区和远东地区，美国，中国华南和青海可可西里地区。

双刺多内椭圆虫 *Copiellintra diploacantha* Nazarov et Ormiston
(Pl. 6, figs. 29—31)

Copiellintra diploacantha Nazarov et Ormiston, 1985, pl. 2, fig. 5.

描述：这个种的特点是海绵外壳椭圆形。2 根三棱形主刺短，相对而生。内球和外壳间发育 10 个以上椭圆形海绵壳。由于没有破损的外壳，在我们的标本中这些海绵壳没有显示。

层位与产地：下中二叠统；俄罗斯南乌拉尔地区，中国青海可可西里地区。

赫格勒虫属 *Hegleria* Nazarov et Ormiston，1985

模式种：*Hegleria mammifera* Nazarov et Ormiston，1985

鉴定要点：壳体由1个海绵外壳和1—2个髓壳组成，外壳上具许多乳头状突起，每个乳头含有许多排列规则的小孔。具有或不发育外刺。

时代与分布：中晚二叠世；世界各地。

乳头状赫格勒虫 *Hegleria mammilla* (Sheng et Wang)

(Pl. 6，figs. 39，44)

Phaenicosphaera mammilla Sheng et Wang，1985，pl. 3，figs. 1—8；Kozur and Krahl，1987，Fig. 7（a）；Wang R J，1993a，pl. 2，figs. 1—6，1993b，pl. 2，figs. 13—16；Xia and Zhang，1998，pl. 1，fig. 19.

Spumellarians gen. et sp. indet.；Blome et al.，1986，pl. 8.1，figs. 20，21.

Hegleria mammifera Nazarov et Ormiston，1985，pl. 6，figs. 3—5；Wang，1991，pl. 3，figs. 5—7；Kuwahara et al.，1997，pl. 3，figs. 17，18；Kuwahara and Yao，2001，pl. 1，fig. 14，pl. 2，fig. 10；Nestell and Nestell，2010，pl. 11，fig. 2.

Hegleria mammilla (Sheng et Wang)；Blome and Reed，1992，Fig. 11 (10，12，13)，1995，pl. 1，figs. 27，28；Wang et al.，1994，pl. 2，figs. 17，18，1998，pl. 5，figs. 16，17，2006，Fig. 11 (JJ)，13 (V)，14 (NN)，2012，pl. 17，fig. 4，pl. 21，fig. 31；Wang and Li，1994，pl. 1，figs. 22，23；Jasin et al.，1995，pl. 1，fig. 4；Wang and Qi，1995，pl. 5，figs. 1—12；Wang and Fan，1997，pl. 1，figs. 1—3；Sashida et al.，1993，pl. 1，fig. 14，1997，Fig. 6 (4，5)，2000a，Fig. 7 (28)，2000b，pl. 3，figs. 1—4，6；Kuwahara et al.，1997，pl. 3，figs. 17，18；Feng et al.，1998，Fig. 3 (F—H，K)，2007，pl. 5，figs. 14，16；Yao and Kuwahara，1999b，pl. 4，fig. 22；Shang et al.，2001，pl. 4，figs. 8，9；Feng and Gu，2002，Fig. 4 (1—3)；Spiller，2002，pl. 6，fig. A；Wang and Yang，2003，pl. 1，figs. 18，19；Suzuki and Kuwahara，2003，pl. 1，fig. 13；Kuwahara et al.，2003，pl. 1，fig. 19，2004，pl. 2，fig. 14，2005，pl. 2，figs. 1，2；Wonganan and Garidroit，2006，pl. 1，fig. 29；Kametaka et al.，2009，Fig. 8 (8，9，non 10)；Maldonado and Noble，2010，pl. 10，figs. 5—7.

Phaeniosphaera mammifera (Nazarov et Ormiston)；Kozur，1989，Fig. 1，1993，pl. 2，fig. 1；Catalano et al.，1991，pl. 6，fig. 6，1992，Fig. 6J.

Hegleria sp. Noble et Renne，1990，pl. 1，figs. 9，10.

gen. et sp. indet. Ujeii et Oba，1991，pl. 4，fig. 10.

Hegleia sp. B，sp. CA，sp. CB，sp. CC Yao and Kuwahara，2002，pl. 1，figs. 12—15.

Hegleria sp. aff. *H. mammilla* (Sheng et Wang)；Kametaka et al.，2009，Fig. 8 (11).

描述：这个种的特点是具有1个海绵状的球形外壳和1—2个髓壳。外壳上发育许多乳头状突起，每个乳头含有许多排列规则的小孔。有些标本外壳上发育外刺。

层位与产地：中上二叠统；世界各地。

古海绵梅虫科 Archaeospongoprunidae Pessagno，1973
古海绵梅虫属 *Archaeospongoprunum* Pessagno，1973

模式种：*Archaeospongoprunum venadoensis* Pessagno，1973

鉴定要点：海绵单壳球形或椭圆形。2根三棱形主刺相对而生，刺的长度相等或不等。刺平直或部分扭曲。辅刺不发育。

时代与分布：二叠纪—白垩纪；世界各地。

清道古海绵梅虫　*Archaeospongoprunum chiangdaoensis*（Sashida et Tonishi）

(Pl. 12，fig. 38)

Pseudospongoprunum? *chiangdaoensis* Sashida et Tonishi，2000a，Fig. 7（20，22，23，non 21）.

Archaeospongoprunum sp. A Wang et Shang，2001，pl. 2，figs. 10—12.

Archaeospongoprunum chiangdaoensis（Sashida）；Feng et al.，2006，pl. 6，figs. 6—13.

描述：这个种的特点是海绵壳椭圆形。2根三棱形极主刺相对而生，主刺短，长度小于壳径。辅刺不发育。

比较：这个种在外形上与*Paroertlispongus fontainei*（Sashida）十分相似，区别在于后者的2根主刺是杆形的。

层位与产地：上二叠统；泰国，中国华南和西藏双湖地区。

左旋刺古海绵梅虫（新种）　*Archaeospongoprunum sinisterispinosum* Wang sp. nov.

(Pl. 6，fig. 28)

词源：sinister，拉丁词，左旋；spin，拉丁词，刺。

描述：海绵壳椭圆形，中等大小，壳孔小。2根形状和大小相似的三棱形主刺细长，长度与壳的长径相近。两刺呈轻微的左旋扭曲。辅刺不发育。

比较：这个新种与产自晚侏罗世—早白垩世的*Archaeospongoprunum patricki* Jud在形状和构造上十分相似，区别在于后者的主刺平直而不扭曲或只有1刺呈轻微左旋扭曲。

层位与产地：中二叠统；中国青海可可西里地区。

罩笼虫目　Nassellaria Ehrenberg, 1875
门孔虫超科　Pylentonemoidea Deflandre, 1963
门孔虫科　Pylentonemidae Deflandre, 1963
门孔虫亚科　Pylentoneminae Deflandre, 1963
门孔虫属　*Pylentonema* Deflandre, 1963

模式种：*Pylentonema antiqua* Deflandre，1963

鉴定要点：壳表发育7根三棱形主刺、1根顶刺、3根侧刺和3根分散足。头大，近球形。门孔大，门孔边缘窄，并轻微翘起。

时代与分布：晚泥盆世—早石炭世；世界各地。

奇异门孔虫　*Pylentonema mira* Cheng

(Pl. 17，figs. 6，7)

Pylentonema mira Cheng，1986，pl. 1，fig. 4，pl. 2，figs. 19，25；Schwartzapfel and Holdsworth，1996，pl. 2，figs. 2，14；Wang et al.，2012，pl. 1，figs. 22，23，pl. 3，figs. 27，32.

描述：这个种的特点是壳大，头近球形，孔多边形。壳表具有7根主刺、1根顶刺、3根侧刺和3根强壮的三棱形分散足。分散足轻微扭曲。门孔圆形，裙边不明显。

比较：这个种与*Pylentonema typica* Cheng的区别为后者的分散足不扭曲。

层位与产地：上泥盆统法门阶；美国荷克拉荷马，中国华南和青海可可西里地区。

门孔虫（未定种 A） *Pylentonema* sp. A Schwartzapfel et Holdsworth

(Pl. 2, fig. 10)

Pylentonema sp. A Schwartzapfel et Holdsworth, 1996, pl. 25, figs. 7, 11, 12, 15, 19.

描述：这个未定种 A 的特点是格状壳小，壳孔大，圆形至椭圆形。具有 7 根细弱的主刺、1 根顶刺、3 根侧刺和 3 根分散足。

层位与产地：上泥盆统法门阶—下石炭统杜内阶；美国何克拉荷马州，中国青海可可西里地区。

原笼虫亚科 Archocyrtiinae Kozur et Mostler, 1981

原笼虫属 *Archocyrtium* Deflandre, 1972

模式种：*Archocyrtium riedeli* Deflandre, 1972, emend. Cheng, 1986

鉴定要点：头大，近球形。有一个大门孔，门孔边缘发育高度不等的裙边。头上发育 4 根三棱形主刺、1 根顶刺和 3 根分散足。

时代与分布：早志留世—早石炭世；世界各地。

卡斯塔利格原笼虫 *Archocyrtium castuligerum* Deflandre

(Pl. 2, figs. 2, 3)

Archocyrtium castuligerum Deflandre, 1972, pl. 4, fig. 6, Gourmelon, 1985, pl. 2, fig. 10, 1987, pl. 18, figs. 4—9; Braun, 1990, pl. 17, fig. 6.

描述：这个种的特点是头近球形，头上具孔，每半圈约有 8 个孔。顶刺长度与头径相近，3 根分散足平直，角度小，其长度比顶刺长，刺呈三棱形。裙边梯形，较高。

比较：这个种与 *Archocyrtium tinnulum* Deflandre 的区别为后者头上具有侧刺，足分散角度较大。

层位与产地：上泥盆统法门阶—下石炭统维宪阶下部；法国，德国，中国青海可可西里地区。

双管原笼虫 *Archocyrtium diductum* Deflandre

(Pl. 2, figs. 4, 6; pl. 3, fig. 2)

Archocyrtium diductum Deflandre, 1973, pl. 1, fig. 4, pl. 2, fig. 4; Gourmelon, 1985, pl. 2, fig. 11, 1987, pl. 18, figs. 10—15; Braun, 1989, pl. 3, fig. 5, 1990, pl. 16, fig. 8, pl. 17, fig. 12; Feng et al., 2004, pl. 1, fig. 10.

Archocyrtium sp. Holdsworth, 1973, pl. 1, fig. 3.

描述：这个种的特点是头长圆形，孔大，圆形，每半圈有 5—6 个孔。4 根三棱形主刺较强壮；1 根顶刺和 3 根分散足，足轻微向内弯曲。裙边较高。

比较：这个种与 *Archocyrtium castaligerum* Deflandre 的区别为后者的裙边更高。

层位与产地：上泥盆统法门阶—下石炭统维宪阶下部；法国，德国，土耳其，中国青海可可西里地区。

铁色原笼虫 *Archocyrtium ferreum* Braun

(Pl. 2, figs. 8, 9)

Archocyrtium ferreum Braun, 1989, pl. 1, fig. 7, 1990, pl. 16, figs. 5, 6, pl. 17, fig. 4; Wang et al., 2012, pl. 6, figs. 13—23, pl. 7, figs. 6—9, 13, 14.

描述：这个种的特点是头近球形，壳孔较大，每半圈有 6—9 个孔。4 根三棱形主刺长度不等，但都比较短，顶刺比 3 根分散足更短。裙边较窄。

比较：这个种与 *Archocyrtium diductum* Deflandre 的区别为后者的头呈长圆形，壳孔较少，裙边不发育。

层位与产地：上泥盆统法门阶—下石炭统维宪阶下部；德国，中国华南和青海可可西里地区。

拉加布赖尔原笼虫　*Archocyrtium lagabriellei* Gourmelon
(Pl. 17, fig. 5)

Archocyrtium lagabriellei Gourmelon, 1987a, pl. 19, figs. 1—4, pl. 22, fig. 2; Braun and Schmidt-Effing, 1988, Fig. 8, 1993, pl. 2, fig. 9; Braun, 1989, pl. 3, fig. 6, 1990, pl. 17, figs. 7, 8; Wang and Kuang, 1993, pl. 2, figs. 9—14, 19; Feng et al., 2004, pl. 2, fig. 8; Noble et al., 2008, Fig. 7 (16, 17).

Archocyrtium? sp. A Gourmelon, 1986, pl. 1, fig. 3.

描述：这个种的特点是头近球形，穿孔，孔稀疏，每半圈有 7—8 个孔。在我们的标本中，只保存了 1 根三棱形顶刺，长度与头径相近。边缘未穿孔，张开成裙边状。3 根分散足未保存。

比较：这个种与 *Archocyrtium riedeli* Deflandre 的区别为后者的头较小，孔排列紧密，特别是 4 根三棱形主刺强壮。

层位与产地：下石炭统；法国，德国，土耳其，泰国，中国华南和青海可可西里地区。

荒谬原笼虫　*Archocyrtium ludicrum* Deflandre
(Pl. 2, fig. 7)

Archocyrtium ludicrum Deflandre, 1973, pl. 1, fig. 5, pl. 2, fig. 5; Kiessling and Tragelehn, 1994, pl. 2, figs. 11—13.

Archocyrtium sp. cf. *A. ludicrum* Deflandre; Aitchison, 1988, Fig. 2 (4).

Archocyrtium sp. B Aitchison, 1990, Fig. 4G.

描述：这个种的特点是头较大，球形，壳孔也大，每半圈有 6—7 个孔。4 根三棱形主刺短小，1 根顶刺略小于 3 根分散足，长度均小于头径。裙边不发育。

比较：这个种与 *Archocyrtium ormistoni* Cheng 的区别为后者的壳孔小，每半圈孔的数目多，4 根主刺长度都大于头径。

层位与产地：上泥盆统法门阶—下石炭统维宪阶下部；法国，德国，澳大利亚，中国青海可可西里地区。

窄原笼虫　*Archocyrtium strictum* Deflandre
(Pl. 2, fig. 5)

Archocyrtium strictum Deflandre, 1973, pl. 1, fig. 6, pl. 2, figs. 1, 2; Gourmelon, 1985, pl. 2, fig. 15, 1987, pl. 19, figs. 5—9.

描述：这个种的特点是头近球形，壳孔粗糙，不甚规则，每半圈有 6—9 个孔。4 根三棱形主刺不等长，顶刺较短，3 根分散足较长。

比较：这个种与 *Archocyrtium castuligerum* Deflandre 的区别为后者的裙边较高。

层位与产地：上泥盆统法门阶—下石炭统杜内阶；法国，中国青海可可西里地区。

旺氏原笼虫　*Archocyrtium wonae* Cheng
(Pl. 3, fig. 1)

Archocyrtium wonae Cheng, 1986, pl. 5, fig. 8, pl. 6, fig. 4; Kiessling and Tragelehn, 1994, pl. 2, fig. 4; Wang et al., 2000, pl. 3, fig. 12, 2003, pl. 4, figs. 20, 25, 2012, pl. 2, fig. 14, pl. 3, fig. 26, pl. 4, fig. 11, pl. 6, figs. 9—12.

Archocyrtium sp. aff. *A. wonae* Cheng; Aitchison and Flood, 1990, Fig. 4c.

Archocyrtium sp. cf. *A. wonae* Cheng; Wu et al., 1989, pl. 1, figs. 1, 7; Feng et al., 1997, pl. 2, fig. 15.

描述：这个种的特点是头小，近球形，孔较大。4根三棱形主刺长度不等，顶刺中等长，平直，3根分散足较长。门孔开放。

比较：这个种与 *Archocyrtium diductum* Deflandre 的区别为后者的主刺纤细，头上孔少，但较大。

层位与产地：上泥盆统弗拉阶—下石炭统维宪阶；美国，德国，澳大利亚，中国华南和青海可可西里地区。

原笼虫（未定种 A） *Archocyrtium* sp. A
(Pl. 3, figs. 3, 4)

描述：头大，近球形。壳孔小，数目多，每半圈有15—20个孔。4根三棱形主刺粗短，长度小于头径；1根顶刺和3根分散足，分散角度较小。

比较：这个未定种 A 与 *Pylentonema racheboeufi* Gourmelon 在外形上十分相似，区别在于后者头上具有3根很小的侧刺。

层位与产地：上泥盆统法门阶；中国青海可可西里地区。

第5章 西藏双湖和青海可可西里地区早中生代放射虫动物群及其时代和对比

5.1 西藏双湖和青海可可西里地区早中生代放射虫动物群及其时代

5.1.1 青海可可西里标本 W5418-2 放射虫动物群及其时代（表 5.1）

这个放射虫动物群共描述了 15 属 27 种，其中新种 3 个（*Paroertlispongus longispinosa* Wang，*Parasepsagon hohxilensis* Wang 和 *Pseudostylosphaera qinghaiensis* Wang），归属于 3 目（泡沫虫目、内射虫目和罩笼虫目）、5 超科（海绵尾虫超科、埃普廷虫超科、六桩虫超科、棘束虫超科和古网冠虫超科）、6 科（奥特里海绵虫科、锥海绵虫科、埃普廷虫科、欣德球虫科、鲁斯特笼虫科和古网冠虫科）。这一放射虫动物群以产 *Spongoserrula rarauana* Dumitrica 为特征，这个种主要出现于中三叠世拉丁期晚期，偶或可以延续到晚三叠世卡尼期早期，甚至到诺利期（Norian）。这个种主要发现于罗马尼亚（Dumitrica，1982）、匈牙利（De Wever，1984）、波斯尼亚-黑塞哥维那（Dosztaly，1991，1994；Kozur and Mostler，1996）、土耳其（Tekin，1999）、泰国（Kamata et al.，2002）。与这个种伴存的放射虫包括 *Annulotriassocampe sulovensis*，*Tritortis kretaensis*，*T. dispiralis*，*Pseudostylosphaera gracilis*，*P. nazarovi*，*Triassospongosphaera latispinosa*。这 6 个种的时限都是中晚三叠世拉丁晚期至卡尼期早期，产出地点为意大利（Kellici and De Wever，1995）、希腊（De Wever et al.，1979）、匈牙利（Kozur and Krahl，1984）、斯洛文尼亚（Gorican and Buser，1990）、奥地利（Kozur et al.，1996）、加拿大（Cordey et al.，1988）、菲律宾（Yeh，1990，1992）、土耳其（Tekin，1999）、俄罗斯远东地区（Bragin，1986，1991）、泰国（Kamata et al.，2002）、日本（Sashida et al.，1993；Sugiyama，1997）、中国西藏（Wang et al.，2002a，2002b；Yang et al.，2000）。*Parasepsagon firmum* 和 *Triassocampe* spp. 2 种在斯洛文尼亚地区产于中三叠世拉丁期晚期地层中（Gorican and Buser，1990）。时限为拉丁期的有 3 个种：*Cryptostephanidium cornigerum*，产于意大利（Dumitrica，1978）、斯洛文尼亚（Gorican and Buser，1990；Ramovs and Gorican，1995）、奥地利（Kozur et al.，1996）、土耳其（Tekin，1999）、菲律宾（Yeh，1990）、日本（Yao，1982）；*Paurinella aequispinosa*，发现于阿尔卑斯山南部和匈牙利（Kozur and Mostler，1981）、意大利（Kellici and De Wever，1995；Martini et al.，1989）、中国云南（Feng et al.，2001）；*Pseudostylosphaera magnispinosa*，见于美国 Oregon（Yeh，1989）、日本（Kojima and Mizutani，1987；Sashida et al.，1993；Sugiyama，1997）。其他一些种时限比较长，但都穿越了拉丁晚期，如 *Parasepsagon variabilis* 和 *Spongostephanidium spongiosum* 2 种时限都从安尼期晚期至卡尼期早期，前种分布于世界各地，如日本（Nakaseko and Nishimura，1979）、希腊（De Wever et al.，1979）、俄罗斯远东地区（Bragin，1991）、美国（Yeh，1989；Blome et al.，1989）、斯洛文尼亚（Ramovs and Gorican，1995）、奥地利（Kozur et al.，1996）、中国云南（Feng et al.，2001）和西藏

（Wang et al.，2002a）；后种发现于意大利和罗马尼亚（Dumitrica，1978）、克罗地亚（Gorican et al.，2005）、约旦安曼和土耳其（Dumitrica and Tekin，2010）。*Eptingium manfredi*、*Pseudostylosphaera japonica* 和 *Triassocampe deweveri* 的时限比较长，从安尼期至诺利期，这 3 个种分布于世界各地。

表 5.1 青海可可西里标本 W5418-2 放射虫动物群属种时代分布表

Table 5.1 Time ranges of the radiolarians of sample W5418-2 from Hoh Xil of Qinghai

属种名称 \ 分布时代	中三叠世		晚三叠世	
	安尼期	拉丁期	卡尼期	诺利期
Annulotriassocampe nova		———		
A. sulovensis		———		
Cryptostephanidium cornigerum		———		
Eptingium manfredi	———————————————			
Hindeosphaera bispinosa		———		
H. spinulosa		———		
Parasepsagon firmum		———		
P. variabilis	—————	———		
Paurinella aequispinosa		———		
Pseudostylosphaera coccostyla	—————	———		
P. compacta	—————	———		
P. gracilis		———		
P. japonica	———————————————			
P. longispinosa	—————	———		
P. magnispinosa		———		
P. nazarovi		———		
Spongoserrula rarauana		———	- - - -	
Spongostephanidium spongiosum	—————	———		
Triassistephanidium laticornis	—————	———		
Triassocampe deweveri	———————————————			
T. spp.		———		
Triassospongosphaera triassica		———		
Tritortis dispiralis		———		
Tiborella florida	—————			

上述放射虫动物群属种的时限分布显示，这个放射虫动物群的时代是属于拉丁期晚期至卡尼期早期，最可能为拉丁期晚期，我们称这个动物群为 *Spongoserrula rarauana* 动物群。

5.1.2 青海可可西里标本 Bb8311-1 放射虫动物群及其时代（表 5.2）

这个放射虫动物群共计包括 13 属 27 种，其中新种 6 个（*Paroertlispongus longispinosa* Wang，*P. opiparus* Wang，*Beturiella variospinosa* Wang，*Triassospongosphaera brevispinosa* Wang，*T. qinghaiensis* Wang 和 *Hozmadia pararotunda* Wang），归属于 3 目（泡沫虫目、内射虫目和罩笼虫

目)、4超科(海绵尾虫超科、埃普廷虫超科、六桩虫超科和棘束虫超科)、7科(古海绵梅虫科、锥海绵虫科、奥特里海绵虫科、埃普廷科、欣德球虫科、多弓形脊虫科和鲁斯特笼虫科)。在这个放射虫动物群中，*Annulotriassocampe multisegmantatus*，*A. sulovensis*，*Pseudostylosphaera gracilis*，*P. helicata*，*P. nazarovi*，*Triassospongosphaera latispinosa* 6种的时限都是从拉丁期晚期至卡尼期早期，它们分布在喀尔巴阡山西部和阿尔卑斯山北部（Kozur and Mostler，1979；Kozur and Mock，1981）、意大利（Lahm，1984；Kellici and De Wever，1995）、美国俄勒冈（Yeh，1989）、斯洛文尼亚（Gorican and Buser，1990）、土耳其（Tekin，1999）、泰国（Kamata et al.，2002）、中国西藏（Wang et al.，2002a，2002b）和云南（Feng et al.，2001）、日本（Nakaseko and Nishimura，1979；Sashida et al.，1993；Sugiyama，1997）、菲律宾（Yeh，1992）、俄罗斯远东地区（Bragin，1991）。*Archaeospongoprunum globosum* 和 *Paratriassocampe brevis* 2种的时限限于拉丁期晚期，前种发现于波斯尼亚-黑塞哥维那（Tekin and Mostler，2005），后种见于匈牙利（Kozur and Mostler，1994）。*Cryptostephanidium cornigerum* 和

表5.2 青海可可西里标本Bb8311-1放射虫动物群属种时代分布表

Table 5.2 Time ranges of the radiolarians of sample Bb8311-1 from Hoh Xil of Qinghai

分布时代 属种名称	中三叠世		晚三叠世	
	安尼期	拉丁期	卡尼期	诺利期
Annulotriassocampe multisegmantatus		——	——	
A. proprium		——	——	
A. sulovensis		——	——	
Archaeospongoprunum globosum		——		
Beturiella robusta	——	——		
Cryptostephanidium cornigerum	——	——		
Cryptostephanidium longispinosum	——	——		
Eptingium manfredi	——	——		
Hindeosphaera spinulosa	——	——		
Parasepsagon variabilis		——	——	
Paratriassocampe brevis		——		
Pseudostylosphaera compacta		——	——	
P. gracilis		——	——	
P. goestlingensis		——	——	
P. helicata		——	——	
P. japonica		——	——	
P. longispinosa		——	——	
P. magnispinosa		——		
Spongostephanidium spongiosum	——	——	——	
Triassospongosphaera latispinosa		——	——	
T. triassica		——	——	

Pseudostylosphaera magnispinosa 2 种的时代为拉丁期，前种见于意大利（Dumitrica，1978）、斯洛文尼亚（Gorican and Buser，1990；Ramovs and Goricam，1995）、菲律宾（Yeh，1990）、奥地利（Kozur et al.，1996）、土耳其（Tekin，1999）、日本（Yao，1982）；后种发现于美国俄勒冈（Yeh，1989）、日本（Kojima and Mizutani，1987；Sashida et al.，1993；Sugiyama，1997）。*Beturiella robusta*，*Cryptostephanidium longispinosum* 和 *Triassospongosphaera triassica* 3 种的时限是从安尼期晚期至拉丁期，第 3 种还可延续至卡尼期早期。它们分布于阿尔卑斯山南部（Dumitrica et al.，1980）、意大利（Lahm，1984）、斯洛文尼亚（Gorican and Buser，1990；Ramovs and Gorican，1995）、匈牙利（Kozur and Mostler，1981）、菲律宾（Cheng，1989）、俄罗斯远东地区（Bragin，1991）、日本（Nakaseko and Nishimura，1979；Sugiyama，1992；Sashida et al.，1999）、中国四川（Feng et al.，1996）和云南（Feng et al.，2001）。*Parasepsagon variabilis*，*Pseudostylosphaera compacta*，*Hindeosphaera bispinosa*，*Spongostephanidium spongiosum* 4 种的时限从安尼期中晚期至卡尼期，它们主要发现于喀尔巴阡山西部和阿尔卑斯山北部（Kozur and Mostler，1979；Kozur and Mock，1981）、意大利（Dumitrica，1978；Lahm，1984）、希腊（De Wever et al.，1979）、斯洛文尼亚（Ramovs and Gorican，1995）、奥地利（Kozur et al.，1996）、克罗地亚（Gorican et al.，2005）、约旦安曼和土耳其（Dumitrica and Tekin，2010；Tekin，1999）、菲律宾（Yeh，1990）、俄罗斯远东地区（Bragin，1991）、日本（Nakaseko and Nishimura，1979）、美国（Yeh，1989；Blome et al.，1989）、中国华南（Feng and Liu，1993）和西藏（Wang et al.，2002a）、云南（Feng et al.，2001）。另外 3 个种 *Eptingium manfredi*，*Pseudostylosphaera japonica* 和 *Hindeosphaera spinulosa* 的时限较长，从安尼期至诺利期，分布于世界各地。只有 *Annulotriassocampe proprium* 1 种的时限是晚三叠世卡尼期最晚期至诺利期，它发现于美国 Oregon（Blome，1984）和土耳其（Tekin，1999）。

在这个动物群中，以 *Annulotriassocampe multisegmantatus* 和众多的 *Pseudostylosphaera* 种为特征，并出现较多的新种，但没有发现拉丁期晚期和卡尼期早期的 *Muelleritortis cochleata* 带和 *Tritortis kretaensis* 带的标准属种，也没有发现代表拉丁期中晚期的 *Ladinocampe multiperforata* 带及其 2 个亚带 *L. annuloperforata* 和 *L. vicentinensis* 的标准属种。根据上述各属种的层位产地，我们仍把这个动物群的时代定成拉丁期中晚期，并称之为 *Annulotriassocampe multisegmantatus* 动物群。

5.1.3 四川理塘大何标本 Z 上部 Z144b1 放射虫动物群及其时代（表 5.3）

这个放射虫动物群共包括 12 属 30 种，其中新种 5 个（*Pseudostylosphaera inaequispinosa* Wang，*P. paragracilis* Wang，*Tubotriassocyrtis fusiformis* Wang，*Tritortis robustispinosa* Wang 和 *Hindeosphaera bella* Wang），归属于 3 目（泡沫虫目、内射虫目和罩笼虫目）、5 超科（光眼虫超科、海绵尾虫超科、埃普廷虫超科、六桩虫超科和棘束虫超科）、6 科（潘坦内尔虫科、锥海绵虫科、埃普廷虫科、欣德球虫科、鲁斯特笼虫科和莫尼卡星虫科）。在这个放射虫动物群中时限为拉丁期晚期至卡尼期早期的种有 10 个，它们是 *Annulotriassocampe sulovensis*，*A. multisegmantatus*，*Muelleritoritis cochleata*，*M. expansa*，*M. koeveskalensis*，*Pseudostylosphaera goestlingensis*，*P. gracilis*，*P. nazarovi*，*Tritortis dispiralis*，*T. kretaensis*。前 2 个种分布在喀尔巴阡山西部和阿尔卑斯山北部（Kozur and Mock，1981）、斯洛文尼亚（Gorican and Buser，1990）、意大利（Kellici and De Wever，1995）、土耳其（Tekin，1999）、美国 Oregon（Yeh，1989）、泰国（Kamata et al.，2002）、中国西藏（Wang et al.，2002a，2002b）；后 3 个种常见于匈牙利和阿尔卑斯山（De Wever，1984；Kozur，1988，Dosztaly，1991）、波斯尼亚-黑塞哥维那（Kozur and Mostler，1996）、土耳其（De Wever，1982，Tekin，1999）、泰国（Kamata et al.，

2002)、俄罗斯远东地区（Bragin，1986，1991）、日本（Nakaseko and Nishimura，1979；Sugiyama，1997）、斯洛文尼亚（Gorican and Buser，1990）、中国四川（Feng et al.，1996）。*Pseudostylosphaera* 属中的 3 个种发现于喀尔巴阡山西部和阿尔卑斯山北部（Kozur and Mostler，1979，1981）、意大利（Lahm，1984）、斯洛文尼亚（Gorican and Buser，1990）、土耳其（Tekin，1999）、泰国（Kamata et al.，2002）、菲律宾（Yeh，1992）、日本（Sashida et al.，1993；Sugiyama，1997）、中国西藏（Yang et al.，2000；Wang et al.，2002a，2002b）。最后 2 个种曾出现于希腊克里特岛（Kozur and Krahl，1984），加拿大不列颠哥伦比亚（Cordey et al.，1988）、阿尔卑斯山和匈牙利（Kozur，1988）、波斯尼亚-黑塞哥维那（Kozur and Mostler，1996）、土耳其（Tekin，1999）、泰国（Kamata et al.，2002）、菲律宾（Yeh，1990）、日本（Sashida et al.，1993；Sugiyama，1997）、俄罗斯远东地区（Bragin，1986，1991）。

表 5.3　四川理塘大何标本 Z 上部 Z144b1 放射虫动物群属种时代分布表

Table 5.3　Time ranges of the radiolarians of sample Z144b1 of upper Z from Dahe, Litang of Sichuan

属种名称 \ 时代	中三叠世		晚三叠世	
	安尼期	拉丁期	卡尼期	诺利期
Annulotriassocampe multisegmantatus		——	——	
A. sulovensis		——		
Cryptostephanidium cornigerum		——————		
Hindeosphaera bispinosa	——	——————		
H. spinulosa	——————	——————		
Muelleritortis cochleata		——		
M. expansa		——		
M. koeveskalensis		——		
Paratriassocampe gaetanii		——		
Paurinella aequispinosa		————		
P. latispinosa		——		
Pseudostylosphaera goestlingensis		——		
P. gracilis		——		
P. imperspinosa		——		
P. inaequispinosa		—		
P. japonica	——————	——————		
P. longispinosa		——		
P. nazarovi		——		
Striatotriassocampe laeviannulata		——		
S. nodosoannulata		——		
Triassospongosphaera triassica	——————	——		
Triassocampe deweveri		——	——	
T. scalaris			——	
Tritortis dispiralis		——		
T. kretaensis		——		

Cryptostephanidium cornigerum，*Paurinella aequispinosa*，*P. latispinosa*，*Pseudostylosphaera fragilis* 和 *P. longispinosa* 这 5 个种的时代限于拉丁期，它们分布在意大利（Lahm，1984；Dumitrica，1978；Kellici and De Wever，1995）、喀尔巴阡山西部和阿尔卑斯山北部（Kozur and Mostler，1981，1994）、斯洛文尼亚（Gorican and Buser，1990；Ramovs and Gorican，1995）、奥地利（Kozur et al.，1996）、土耳其（Tekin，1999）、泰国（Kamata et al.，2002）、菲律宾（Yeh，1990）、俄罗斯远东地区（Bragin，1991）、日本（Yao，1982；Sugiyama，1997）、中国云南（Feng et al.，2001）和西藏（Wang et al.，2002a）。

时代分布为拉丁期中晚期的种有 *Paratriassocampe gaetanii*，*Pseudostylosphaera imperspicua*，*Striatotriassocampe laeviannulata*，*S. nodosoannulata*，*Triassocampe scalaris*。这 5 个种产出的地点有阿尔卑斯山南部（Dumitrica et al.，1980）、斯洛文尼亚（Gorican and Buser，1990）、土耳其（Tekin，1999）、意大利（Martini et al.，1989；Kozur and Mostler，1994）、泰国（Sashida et al.，2000）、菲律宾（Yeh，1990）、日本（Yao，1982；Mizutani and Koike，1982）、中国云南（Feng et al.，2001）。

Hindeosphaera spinulosa 和 *Pseudostylosphaera japonica* 2 种的时限较长，从安尼期至诺利期，分布于世界各地。

这个动物群以出现 *Muelleritortis cochleata* 和 *Tritortis kretaensis* 2 种为特点，我们称这个动物群为 *Muelleritortis cochleata* 动物群，它的时代为中三叠世拉丁期晚期。

5.1.4 青海标本 8PFS6 和 8PFS7 放射虫动物群及其时代（表 5.4）

这个放射虫动物群的属种组成比较单调，共描述了 5 属 8 种，归属于 2 目（内射虫目和罩笼虫目）、3 超科（埃普廷虫超科、六桩虫超科和棘束虫超科）、4 科（埃普廷虫科、欣德球虫科、多弓形脊虫科和鲁斯特笼虫科）。

表 5.4 青海标本 8PFS6 和 8PFS7 放射虫动物群属种时代分布表
Table 5.4 Time ranges of the radiolarians of sample 8PFS6 and 8PFS7 from Qinghai

属种名称 \ 分布时代	中三叠世		晚三叠世	
	安尼期	拉丁期	卡尼期	诺利期
Eptingium manfredi	—	—	—	—
Hindeosphaera bispinosa	—	—		
Pseudostylosphaera compacta	—	—	—	
P. longispinosa		—		
P. magnispinosa	—	—		
P. nazarovi		—		
Parasepsagon variabilis	—	—		
Tiborella florida		—		

这个放射虫动物群以产 *Tiborella florida* 为特点，这个种的时代为安尼期晚期，主要见于奥地利（Kozur et al.，1996）、斯洛文尼亚（Ramovs and Gorican，1995）、俄罗斯远东地区（Bragin，1991）、日本（Nakaseko and Nishimura，1979）、中国华南（Feng and Liu，1993）和云南（Feng et al.，2001）。时限比较长的种，如 *Eptingium manfredi* 和 *Hindeosphaera spinulosa* 2 种的时限从安尼期至诺利期，分

布于世界各地；*Pseudostylosphaera compacta* 和 *P. longispinosa* 2 种的时限也为安尼期至诺利期；还有少量的种从拉丁期才开始出现，如 *Hindeosphaera bispinosa* 的时限为拉丁期至卡尼期，*Pseudostylosphaera magnispinosa* 的时限为拉丁期，*P. nazarovi* 的时限为拉丁期晚期至卡尼期中期。由于 *Tiborella florida* 特征种的出现，我们把这个动物群的时代定为中三叠世安尼期晚期，称之为 *Tiborella florida* 动物群。

5.1.5 西藏双湖角木日地区标本 D1339WF1 放射虫动物群及其时代（表5.5）

这个放射虫动物群的属种比较稀少，仅包括 6 属 8 种，归属于 3 目（泡沫虫目、内射虫目和罩笼虫目）、4 超科（海绵尾虫超科、埃普廷虫超科、六桩虫超科和棘束虫超科）、5 科（锥海绵虫科、埃普廷虫科、欣德球虫科、多弓形脊虫科和鲁斯特笼虫科）。

表 5.5 西藏双湖角木日标本 D1339WF1 放射虫动物群属种时代分布表
Table 5.5 Time ranges of the radiolarians of sample D1339WF1 from Jiaomuri, Shuanghu of Tibet

分布 时代 属种名称	中三叠世		晚三叠世	
	安尼期	拉丁期	卡尼期	诺利期
Eptingium nakasekoi	—			
Parasepsagon praetetracanthus	——	—		
Pseudostylosphaera compacta	————	————	————	
P. japonica	————	————	————	
Tiborella florida	——			
Triassocampe coronata	——	—		
T. nanpanensis	——			
Triassospongosphaera triassica	—	————	————	

这个放射虫动物群的特征种是 *Eptingium nakasekoi*，它的时代为安尼期早期，曾在匈牙利（Kozur and Mostler，1994）、斯洛文尼亚（Ramovs and Gorican，1995）、泰国（Kamata et al.，2002）、马来西亚（Spiller，2002）、菲律宾（Cheng，1989）、日本（Nakaseko and Nishimura，1979；Sugiyama，1997）、中国云南（Feng et al.，2001）等地找到过。伴存放射虫 *Triassocampe coronata* 的时限为安尼期中晚期至拉丁期早期，主要分布在希腊（De Wever et al.，1979）、马来西亚（Spiller，2002）、泰国（Kamata et al.，2002）、俄罗斯远东地区（Bragin，1991）、日本（Mizutani and Koike，1982；Sugiyama，1992，1997）、中国华南（Feng and Ye，1996；Feng et al.，1996）和云南（Feng et al.，2001）；*Parasepsagon praetetracanthus* 的时限为安尼期中晚期，这个种只见于匈牙利（Kozur and Mostler，1994）、阿尔卑斯山北部（Kozur and Mostler，1981）、中国云南（Feng et al.，2001）；*Triassocampe nanpanensis* 的时代为安尼期中晚期，这个种仅见于中国华南（Feng，1992；Feng and Liu，1993；Fang and Feng，1996）和云南（Feng et al.，2001）；*Pseudostylosphaera compacta*，*P. japonica* 和 *Triassospongosphaera triassica* 3 种的时限较长，前 2 种的时限从安尼期开始至卡尼期，有的还可延续到诺利期；第 3 种也是从安尼期晚期至卡尼期早期，它们分布于世界各地。这个动物群的特征种 *Eptingium nakasekoi* 的时代为安尼期早期，尽管许多伴存放射虫的时限为安尼期中晚期，我们仍然把这个动物群称

为 *Eptingium nakasekoi* 动物群，时代为安尼期早期，这个意见是否正确，有待今后进一步采集和研究。

5.2 西藏双湖和青海可可西里地区早中生代放射虫化石带及其对比（表5.6）

笔者在北羌塘地区发现了5个早中生代放射虫动物群，它们分别是 *Eptingium nakasekoi* 动物群、*Tiborella florida* 动物群、*Annulotriassocampe multisegmantatus* 动物群、*Muelleritortis cochleata* 动物群和 *Spongoserrula rarauana* 动物群，这些放射虫动物群的时限属于中三叠世安尼期早期至拉丁期晚期。Kozur 和 Mostler（1994，1996）、Kozur 等（1996）、Sugiyama（1997）、Feng 等（2001）分别建立了欧洲、日本和中国云南中三叠世安尼期至拉丁期放射虫化石带（表5.6），他们将安尼期早期的放射虫化石带分别称作 *Parasepsagon robustus* 带、*Eptingium nakasekoi* 带和 *Triassocampe dumitricai* 带。欧洲的 *Parasepsagon robustus* 带主要特征种有 *P. asymmetricus asymmetricus*、*Plafkerium anisicum*，伴存放射虫包括 *P. asymmetricus praetetracanthus*、*Tiborella anisica* 和 *Triassocampe* 属的最原始种。日本的 *Eptingium nakasekoi* 带包括 *Hozmadia gifuensis*、*Katorella bifurcata*、*Neopylentonema nakasekoi*、*Oertlispongus diacanthus*、*Parasepsagon*（?）*antiquus*、*Spongosilicormiger gigantoceras*、*Triassocampe eruca* 等放射虫。中国云南的 *Triassocampe dumitricai* 带含有 *Paroertlispongus multispinosus*、*P.* cf. *diacanthus*、*P. chinensis*、*Triasocampe solida*、*T. exilis*、*T. goricani*、*T. nanpanensis*、*Eptingium* cf. *nakasekoi* 等属种。我国羌塘的 *Eptingium nakasekoi* 带，除带种外，还有特征种 *Parasepsagon praetetracanthus*、*Triassocampe coronata*、*T. nanpanentsis*、*Tiborella florida* 和一些时限较长的属种 *Pseudostylosphaera compacta*、*P. japonica*、*Triassospongosphaera triassica* 等。羌塘的这个带与日本的 *E. nakasekoi* 带有着相同的指引种，同欧洲特提斯地区的 *Parasepsagon robustus* 带有相同的特征种 *P. praetetracanthus* 和一些相似的属种；和中国云南的 *Triassocampe dumitricai* 带有相同的特征种，如 *Eptingium* cf. *nakasekoi*，*Triassocampe nanpanensis* 和其他一些相似的属种群。因此，羌塘的 *E. nakasekoi* 带完全可以同日本的同名带和中国云南的 *T. dumitricai* 带对比，也可同欧洲的 *P. robustus* 带大致对比。

对安尼期晚期的放射虫化石带，Sugiyama（1997）建立了Tr3A：spineA2带，日本的这个带主要放射虫包括 *Yeharaia elegans*、*Y. cornigera*、*Oertlispongus inaequispinosus*、*Pentactinocarpus tetracanthus*、*Spinotriassocampe hungarica*、*S. annulata*、*Baumgartneria bifurcata*、*Hinedorcus alatus*、*Ladinocampe* sp.、*Cryptostephanidium cornigerum*、*C. japonicum*、*Eptingium manfredi*、*Pseudostylosphaera japonica*、*Triassocampe deweveri* 和 *Hozmadia* 属等；Kozur 等（1996）建立的 *Spongosilicarmiger transitus* 带分为2个亚带，下亚带为 *Tiborella florida*，上亚带是 *Yeharaia annulata*。*Tiborella florida* 下亚带的伴存放射虫特别丰富，包括 *Eptingium manfredi japonicum*、*E. nakasekoi*、*E. ramovsi*、*Hindeosphaera spinulosa*、*Hozmadia costata*、*Parasepsagon variabilis*、*Spongostephanidium japonicum*、*S. longispinosum*、*Tiborella florida austriaca*、*T. florida florida*、*Hozmadia rotunda*、*H. latispinosa*、*H. rotundispina*、*Poulpus illyricus*、*Triassistephanidium anisicum*、*Paurinella fusina*、*P. sinensis*、*Triassocampe scalaris baloghi* 等。我国青海的 *Tiborella florida* 带以出现 *Tiborella florida* 为特征，此带还包含 *Eptingium manfredi*、*Hindeosphaera bispinosa*、*H. spinulosa*、*Paroertlispongus longispinosa*、*Pseudostylosphaera compacta*、*P. longis-pinosa*、*P. nazarovi*、

表 5.6 羌塘北部与其他地区中、晚三叠世放射虫动物群对比表

Table 5.6 Correlation chart of the Middle and Late Triassic radiolarian faunas from northern Qiangtang with other regions

时代	地区	中国羌塘北部	日本 (Sugiyama, 1997)	俄罗斯远东地区 (Bragin, 1991)	欧洲 (Kozur et al., 1996)	马来西亚 (Spiller, 2002)	泰国 (Kamata et al., 2002; Sashida et al., 1993)	中国华南 (Feng et al., 1996, 2001)
晚三叠世	卡尼期 Carnian		Capnodoce–Trialatus	Triassocampe nova	Nakasekoellus inkensis / Tetraporobrrachia haeckeli		Capnuchosphaera sp.	
			Poulpus carcharus					
		Spongoserrula rarauana	Capnuchosphaera		Tritortis kretaensis			
		Muelleritortis cochleata	Spongoserrula dehli	Sarla dispiralis	Muelleritortis cochleata / Pterospongus priscus / Spongoserrula rarauana / S. fluegeli	Triassocampe deweveri	Cryptostephanidium sp.	Muelleritortis cochleata
中三叠世	拉丁期 Ladinian	Annulotriassocampe multisegmantatus	Yeharaia elegans	Triassocampe deweveri	Ladinocampe multiperforata / Spongosilicarniger italicus			?
			Spine A2		S. transitus		Pseudostylosphaera helicata	
	安尼期 Anisian	Tiborella florida	Triassocampe deweveri	Tetraspinocyrtis laevis		Triassocampe coronata	Pseudostylosphaera japonica	Triassocampe deweveri
			Triassocampe coronata	Triassocampe diordinis				T. coronata coronata / T. coronata inflata
		Eptingium nakasekoi	Eptingium nakasekoi	Hozmadia	Parasepsagon robustus			T. dumitricai
早三叠世	奥伦尼克期 Olenekian		Parentactinia nakatsugawaensis	"Stylosphaera" fragilis			Parentactinia nakatsugawaensis	Eptingium sp.

P. magnispinosa、*Triassocampe* sp. A 等。我国的这个带可以同欧洲特提斯地区的同名带相比较。尽管这2个动物群在丰度和分异度上有明显的不同，前者放射虫的丰度和分异度很高，后者则属种单调，但这种现象是否与这2个动物群所处的环境有关，因为前者赋存于灰岩相地层中，而后者发现于硅质岩相地层中，值得进一步探讨。Kozur 等（1996）在讨论 Feng（1992）及 Feng 和 Liu（1993）在我国西南地区建立的 *Shengia yini* 组合和 *Pseudoeucyrtis liui* 组合（早三叠世早期和晚期）时，根据这2个放射虫组合的面貌与他们建立的安尼期晚期 *Spongosilicarmiger transitus* 带相似的特点而把 *Shengia yini* 组合与他们的 *Tiborella florida* 亚带对比。如果这一对比可靠的话，那么 *T. florida* 带也可以同我国华南的 *Shengia yini*（= *Archaeospongoprumum mesotriassicum*）组合比对。

Kozur 和 Mostler（1994）根据匈牙利的材料，建立了拉丁期晚期的放射虫化石 *Muelleritortis cochleata* 带。2年之后（1996），他们又根据波斯尼亚-黑塞哥维那地区同一时期 Oertlispongidae 科属的演化特征，把这个带细分为3个亚带，自下而上为 *Pterospongus priscus* 亚带、*Spongoserrula rarauana* 亚带和 *S. fluegeli* 亚带。Sugiyama（1997）根据日本的材料同样建立了拉丁期晚期的 *Muelleritortis cochleata* 带，日本以 *M. cochleata* 的出现作为这个带的开始，在这个带的下部这个种异常丰富，但至上部相对减少。*Tritortis kretaensis* 在上部开始出现，并向上显示出越来越多的趋势。这个带还包括 *Yeharaia compsa*、*Falcispongus dumitricae*、*Hozmadia* sp. A、*Pseudostylosphaera goestlingensis*、*P. nazarovi* 等伴存放射虫分子。Kozur 和 Mostler（1994）建立的 *Muelleritortis cochleata* 带是以带的指引种占优势作为这个带的特点。此带的常见种还包括 *Hungarosaturnalis* 诸种。在此带上部，*Tritortis kretaensis* 开始出现并逐渐趋向繁盛。我国四川理塘的 *Muelleritortis cochleata* 带以出现 *M. cochleata*、*M. expansa*、*M. koeveskalensis* 和 *Tritortis kretaensis*、*T. dispiralis* 为特征种，伴存放射虫较多，包括 *Annulotriassocampe multisegmantatus*、*A. sulovensis*、*Hindeosphaera bella*、*H. bispinosa*、*H. spinulosa*、*Triassocampe deweveri*、*T. scalaris*、*Pseudostylosphaera fragilis*、*P. goestlingensis*、*P. gracilis*、*P. imperspicua*、*P. inaequispinosa*、*P. longispinosa*、*P. japonica*、*P. nazarovi*、*P. paragracilis*、*Cryptostephanidium cornigerum*、*Paurinella aequispinosa*、*P. latispinosa*、*Tubotriassocyrtis fusiformis*、*Triassospongosphaera triassica*、*Striatotriassocampe laeviannulata*、*S. nodosoannulata*、*Paratriassocampe gaetanii* 等。我国的 *M. cochleata* 动物群完全可以同日本（Sugiyama，1997）和欧洲（Kozur and Mostler，1994）同名放射虫 *Muelleritortis cochleata* 带对比。Kozur 和 Mostler（1996）的 *Spongoserrula rarauana* 亚带以丰富的带种作为这个亚带的特征，时代为晚拉丁期中期，其中伴存射虫包括 *Spongoserrula rarauana trinodosa*、*S. bidentata*、*Steigerispongus cristagalli*、*Scutispongus undulatus*、*S. rostratus rostratus*、*Pteraspongus patrulii*、*P. incissus*、*P. aquilus*、*P. alatus*、*Falcispongus hamatus*、*F. falciformis minor*、*Bogdanella praecursor*、*Baumgartneria dumitricae*、*B. curvispina* 等。青海标本 W5418-2 动物群以 *Spongoserrula rarauana* 十分丰富为特点，其他特征种还有 *Parasepsagon firmum*、*Tritortis kretaensis*、*T. dispiralis*。这一动物群还包括伴存放射虫 *Annulotriassocampe nova*、*A. sulovensis*、*Hindeosphaera bispinosa*、*Triassocampe deweveri*、*Cryptostephanidium cornigerum*、*Paurinella aequispinosa*、*Spongostephanidium spongiosum*、*Triassistephanidium laticornis*、*Triassospongosphaera triassica*、*T. latispinosa*、*T. qinghaiensis*、*Pseudostylosphaera coccostyla*、*P. compacta*、*P. gracilis*、*P. japonica*、*P. magnispinosa*、*P. nazarovi*、*P. longispinosa* 等。虽然我国这个动物群与欧洲同名放射虫带（Kozur and Mostler，1996）在放射虫属种组成上有较大的差异，但在主要属种的特征上相似，所以我国的 *S. rarauana* 动物群与欧洲的同名亚带 *S. rarauana* 带完全可以对比。

青海标本 Bb8311-1 动物群以出现 *Annulotriassocampe multisegmantatus*、*A. proprium*、*A. sulo-*

vensi 和众多的 *Pseudostylosphaera* 属的分子，如 *P. compacta*、*P. gracilis*、*P. helicata*、*P. japonica*、*P. magnispinosa*、*P. nazarovi*、*P. longispinosa*、*Hindeosphaera bispinosa*、*H. spinulosa* 为特征，共存放射虫还包括 *Archaeospongoprunum globosa*、*Beturiella robusta*、*Cryptostephanidium longispinosa*、*C. cornigerum*、*Eptingium manfredi*、*Parasepsagon variabilis*、*Paratriassocampe brevis*、*Spongostephanidium spongiosum*、*Triassospongosphaera brevispinosa*、*T. latispinosa*、*T. qinghaiensis*、*T. triassica* 等，在这个放射虫动物群中没有发现拉丁期晚期 *Muelleritortis cochleata* 带和卡尼期早期 *Tritortis kretaensis* 带种。Kozur 和 Mostler（1994，1996）把拉丁期中晚期的放射虫 *Ladinocampe multiperforata* 带分为 2 个亚带，下亚带为 *L. annuloperforata*，上亚带为 *L. vicentinensis*。*L. vicentinensis* 亚带以 *L. vicentinensis*、*L. multiperforata* 和 *Conospongocyrtis cephaloconica* 为特征种，共存放射虫包括 *Triassocampe longicephalis*、*Triassospongocyrtis ruesti*、*Spongolophopaena longa*、*Anisicyrtis trettoensis postera*、*Planispinocyrtis thoraciglobulosa*、*Pararuesticyrtium constrictum*、*Silicarmiger costatus costatus*、*S. costatus magnicornus*、*S. latus mediospinosus*、*Annulotriassocampe campanilis longiporata* 等。在我国的动物群中也没有发现 *Ladinocampe multiperforata* 带及其 2 个亚带的指引种。但是，我国的 *Annulotriassocampe multisegmantatus* 放射虫动物群的主要属种和共存放射虫的时代大多数是拉丁期中晚期的。因此，我们认为 *A. multisegmantatus* 动物群最有可能是居于 *Ladinocampe multiperforata* 带和 *Muelleritortis cochleata* 带之间的一个独立的放射虫带，其位置上与 Kozur 和 Mostler（1994，1996）尚未命名的放射虫带大致相当。

第6章 早中生代放射虫分类描述

放射虫亚纲　Radiolaria Müller，1858
　多囊虫超目　Polycystina Ehrenberg，1838，emend. Riedel，1967
　　泡沫虫目　Spumellaria Ehrenberg，1875，emend. De Wever et al.，2001
　　　光眼虫超科　Actinommacea Haeckel，1862，emend. De Wever et al.，2001
　　　　潘坦内尔虫科　Pantanelliidae Pessagno，1977
　　　　　潘坦内尔虫亚科　Pantanellinae Pessagno，1977
　　　　　　潘坦内尔虫属　*Pantanellium* Pessagno，1977

模式种：*Pantanellium riedeli* Pessagno，1977

鉴定要点：壳近球形，壳孔大，缺失门孔，具有2根结实的三棱形主刺。

时代与分布：中三叠世拉丁期晚期—早白垩世阿普特期；世界各地。

多孔潘坦内尔虫（新种）　*Pantanellium multiporum* Wang sp. nov.
(Pl. 23，figs. 29，30)

词源：multi，拉丁词，多；porus，拉丁词，孔。

描述：格状外壳近球形，壳孔较大，壳表半圈可见8—9个孔。具2根三棱形主刺，长度不相等，3根纵脊和3条纵沟相间，在脊的基部发育次生沟，短刺平直，长刺在末端呈轻微左旋扭曲。

比较：这个新种以其壳孔较多、主刺中长刺末端呈轻微左旋扭曲区别于这个属的其他种。

层位与产地：中三叠统拉丁阶上部；中国青海可可西里地区。

潘坦内尔虫（未定种A）　*Pantanellium* sp. A
(Pl. 24，figs. 22，24)

描述：壳近球形，壳孔大，数目少，半圈有7—8个。孔的接点发育瘤状构造。2根三棱形主刺平直，1根较长，另一根略短。3根纵脊和3条纵沟相间，脊较窄，沟较宽，无次生沟。

比较：这个未定种A与*P. multiporum* Wang sp. nov. 的区别为后者的1根主刺末端呈轻微左旋扭曲。

层位与产地：中三叠统拉丁阶上部；中国四川理塘地区。

　　　海绵尾虫超科　Sponguracea Ehrenberg，1875，emend. De Wever et al.，2001
　　　　古海绵梅虫科　Archeospongoprunidae Pessagno，1973
　　　　　古海绵梅虫亚科　Archeospongopruninae Pessagno，1973
　　　　　　古海绵梅虫属　*Archeospongoprunum* Pessagno，1973

模式种：*Archeospongoprunum venadoensis* Pessagno，1973

鉴定要点：海绵壳近球形或椭圆形，具2根三棱形主刺，刺平直或轻微扭曲，3根纵脊和3条纵沟相间。

时代与分布：中三叠世拉丁期晚期—早白垩世阿普特期；世界各地。

球形古海绵梅虫 *Archaeospongoprunum globosum* Tekin et Mostler

(Pl. 23, figs. 33, 34)

Archaeospongoprunum globosum Tekin et Mostler, 2005, pl. 3, figs. 9, 10.

描述：这个种的特点是海绵壳大，近球形，2根三棱形主刺平直，细，长度不相等，但长度均小于壳径。

比较：这个种与 *A. bispinosum* Kozur et Mostler 的区别为后者海绵壳为长椭圆形，2根三棱形主刺长度相近。

层位与产地：中三叠统拉丁阶上部；波斯尼亚、黑塞哥维那，中国青海可可西里地区。

锥海绵虫科 Pyramispongiidae Kozur et Mostler, 1978, emend. De Wever et al., 2001
保林虫属 *Paurinella* Kozur et Mostler, 1981

模式种：*Paurinella curvata* Kozur et Mostler, 1981

鉴定要点：海绵壳近球形，具有3根杆形主刺，刺平直或弯曲，有时具辅刺。

时代与分布：中三叠世拉丁期晚期—卡尼期早期；欧洲特提斯区，中国云南、四川和青海。

等刺保林虫 *Paurinella aequispinosa* Kozur et Mostler

(Pl. 19, figs. 13, 14; pl. 20, figs. 23—25; pl. 25, fig. 31)

Paurinella aequispinosa Kozur et Mostler, 1981, pl. 42, fig. 1, pl. 43, fig. 1, 1994, pl. 15, figs. 9, 11; Kellici and De Wever, 1995, pl. 3, fig. 15; Feng et al., 2001, pl. 7, figs. 1—4.

Eptingiidae? gen. et sp. indet. Martini; De Wever et al., 1989, pl. 1, fig. 6.

描述：这个种的特点是海绵壳小，近球形，具有3根杆形主刺，刺细长，刺的长度大于壳径。

比较：这个种与 *P. latispinosa* Kozur et Mostler 的区别为后者刺较短，粗壮，中部膨大，长度小于或等于壳径。

层位与产地：中三叠统拉丁阶；南阿尔卑斯，匈牙利，意大利，中国云南、四川和青海。

宽刺保林虫 *Paurinella latispinosa* Kozur et Mostler

(Pl. 25, figs. 29, 30)

Paurinella latispinosa Kozur et Mostler, 1994, pl. 15, fig. 4; Tekin, 1999, pl. 14, fig. 9.

描述：这个种的特点是海绵壳近球形，3根杆形主刺较粗壮，其形状、大小和夹角都相近，主刺中部膨大，向远端变锥形，主刺的长度等于或小于壳径。

层位与产地：中三叠统拉丁阶；匈牙利，土耳其，中国四川理塘地区。

保林虫（未定种A） *Paurinella* sp. A

(Pl. 19, fig. 15)

描述：海绵壳近球形，壳孔小。具有3根杆形短刺，大小和形状相似，长度小于壳径。

比较：这个未定种A与 *P. aequispinosa* Kozur et Mostler 的区别为后者的主刺细长，长度大于壳径，其中2根很长，宽针形，第3根较小，细针形。

层位与产地：中三叠统拉丁阶上部；中国青海可可西里地区。

三叠海绵球虫属 *Triassospongosphaera* (Kozur et Mostler), 1981

模式种：*Spongechinus triassicus* Kozur et Mostler, 1979

鉴定要点：海绵壳近球形，壳上发育很多放射状排列的杆形或角锥形主刺。

时代与分布：中晚三叠世拉丁期晚期—卡尼期早期；欧洲特提斯区，日本，俄罗斯远东地区，中国云南、四川和青海。

短刺三叠海绵球虫（新种） *Triassospongosphaera brevispinosa* Wang sp. nov.
(Pl. 22, fig. 33)

词源：brev, 拉丁词, 短；spinos, 拉丁词, 刺。

描述：海绵壳近球形，壳较大，孔细小，具有很多短小的角锥形主刺，刺的数目每半圈在40根以上。

比较：这个新种与 *T. latispinosa* (Kozur et Mostler) 的区别为后者的壳体小，角锥形主刺较粗壮，刺的数目较少，每半圈有14—18根。

层位与产地：中三叠统拉丁阶上部；中国青海可可西里地区。

宽刺三叠海绵球虫 *Triassospongosphaera latispinosa* (Kozur et Mostler)
(Pl. 22, fig. 37)

Spongechinus? latispinosus Kozur et Mostler, 1979, pl. 5, fig. 4.

Triassospongosphaera latispinosa (Kozur et Mostler), Kozur and Mostler, 1981, pl. 3, fig. 6.

描述：这个种的特点是壳小，壳上具有14—18根角锥形宽刺，刺的长度不及壳径的一半。

层位与产地：中上三叠统拉丁阶上部—卡尼阶下部；欧洲特提斯区，中国青海可可西里地区。

青海三叠海绵球虫（新种） *Triassospongosphaera qinghaiensis* Wang sp. nov.
(Pl. 22, figs. 34—36)

词源：qinghai, 青海, 为标本的原产地。

描述：海绵壳较小，近球形，壳孔很小，壳上发育放射状排列的杆形主刺，刺中等长度，每个刺的长度约为壳半径的1/3，每半圈有16—18根刺，每个刺的基部宽，向远端宽度逐渐减少，至末端锐尖。

比较：这个新种与 *T. triassica* (Kozur et Mostler) 的区别为后者的主刺较长，壳体较大，刺的长度约为壳径的1/2。

层位与产地：中三叠统拉丁阶上部；中国青海可可西里地区。

三叠海绵球虫（未定种 A） *Triassospongosphaera* sp. A
(Pl. 12, figs. 8—11)

描述：海绵壳近球形，壳孔较大，壳表发育杆形主刺，数目少，刺圆锥形，中等粗细，刺的长度约为壳径的一半。

比较：这个未定种 A 与 *T. qinghaiensis* Wang sp. nov. 的区别为后者的壳孔较小，主刺较短，长度约为壳径的1/3。

层位与产地：中三叠统安尼阶下部；中国西藏双湖地区。

三叠三叠海绵球虫 *Triassospongosphaera triassica* (Kozur et Mostler)
(Pl. 12, fig. 12; pl. 19, figs. 17—19; pl. 20, figs. 11—22; pl. 22, figs. 38—40; pl. 25, figs. 27, 28)

Spongechinus triassicus Kozur et Mostler, 1979, pl. 13, figs. 6, 7.

Triassospongosphaera triassica (Kozur et Mostler), Kozur and Mostler, 1981, pl. 58, fig. 4, pl. 59, fig. 4; Lahm, 1984, pl. 11, fig. 9.

Triassospongosphaera sp. Bragin, 1991, pl. 4, fig. 2.

Astrocentrus sp. cf. *A. pulcher* Kozur et Mostler, Nakaseko and Nishimura, 1979, pl. 6, figs. 6, 7; Feng et al., 2001, pl. 9, figs. 1—5.

Astrocentrus cf. *pulcher* Kozur et Mostler; Feng et al., 1996, pl. 3, fig. 14 (non 11).

描述：这个种的特点是海绵壳近球形，壳上具有较多的放射状排列的杆形刺，刺的宽度自基部向远端逐渐变窄，末端锐尖。过去曾被有些作者鉴定成 *Astrocentrus pulcher* 的标本，由于这些种的壳上发育较多的杆形刺而被笔者作为这个种的同义名。

比较：这个种与 *T. multispinosa* (Kozur et Mostler) 的区别为后者杆形刺的数目更多，刺更长。

层位与产地：中上三叠统安尼阶上部—卡尼阶下部；匈牙利，日本，俄罗斯远东地区，中国云南、四川理塘地区和青海可可西里地区。

奥特尔海绵虫科　Oertlispongidae Kozur et Mostler, 1980

拟奥特尔海绵虫属　*Paroertlispongus* Kozur et Mostler, 1981

模式种：*Paroertlispongus multispinosus* Kozur et Mostler, 1981

鉴定要点：海绵壳近球形，壳孔小，具 2 根杆形主刺，有时还发育许多细小的辅刺，刺的长度相等或不相等。刺平直或弯曲。

时代与分布：中三叠世拉丁期；欧洲特提斯区，中国华南和青海。

长刺拟奥特尔海绵虫（新种）　*Paroertlispongus longispinosus* Wang sp. nov.

(Pl. 19, figs. 1, 2, 21, 22; pl. 22, figs. 16—20)

词源：long，拉丁词，长；spinos，拉丁词，刺。

描述：海绵壳较小，近球形，壳孔也小，2 根杆形主刺平直，形状和大小相似，长度是壳径的 2 倍以上。

比较：这个新种与 *P. chinensis* (Feng) 的区别为后者的主刺较短，长度与壳径相近。

层位与产地：中三叠统拉丁阶上部；中国青海可可西里地区。

美丽拟奥特尔海绵虫（新种）　*Paroertlispongus opiparus* Wang sp. nov

(Pl. 22, figs. 22, 23)

词源：opipar，拉丁词，美丽。

描述：海绵壳较大，近球形，壳孔小，具有 2 根长度不等的杆形主刺，刺粗壮，相对而生，位于同一直线上，长刺的长度是壳径的 2 倍，而短刺的长度略大于壳径。

比较：这个新种与 *P. hermi* (Lahm) 的区别为后者的主刺细弱，长刺和短刺长度分别大于和小于壳径。

层位与产地：中三叠统拉丁阶上部；中国青海可可西里地区。

拟奥特尔海绵虫（未定种 A）　*Paroertlispongus* sp. A

(Pl. 22, fig. 21)

描述：海绵壳小，椭圆形，具有 2 根细弱的杆形主刺，大小和形状相似，2 根主刺不位于同一直线上。刺的长度为壳径的 2 倍以上。

比较：这个未定种 A 与 *P. longispinosus* Wang sp. nov. 的区别为后者海绵壳近球形，2 根杆形主刺位于同一直线上；与 *P. obliqus* Kozur et Mostler 的区别为后者的 2 根杆形主刺长度不相等，主刺较短，壳近球形。

层位与产地：中三叠统拉丁阶上部；中国青海可可西里地区。

海绵锯齿虫属 *Spongoserrula* Dumitrica，1982

模式种：*Spongoserrula rarauana* Dumitrica，1982

鉴定要点：刺层状，不对称，弯曲，外边缘的齿数目不定。

时代与分布：中晚三叠世拉丁期晚期—卡尼期；欧洲特提斯地区，泰国，土耳其，中国青海。

拉劳海绵锯齿虫 *Spongoserrula rarauana* Dumitrica

(Pl. 19，figs. 10，11；pl. 20. figs. 1—10)

Spongoserrula rarauana Dumitrica，1982，pl. 5，figs. 5—7，pl. 6，figs. 1—5，pl. 12，figs. 10—13；De Wever，1984，pl. 1，figs. 2，3，5，pl. 2，figs. 1，2，4；Cordey et al.，1988，pl. 1，figs. 6，7；Dorztaly，1991，pl. 1，fig. 2，1993，pl. 1，fig. 6，1994，pl. 1，fig. 6；Kozur and Mostler，1996，pl. 5，figs. 8，10，11，13—15，pl. 6，figs. 1—3，6，9，pl. 8，fig. 9；Kamata et al.，2002，Fig. 6（N，O）.

Spongoserrula rarauana rarauana Dumitrica；Tekin，1999，pl. 15，figs. 7，8.

Spongoserrula sp. Kamata et al.，2002，Fig. 6（M）.

描述：这个种的特点是具有宽大、强壮和弯曲的层状齿，在外边缘有 5—6 个齿，末端齿短。在茎和第 1 齿间外边缘上发育一个肿状物。茎短，宽度从根到顶逐渐增大。

层位与产地：中上三叠统拉丁阶上部—卡尼阶；加拿大不列颠哥伦比亚，罗马尼亚，匈牙利，波斯尼亚-黑塞哥维那，泰国，土耳其，中国青海可可西里地区。

内射虫目 Entactinaria Kozur et Mostler，1982
埃普廷虫超科 Eptingiacea Dumitrica，1978
埃普廷虫科 Eptingiidae Dumitrica，1978
埃普廷虫亚科 Eptingiinae Dumitrica，1978
埃普廷虫属 *Eptingium* Dumitrica，1978

模式种：*Eptingium manfredi* Dumitrica，1978

鉴定要点：主刺和弓发育在头腔中，由若干榍同头壁相连。具口，位于侧面。3 根主刺三棱形。由 3 根纵脊与 3 条纵沟相间，有或缺失次生沟。

时代与分布：中三叠世安尼期中期—拉丁期；世界各地。

曼弗雷德埃普廷格虫 *Eptingium manfredi* Dumitrica

(Pl. 19，figs. 34—39；pl. 23，figs. 15，20—26)

Eptingium manfredi Dumitrica，1978，pl. 3，figs. 3，4，pl. 4，figs. 1，2，6，7（non 5）；Pesssagno et al.，1979，pl. 6，figs. 9—11；De Wever，1982，pl. 35，fig. 5；Sato et al.，1982，pl. 2，fig. 13；Gorican and Buser，1990，pl. 8，figs. 7，8；Jurkovsek，1990，pl. 10，fig. 4a，4b；Bragin，1991，pl. 2，figs. 12，13；Feng and Liu，1993，pl. 1，figs. 14，15；Ramovs and Gorican，1995，pl. 5，figs. 6—8；Sashida et al.，1999，Fig. 6（16，17）；Kozur et al.，1996，pl. 10，figs. 1—4，6，10；Feng et al.，2001，pl. 4，figs. 14—18；Sashida et al.，2000，Fig. 7（22—24）.

Eptingium manfredi manfredi Dumitrica; Dumitrica et al., 1980, pl. 3, figs. 1—3, pl. 6, figs. 5—7; Kozur and Mostler, 1994, pl. 1, fig. 3; Spiller and Metcalfe, 1995, Fig. 6 (14); Spiller, 2002, pl. 4, figs. N, O.

Eptingium manfredi robustum Kozur et Mostler, 1980, pl. 6, figs. 1—4, 8; Kozur and Mostler, 1994, pl. 1, figs. 1, 2.

Eptingium cf. *manfredi* Dumitrica; Matsuda and Isozaki, 1982, pl. 3, fig. 25; Yao, 1982, pl. 1, fig. 17; Yao et al., 1982, pl. 1, fig. 12.

Eptingium cf. *manfredi manfredi* Dumitrica; Takashima and Koike, 1982, pl. 1, figs. 7, 8.

描述：这个种的特点是壳近球形，壳表发育瘤状构造。3根三棱形主刺粗壮，形态、大小和弯曲度不一，每个刺由3根纵脊和3条纵沟相间，在纵脊上发育次生沟，相邻脊间发育横脊。

层位与产地：中上三叠统；世界各地。

中世古幸次郎埃普廷格虫 *Eptingium nakasekoi* Kozur et Mostler
（Pl. 12, figs. 16—23）

Eptingium nakasekoi Kozur et Mostler, 1994, pl. 1, fig. 5; Ramovs and Gorican, 1995, pl. 5, figs. 9, 10; Sugiyama, 1997, Fig. 27 (4, 5); Spiller, 2002, pl. 4, fig. q.

Tripocyclia cf. *acythus* De Wever; Nakaseko and Nishimura, 1979, pl. 4, figs. 1—3.

Tripocyclia sp. cf. *T. acythus* De Wever; Kido, 1982, pl. 1, fig. 7.

Eptingium (?) sp. A Cheng, 1989, pl. 6, fig. 9.

Eptingium? sp. Sashida et al., 1993, Fig. 6 (3, 4, non 5).

Eptingium sp. cf. *E. nakasekoi* Kozur et Mostler; Feng et al., 2001, pl. 4, fig. 9.

Eptingium cf. *nakasekoi* Kozur et Mostler; Kamata et al., 2002, Fig. 7 (M).

描述：这个种的特点是壳近球形，二层结构，内层孔小，外层孔大，孔的接点发育瘤状构造。3根三棱形主刺较粗壮，长度略大于壳径。刺平直，基部最宽，向远端逐渐减小。3根纵脊和3条纵沟相间。

比较：这个种与 *E. manfredi* Dumitrica 的区别为后者主刺脊上发育次生沟，在相邻脊间有横脊。

层位与产地：中三叠统安尼阶下部；奥地利，匈牙利，斯洛文尼亚，马来西亚，日本，菲律宾，泰国，中国云南和西藏双湖地区。

隐王冠虫属 *Cryptostephanidium* Dumitrica, 1978

模式种：*Cryptostephanidium cornigerum* Dumitrica, 1978

鉴定要点：矢环三角形，位于头腔中，弓插入头壁中，背刺很短，口缺失。

时代与分布：中晚三叠世安尼期中期—诺利期早期；欧洲特提斯地区，日本，菲律宾，土耳其，约旦安曼，中国青海。

带角隐王冠虫 *Cryptostephanidium cornigerum* Dumitrica
（Pl. 18, figs. 22, 23, 25; pl. 21, figs. 33—36; pl. 23, fig. 13; pl. 25, fig. 41）

Cryptostephanidium cornigerum Dumitrica, 1978, pl. 1, figs. 1—4, pl. 4, fig. 4; Gorican and Buser, 1990, pl. 8, figs. 1—3; Yeh, 1990, pl. 5, figs. 11, 15, pl. 11, fig. 5; Ramovs and Gorican, 1995, pl. 5, fig. 3; Kozur, Krainer and Mostler, 1996, pl. 10, fig. 12; Tekin, 1999, pl. 24, fig. 6.

Cryptostephanidium cf. *cornigerum* Dumitrica, Yao, 1982, pl. 1, fig. 16.

描述：这个种的特点是头球形，具三角形、四边形或不规则的大孔，3根三棱形主刺较粗壮，平直，长度不相等，其中有1根较短。3根纵脊和3条纵沟相间，在脊上无次生沟。

比较：这个种与 *C. verrucosum* Dumitrica 的区别为后者的刺较长，刺的末端针形，壳表多瘤。

层位与产地：中三叠统安尼阶上部—拉丁阶上部；意大利，斯洛文尼亚，匈牙利，罗马尼亚，奥地利，菲律宾，日本，土耳其，中国青海可可西里地区。

长刺隐王冠虫 *Cryptostephanidium longispinosum* (Sashida)

(Pl. 23, figs. 16, 28)

Spongostephanidium longispinosum Sashida, 1991, Fig. 7 (1—8).

Cryptostephanidium longispinosum (Sashida); Sugiyama, 1992, Fig. 13 (1, 2); Ramovs and Gorican, 1995, pl. 5, fig. 2.

Cryptostephanidium sp. A Cheng, 1989, pl. 6, fig. 3.

Cryptostephanidium sp. B Cheng, 1989, pl. 6, fig. 4.

Cryptostephanidium sp. C Cheng, 1989, pl. 6, fig. 11.

Cryptostephanidium sp. D Cheng, 1989, pl. 7, fig. 5.

Cryptostephanidium sp. E Cheng, 1989, pl. 7, fig. 6.

描述：这个种的特点是头近球形，壳孔较大，壳表瘤小。3根三棱形主刺欠粗壮，长度不相等，长度通常不超过壳径。

比较：这个种与 *C. japonicum* (Nakaseko et Nishimura) 的区别为后者壳表的瘤较粗壮，主刺也较长。

层位与产地：中三叠统安尼阶上部—拉丁阶；斯洛文尼亚，日本，菲律宾，中国青海。

隐王冠虫（未定种A） *Cryptostephanidium* sp. A

(Pl. 21, figs. 11, 37)

描述：头近球形，壳孔大，孔的接点发育瘤状构造。3根三棱形主刺大小和形状相似，较粗壮，平直，以120°角度相交，刺的长度大于壳径。3根纵脊和3条纵沟相间，脊上无次生沟。

比较：这个未定种A与 *C. cornigerum* Dumitrica 的区别为后者主刺长度不相等，其中有1根较短，另外2根较长，其长度都小于壳径。

层位与产地：中三叠统拉丁阶上部；中国青海可可西里地区。

海绵王冠虫属 *Spongostephanidium* Dumitrica, 1978

模式种：*Spongostephanidium spongiosum* Dumitrica, 1978

鉴定要点：海绵壳近球形，二层结构，3根主刺杆形，长度不相等，不对称排列。背刺缺失，无口或门孔。

时代与分布：中晚三叠世安尼期—卡尼期；欧洲特提斯地区，中国青海。

海绵质海绵王冠虫 *Spongostephanidium spongiosum* Dumitrica

(Pl. 18, fig. 27)

Spongostephanidium spongiosum Dumitrica, 1978, pl. 2, figs. 2—5; Gorican et al., 2005, pl. 2, fig. 8; Dumitrica et al., 2010, Fig. 10 (U, V).

描述：这个种的特点是海绵壳近球形，二层结构，壳上发育瘤状构造，3根主刺杆形，长度不相等，但长度一般都比壳径小，排列不对称。

层位与产地：中上三叠统安尼阶中部—卡尼阶；罗马尼亚，意大利，约旦安曼，土耳其，中国青海可可西里地区。

三叠王冠虫属 *Triassistephanidium* Dumitrica, 1978

模式种：*Triassistephanidium laticornis* Dumitrica, 1978

鉴定要点：矢环的弓与头壁相接，背刺短，无口或无门孔。

时代与分布：中三叠世拉丁期早期；欧洲特提斯地区，中国青海。

宽角三叠王冠虫 *Triassistephanidium laticornis* Dumitrica
(Pl. 18, figs. 24, 26, 41)

Triassistephanidium laticornis Dumitrica, 1978, pl. 1, figs. 5, 6, pl. 2, fig. 1, pl. 4, fig. 3; Dumitrica et al., 1980, pl. 6, fig. 9; Kozur and Reti, 1986, Fig. 6 (F); Gorican and Buser, 1990, pl. 8, fig. 6.

Triassistephanidium sp. Bragin, 1986, pl. 2, fig. 10.

Triassistephanidium cf. *laticornis* Dumitrica; Bragin, 1991, pl. 3, fig. 2.

描述：这个种的特点是壳近三角形，孔的接点发育瘤状构造。3根主刺三棱形，平直，粗壮，刺的中部膨大，末端变钝，由3根纵脊和3条纵沟相间，在纵脊上发育次生沟。

层位与产地：中三叠统安尼阶上部—拉丁阶；罗马尼亚，斯洛文尼亚，匈牙利，俄罗斯远东地区，中国青海可可西里地区。

六桩虫超科 Hexastylacea Haeckel, 1882, emend. Petrushevskaya, 1970
欣德球虫科 Hindeosphaeridae Kozur et Mostler, 1981
欣德球虫属 *Hindeosphaera* Kozur et Mostler, 1979

模式种：*Hindeosphaera foremanae* Kozur et Mostler, 1979

鉴定要点：壳近球形，二层结构。2根主刺长度不相等，刺平直或弯曲，三棱形或杆形，锥形。

时代与分布：中晚三叠世拉丁期—卡尼期；世界各地。

华丽欣德球虫（新种） *Hindeosphaera bella* Wang sp. nov.
(Pl. 24, fig. 25)

词源：bell，拉丁词，华丽。

描述：壳近球形，二层结构，内层孔小，外层孔大，孔的接点发育瘤状构造。2根三棱形极主刺平直，长度不相等，长刺细长，长度大于壳径，短刺小，长度小于壳径。纵沟宽于纵脊，在纵脊上未发现次生沟。

比较：这一新种与 *H. spinulosa* (Nakaseko et Nishimura) 的区别为后者长刺的长度小于或等于壳径，长刺更强壮。

层位与产地：中三叠统拉丁阶上部；中国四川理塘地区。

双刺欣德球虫 *Hindeosphaera bispinosa* Kozur et Mostler
(Pl. 21, figs. 20—32; pl. 24, figs. 27, 28, 31)

Hindeosphaera bispinosa Kozur et Mostler, 1979, pl. 2, fig. 6, 1981, pl. 67, fig. 1; Lahm, 1984, pl. 6, figs. 1, 2; Tekin, 1999, pl. 25, figs. 6, 7.

Pseudostylosphaera sp. B Wang et al., 2002a, pl. 1, figs. 41—43.

Hindeosphaera sp. B Wang et al., 2005, pl. 1, figs. 5—7.

描述：这个种的特点是壳近球形或椭圆形，二层结构，孔的接点发育瘤状构造。2根极主刺长度不一，长刺粗壮，轻微左旋扭曲，长度小于或大于壳径，短刺很小，杆形，大部分未保存。

比较：这个种与 *H. spinulosa*（Nakaseko et Nishimura）的区别为后者的长刺比较短，平直而不扭曲。

层位与产地：中上三叠统拉丁阶—卡尼阶；欧洲特提斯地区，土耳其，中国西藏双湖地区、四川理塘地区和青海可可西里地区。

小刺欣德球虫 *Hindeosphaera spinulosa* (Nakaseko et Nishimura)

(Pl. 19, fig. 27; pl. 23, figs. 10, 12; pl. 24, figs. 26, 29, 30)

Archaeospongoprunum spinulosa Nakaseko et Nishimura, 1979, pl. 2, figs. 3, 4, 6; Mizutani and Koike, 1982, pl. 3, fig. 4; Takashima and Koike, 1982, pl. 2, fig. 4; Matsuda and Isozaki, 1982, pl. 3, figs. 27, 28; Sato et al., 1982, pl. 2, fig. 4.

Pseudostylosphaera spinulosa (Nakaseko et Nishimura); Yeh, 1990, pl. 4, fig. 14; Wang et al., 2002a, pl. 1, fig. 44, 2002b, pl. 2, fig. 37.

Hindeosphaera spinulosa (Nakaseko et Nishimura); Martini et al., 1989, pl. 3, fig. 8; De Wever et al., 1990, pl. 1, fig. 1, Kozur and Mostler, 1996, pl. 4, figs. 4, 8; Feng et al., 2001, pl. 5, figs. 15, 21; Yang et al., 2000, pl. 1, fig. 7.

Hindeosphaera? spinulosa (Nakaseko et Nishimura); Ramovs and Gorican, 1995, pl. 3, figs. 6—8.

Pseudostylosphaera sp. cf. *P. spinulosa* (Nakaseko et Nishimura); Cheng, 1989, pl. 6, fig. 2, pl. 7, figs. 8, 9; Wang et al., 2002b, pl. 2, figs. 38—40.

"*Stylosphaera*" *spinulosa* (Nakaseko et Nishimura); Bragin, 1991, pl. 3, figs. 4, 5, 9.

Stylosphaera (?) *spinulosa* (Nakaseko et Nishimura); Kishida and Sugano, 1982, pl. 1, figs. 16, 17.

描述：这个种的特点是壳近球形，二层结构，孔的接点发育瘤状构造。具有数根不等长的极主刺，较长的一根粗壮，平直，三棱形，另几根特短。

层位与产地：中上三叠统；世界各地。

缪勒旋扭虫属 *Muelleritortis* Kozur, 1988

模式种：*Emiluvia* (?) *cochleata* Nakaseko et Nishimura, 1979

鉴定要点：壳近球形，壳孔大，孔的接点发育瘤状构造。4根主刺三棱形，相互间以直角相间，其中3根主刺扭曲，另一根主刺平直。

时代与分布：中晚三叠世拉丁期晚期—卡尼期早期；欧洲特提斯地区，日本，土耳其，中国云南和四川。

匙形缪勒旋扭虫 *Muelleritortis cochleata* (Nakaseko et Nishimura)

(Pl. 25, figs. 38, 40)

Emiluvia (?) *cochleata* Nakaseko et Nishimura, 1979, pl. 3, figs. 2—4, 6; Nishizono et al., 1982, pl. 1, fig. 19; Kishida and Sugano, 1982, pl. 1, figs. 1—3.

Muelleritortis cochleata cochleata (Nakaseko et Nishimura); Kozur, 1988, pl. 1, figs. 1—8, pl. 2, figs. 1, 2, pl. 3, fig. 1; Dosztaly, 1991, pl. 5, fig. 1; Kozur and Mostler, 1996, pl. 1, fig. 9; Tekin, 1999, pl. 26, figs. 4, 5; Kamata et al., 2002, Fig. 6 (D).

Plafkerium cochleata (Nakaseko et Nishimura); Bragin, 1991, pl. 4, fig. 12 (non 10).

Muelleritortis cochleata (Nakaseko et Nishimura); Sugiyama, 1997, Fig. 27 (11).

Plafkerium sp. De Wever, 1982, pl. 9, fig. 4.

Emiluvia (?) aff. *cochleata* Nakaseko et Nishimura, Bragin, 1986, pl. 2, fig. 8.

描述：这个种的特点是壳近球形，二层结构，内层孔小，外层孔大，孔的接点发育瘤状构造。4根主刺三棱形，细长，由3根纵脊和3条纵沟相间，纵脊上发育次生沟，其中3根形状和大小相似，呈左旋扭曲，另一根主刺平直，较长，长度大于壳径。

比较：这个种与 *M. tumidospina* Kozur 的区别为后者的3根左旋扭曲主刺短小，长度小于壳径。

层位与产地：中上三叠统拉丁阶上部—卡尼阶上部；匈牙利，波斯尼亚-黑塞哥维那，土耳其，泰国，俄罗斯远东地区，日本，中国四川理塘地区。

膨大缪勒旋扭虫 *Muelleritortis expansa* Kozur et Mostler

(Pl. 24, figs. 33, 34)

Muelleritortis expansa Kozur et Mostler, 1996, pl. 1, figs. 1—5, 8; Tekin, 1999, pl. 26, fig. 8.

Plafkerium sp. De Wever, 1984, pl. 3, fig. 10; Gorican and Buser, 1990, pl. 6, fig. 9.

Emiluvia? cochleata Nakaseko et Nishimura; Nishizono and Murata, 1983, pl. 2, fig. 7.

Muelleritortis cochleata cf. *tumidospina* Kozur; Feng et al., 1996, pl. 3, figs. 1—3.

描述：这个种的特点是壳近球形，二层结构，内层孔小，外层孔大，孔的接点发育瘤状构造。4根主刺特别粗壮，每根刺由3根窄脊和3条宽沟相间。主刺末端膨大，其中3根主刺形状和大小相似，呈明显的左旋扭曲，另一根主刺平直。

比较：这个种和 *M. tumidospina* Kozur 的区别为后者的4根主刺末端锐尖。

层位与产地：中上三叠统拉丁阶上部—卡尼阶下部，匈牙利，斯洛文尼亚，波斯尼亚-黑塞哥维那，土耳其，日本，中国四川理塘地区。

凯维斯卡尔缪勒旋扭虫 *Muelleritortis koeveskalensis* Kozur

(Pl. 24, fig. 35; pl. 25, figs. 37, 39)

Muelleritortis cochleata koeveskalensis Kozur, 1988, pl. 3, fig. 3, Kozur and Mostler, 1996, pl. 2, figs. 1, 8; Tekin, 1999, pl. 26, fig. 6.

Muelleritortis cf. *cochleata koeveskalensis* Kozur; Kozur and Mostler, 1996, pl. 2, fig. 4.

Muelleritortis cochleata cochleata; Kamata et al., 2002, Fig. 6 (D).

描述：这个种的特点是壳近球形，二层结构，内层孔小，外层孔大，孔的接点发育瘤状构造。4根三棱形主刺中等强壮，相互间以直角相间，其中3根主刺呈左旋扭曲，刺的中部膨大，至末端又变锐尖，另一根主刺平直，比扭曲刺略长，长度大于壳径。

比较：这个种与 *M. cochleata* (Nakaseko et Mishimura) 的区别为后者的3根左旋扭曲主刺中部没有膨大，刺的近端向远端平缓地变细。由于 Kamata 等（2002）鉴定的 *M. cochleata cochleata* (Nakaseko et Nishimura) 的3根左旋扭曲主刺中上部膨大而改归 *Muelleritortis koeveskalensis* Kozur。

层位与产地：中上三叠统拉丁阶上部—卡尼阶下部；匈牙利，土耳其，泰国，俄罗斯远东地区，中国四川理塘地区。

拟彼萨格诺虫属 *Parasepsagon* Dumitrica, Kozur et Mostler, 1980

模式种：*Parasepsagon tetracanthus* Dumitrica, Kozur et Mostler, 1980

鉴定要点：外壳球形或椭圆形，二层结构，外层发育瘤状构造。4根长度不等或相等的三棱形主刺平直，刺间大部分以直角相间。

时代与分布：中晚三叠世；世界各地。

坚固拟彼萨格诺虫 *Parasepsagon firmum* (Gorican)

(Pl. 18, figs. 30, 31, 33, 35—39, 42—45; pl. 21, figs. 1—7, 14)

Plafkerium? firmum Gorican, 1990, pl. 6, figs. 3—5 (non 6).

Parasepsagon firmum (Gorican); Wang et al., 2005, pl. 1, figs. 10—14.

描述：这个种的特点是外壳近方形，壳孔较大，壳上发育瘤状构造。4根三棱形主刺粗壮，平直，其中有1根比其余3根长。每个刺由3根宽纵脊和3条窄纵沟相间，脊上发育次生沟。刺的中上部膨大，末端变尖。

比较：这个种与 *Muelleritortis expansa* Kozur et Mostler 在外壳特征和主刺形状上比较相似，区别在于后者的主刺欠粗壮，纵脊上不发育次生沟，刺的宽度从基部向远端逐渐变小。

层位与产地：中三叠统拉丁阶上部；斯洛文尼亚，中国青海可可西里地区。

可可西里拟彼萨格诺虫（新种） *Parasepsagon hohxiliensis* Wang sp. nov.

(Pl. 18, figs. 32, 34, 40, 46)

Plafkerium? firmum Gorican, 1990, pl. 6, fig. 6 (non 3—5).

词源：hoh xil，青海可可西里。

描述：外壳近球形，二层结构，内层孔小，外层孔大，壳孔数目少，孔的接点发育瘤状构造。4根三棱形主刺较粗壮，平直。刺的两边平行，至末端变锐尖。每个刺由3根纵脊和3条纵沟相间，脊上没有次生沟。刺的长度不相同，其中3根主刺长度和形状相似，长度与壳径相近，另一根主刺较长，长度大于壳径。

比较：这个新种与 *P. variabilis* (Nakaseko et Nishimura) 的区别为后者的主刺从基部向远端宽度逐渐变小，长刺和短刺长度均小于壳径。新种与 *P. firmum* (Gorican) 的区别为后者的主刺长度较小，短刺长度小于壳径，长刺长度与壳径相近。另外，主刺在中上部膨大。

层位与产地：中三叠统拉丁阶上部；斯洛文尼亚，中国青海可可西里地区。

前四刺拟彼萨格诺虫 *Parasepsagon praetetracanthus* Kozur et Mostler

(Pl. 12, fig. 7)

Parasepsagon asymmetrica praetetracanthus Kozur et Mostler, 1994, pl. 5, fig. 3; Feng et al., 2001, pl. 8, figs. 4, 5.

描述：这个种的特点是外壳椭圆形，二层结构，内层孔小，外层孔大，孔的接点发育瘤状构造。4根三棱形主刺形状和大小相近，位于同一平面中。刺长略大于壳径。每个刺由3根纵脊和3条纵沟相间，脊上无次生沟。

比较：这个种与 *P. tetracanthus* Dumitrica, Kozur et Mostler 的区别为后者4根主刺的长度不相等，2根极主刺长，而另外2根则较短。

层位与产地：中三叠统安尼阶上部；匈牙利，中国云南、四川理塘地区和西藏双湖地区。

拟彼萨格诺虫（未定种 A） *Parasepsagon* sp. A

(Pl. 19, fig. 9)

描述：外壳近球形，二层结构，外层孔较大，孔的接点发育瘤状构造。4 根三棱形主刺粗壮，平直。刺的中部轻微膨大。3 根纵脊和 3 条纵沟相间，其中 2 根极主刺对称排列，而其余 2 根主刺不对称。纵脊较宽，其上发育次生沟。

比较：这个未定种 A 十分相似于 *P. asymmetricus* Kozur et Mostler，与后者的区别为壳孔较大，具有更粗壮的主刺，纵脊较宽，其上发育次生沟。

层位与产地：中三叠统拉丁阶上部；中国青海可可西里地区。

拟彼萨格诺虫（未定种 B） *Parasepsagon* sp. B

(Pl. 21, figs. 9, 10)

描述：外壳近球形，二层结构，外层孔大，每半圈有 6—7 个，孔的接点发育瘤状构造。4 根三棱形主刺较粗壮，平直，排列不太对称。3 根纵脊和 3 条纵沟相间，脊上无次生沟。

比较：这个未定种 B 与 *P.* sp. A 的区别为后者壳孔更大，纵脊上发育次生沟。

层位与产地：中三叠统拉丁阶上部；中国青海可可西里地区。

变异拟彼萨格诺虫 *Parasepsagon variabilis* (Nakaseko et Nishimura)

(Pl. 19, fig. 33; pl. 23, fig. 14)

Staurodoras variabilis Nakaseko et Nishimura, 1979, pl. 3, figs. 5, 8; Kido, 1982, pl. 1, fig. 5; Kishida and Sugano, 1982, pl. 1, fig. 20.

Staurolonche variabilis (Nakaseko et Nishimura); Bragin, 1991, pl. 1, fig. 16; Wang et al., 2002a, pl. 1, fig. 40.

Parasepsagon variabilis (Nakaseko et Nishimura); Yeh, 1989, pl. 1, figs. 16, 19; Ramovs and Gorican, 1995, pl. 3, fig. 5; Kozur, Krainer and Mostler, 1996, pl. 4, figs. 2, 3, 7, 9; Feng et al., 2001, pl. 8, fig. 7; Wang et al., 2005, pl. 1, figs. 18—21.

Staurolonche sp. aff. *S. variabilis* (Nakaseko et Nishirmura); Blome et al., 1989, pl. 33.1, fig. 11.

描述：这个种的特点是外壳近球形，壳上发育瘤状构造。4 根三棱形主刺平直，长度和大小不相等，其中 1 根较长，另一根次长，其余 2 根长度相似，但长度都小于壳径。每个刺由 3 根纵脊和 3 条纵沟相间，脊上无次生沟。

层位与产地：中上三叠统安尼阶上部—卡尼阶；世界各地。

三旋扭虫属 *Tritortis* Kozur, 1988

模式种：*Sarla* (?) *kretaensis* Kozur et Krahl, 1984

鉴定要点：外壳近球形，壳孔较大，壳上发育瘤状构造。具 3 根三棱形主刺，以 120°角相间，其中 1 根主刺平直，另外 2 根主刺呈轻微或强烈扭曲。

时代与分布：中晚三叠世拉丁期—卡尼期；欧洲特提斯地区，土耳其，加拿大，日本，俄罗斯远东地区，中国青海和四川。

双旋三旋扭虫 *Tritortis dispiralis* (Bragin)

(Pl. 18, figs. 28, 29; pl. 25, figs. 32, 33)

Sarla dispiralis Bragin, 1986, pl. 1, fig. 12, 1991, pl. 4, figs. 6, 11, pl. 5, fig. 8.

Tripocyclia sp. Nishizono et al.，1982，pl. 1，fig. 15.
Tripocyclia sp. A Nishizono et Murata，1983，pl. 2，fig. 6.
Sarka? *kretaensis* Kozur et Krahl；Cordey et al.，1988，pl. 2，fig. 7（non 9—11）.
Tritortis dispiralis（Bragin）；Kozur and Mostler，1996，pl. 3，fig. 11；Tekin，1999，pl. 27，fig. 3.

描述：这个种的特点是壳近球形，二层结构，内层孔小，外层孔大，孔的接点发育瘤状构造。3根三棱形主刺较粗壮，形状不一，其中2根作轻微左旋扭曲，另一根平直。纵脊上未见次生沟。主刺长度等于或小于壳径。

比较：这个种与 *T. kretaensis*（Kozur et Krahl）的区别为后者的壳较小，2根主刺呈强烈的左旋扭曲。主刺长，长度约为壳径的2倍。

层位与产地：中上三叠统拉丁阶上部—卡尼阶下部；匈牙利，波斯尼亚，黑塞哥维那，土耳其，加拿大，俄罗斯远东地区，中国青海可可西里地区和四川理塘地区。

克兰特三旋扭虫 *Tritortis kretaensis*（Kozur et Krahl）

(Pl. 25，figs. 35，36，42)

Sarla（?）*kretaensis* Kozur et Krahl，1984，pl. 1，figs. 3，4；Cordey et al.，1988，pl. 2，figs. 9—11.
Tritortis kretaensis（Kozur et Krahl）；Sugiyama，1997，Fig. 48（21）.
Tritortis kretaensis kretaensis（Kozur et Krahl）；Kozur，1988a，pl. 4，figs. 3—5；Tekin 1999，pl. 27，figs. 4，5；Kamata et al.，2002，Fig. 6（F）.
Tritortis cf. *kretaensis kretaensis*（Kozur et Krahl）；Kamata et al.，2002，Fig. 6（K，L）.
Tritortis kretaensis subcylindrica Kozur，1988a，pl. 4，figs. 6，8.
Spumellaria gen. et sp. indet.；Sashida et al.，1993，Fig. 8（9，10，non 7，11，12）.
Sarla（?）sp. Yeh，1990，pl. 9，fig. 5.

描述：这个种的特点是壳近球形，壳孔大，壳面上发育瘤状构造。3根三棱形主刺形状不一，长度相近，其中1根平直，另外2根较强烈扭曲。主刺长度约为壳径的2倍。

层位与产地：中上三叠统拉丁阶上部—卡尼阶下部；希腊，加拿大，匈牙利，日本，土耳其，中国四川理塘地区。

粗刺三旋扭虫（新种） *Tritortis robustispinosa* Wang sp. nov.

(Pl. 25，fig. 34)

词源：robust，拉丁词，强壮；spines，拉丁词，刺。

描述：外壳小，亚球形，二层结构，外层孔较大，孔的结点发育小瘤。只保存2根粗壮的、轻微左旋的三棱形主刺。纵脊较窄，而纵沟较宽；主刺宽度与壳径相近，而长度大于壳径。

比较：这个新种与 *Tritortis dispiralis*（Bragin）的区别为后者的壳体较大，主刺细长，宽度较小。

层位与产地：中三叠统拉丁阶上部；中国四川理塘地区。

假桩球虫属 *Pseudostylosphaera* Kozur et Mock，1981

模式种：*Pseudostylosphaera gracilis* Kozur et Mock，1981

鉴定要点：壳近球形至椭圆形，壳表发育瘤状构造。2根三棱形主刺平直或扭曲，刺由3根纵脊和3条纵沟相间组成。

时代与分布：中晚三叠世；世界各地。

颗石柱假桩球虫 *Pseudostylosphaera coccostyla* (Rüst)

(Pl. 18, fig. 19)

Spongotractus coccostyla (Rüst), 1892, pl. 21, fig. 8.

Pseudostylosphaera coccostyla (Rüst); Kozur and Mostler, 1981, pl. 15, fig. 3, pl. 46, fig. 5, 1994, pl. 1, fig. 18; Hahm, 1984, pl. 4, figs. 7, 8; Kozur and Reti, 1988, Fig. 6 (E); Gorican and Buser, 1990, pl. 5, fig. 1; Chiari et al., 1995, pl. 2, fig. 13; Ramovs and Gorican, 1995, pl. 3, figs. 3, 4; Sashida et al., 1999, Fig. 8 (10); Feng et al., 2001, pl. 5, figs. 13, 14; Wang et al., 2002a, pl. 1, fig. 23, 2005, pl. 1, fig. 9.

描述：这个种的特点是三棱形主刺较粗壮，在纵脊上发育次生沟。

比较：这个种与 *P. compacta* (Nakaseko et Nishimura) 的区别为后者 2 根主刺的纵脊上没有次生沟。

层位与产地：中上三叠统拉丁阶—卡尼阶；匈牙利，奥地利，南斯拉夫，斯洛文尼亚，土耳其，泰国，中国青海可可西里地区。

致密假桩球虫 *Pseudostylosphaera compacta* (Nakaseko et Nishimura)

(Pl. 12, figs. 24, 26—30; pl. 18, figs. 20, 21; pl. 19, figs. 12, 24, 25, 47, 48; pl. 21, fig. 16; pl. 23, figs. 2, 4—7)

Archaeospongoprunum compactum Nakaseko et Nishimura, 1979, pl. 1, figs. 3, 7.

Pseudostylosphaera compacta (Nakaseko et Nishimura); Yeh, 1990, pl. 4, figs. 3, 4, 20; Feng and Liu, 1993, pl. 1, figs. 1, 2; Feng et al., 2001, pl. 5, figs. 5—12.

Pseudostylosphaera coccostyla compacta (Nakaseko et Nishimura); Kozur and Mostler, 1994, pl. 1, fig. 8.

"*Stylosphaera*" *compacta* (Nakaseko et Nishimura); Bragin, 1991, pl. 10, figs. 1, 2.

描述：这个种的特点是外壳近球形，壳上发育瘤状构造。2 根三棱形主刺平直不扭曲，形状和大小相似，3 根纵脊和 3 条纵沟相间，脊上无次生沟。刺的基部最宽，向远端变窄，至末端呈尖锐形。

层位与产地：中上三叠统拉丁阶—卡尼阶；日本，菲律宾，俄罗斯远东地区，中国华南和青海可可西里地区。

纤细假桩球虫 *Pseudostylosphaera gracilis* Kozur et Mock

(Pl. 18, figs. 11—13, 15—18; pl. 21, figs. 17, 18; pl. 23, figs. 17, 18; pl. 24, figs. 16—18)

Pseudostylosphaera gracilis Kozur et Mock, 1981, pl. 66, fig. 1; Lahm, 1984, pl. 5, figs. 1, 2; Jurkovsek, 1990, pl. 5, fig. 1; Sugiyama, 1997, Fig. 48 (18); Tekin, 1999, pl. 25, figs. 10, 11; Wang et al., 2002a, pl. 2, fig. 29, 2002b, pl. 1, figs. 27, 28.

Stylosphaera (?) *hellenica* (De Wever); Kishida and Sugano, 1982, pl. 1, fig. 1.

Pseudostylosphaera sp. Sashida et al., 1993, Fig. 7 (19).

Stylosphaera hellenica (De Wever); Kozur and Mostler, 1979, pl. 1, fig. 8.

Pseudostylosphaera cf. *hellenica* (De Wever); Gorican and Buser, 1990, pl. 5, fig. 8.

Archaeospongoprunum hellenica De Wever, 1979, pl. 1, fig. 8.

Pseudostylosphaera hellenica (De Wever); Kamata et al., 2002, Fig. 5 (J).

描述：这个种的特点是壳椭圆形，壳上发育瘤状构造。2 根三棱形主刺在末端膨大，呈轻微的左旋扭曲。

比较：这个种与 *P. nazarovi* (Kozur et Mostler) 的区别为后者的 2 根主刺末端不膨大，刺自基部向远端逐渐变细。

层位与产地：中上三叠统拉丁阶上部—卡尼阶下部；奥地利，斯洛文尼亚，土耳其，日本，中国青海可可西里地区和四川理塘地区。

戈斯特林假桩球虫 *Pseudostylosphaera goestlingensis* （Kozur et Mostler）

(Pl. 23, fig. 11; pl. 24, figs. 21, 23)

Stylosphaera? goestlingensis Kozur et Mostler, 1979, pl. 17, fig. 5, pl. 18, fig. 1.

Pseudostylosphaera goestlingensis (Kozur et Mostler); Kozur and Mostler, 1981, p. 31; Lahm, 1984, pl. 5, figs. 3, 4; Gorican and Buser, 1990, pl. 5, fig. 7; Sugiyama, 1997, Fig. 48 (19); Tekin, 1999, pl. 25, fig. 9; Yang et al., 2000, pl. 1, fig. 5.

描述：这个种的特点是壳近球形，壳表发育瘤状构造。2根三棱形主刺呈轻微的右旋扭曲。

比较：这个种与 *P. nazarovi* （Kozur et Mostler）的区别为后者的2根主刺呈轻微的左旋扭曲。

层位与产地：中上三叠统拉丁阶上部—卡尼阶下部；奥地利，斯洛文尼亚，土耳其，日本，中国西藏双湖地区和四川理塘地区。

螺旋状假桩球虫 *Pseudostylosphaera helicata* （Nakaseko et Nishimura）

(Pl. 23, fig. 9)

Archaeospongoprunum helicata Nakaseko et Nishimura, 1979, pl. 2, figs. 1, 2, pl. 12, fig. 3.

Pseudostylosphaera helicata (Nakaseko et Nishimura), Yeh, 1992, pl. 7, fig. 1; Sashida et al., 1993, Fig. 7 (11).

P. sp. cf. *P. helicata* (Nakaseko et Nishimura); Yeh, 1992, pl. 7, fig. 2.

描述：这个种的特点是外壳椭圆形，二层结构，孔的接点发育瘤状构造。2根三棱形主刺较长，形状和大小大致相近，3根纵脊和3条纵沟相间，在远端刺呈轻微的扭曲。

比较：这个种与 *P. compacta* （Nakaseko et Nishimura）的区别为后者的2根主刺平直而不扭曲，外壳近球形。

层位与产地：中上三叠统拉丁阶上部—卡尼阶下部；日本，菲律宾，中国青海可可西里地区。

不明显假桩球虫 *Pseudostylosphaera imperspicua* （Bragin）

(Pl. 24, figs. 7—9)

Archaeospongoprunum imperspicua Bragin, 1986, pl. 2, fig. 9.

"*Stylosphaera*" *imperspicua* (Bragin); Bragin, 1991, pl. 5, fig. 4.

Pseudostylosphaera imperspicua (Bragin); Tekin, 1999, pl. 25, fig. 13.

Archaeospongoprunum cf. *hellenicum* De Wever; Takashima and Koike, 1982, pl. 2, figs. 7, 8.

Pseudostylosphaera nazarovi (Kozur et Mostler); Feng et al., 1996, pl. 1, fig. 10.

Pseudostylosphaera cf. *nazarovi* (Kozur et Mostler); Feng et al., 1996, pl. 1, figs. 14, 15.

描述：这个种的特点是壳近球形或椭圆形，二层结构，内层孔小，外层孔大，孔的接点发育瘤状构造。2根主刺粗壮，三棱形，形状和大小相似，3根纵脊和3条纵沟相间，沟较深。刺呈不明显的轻微扭曲。

比较：这个种与 *P. nazarovi* （Kozur et Mostler）的区别为后者的2根主刺呈明显的左旋扭曲。

层位与产地：中三叠统拉丁阶上部；日本，俄罗斯远东地区，土耳其，中国四川理塘地区。

不等刺假桩球虫（新种） *Pseudostylosphaera inaequispinosa* Wang sp. nov.

(Pl. 24, figs. 1—3, 6, 12, 13)

词源：inaequi, 拉丁词，不等；spinosus, 拉丁词，多刺。

描述：壳球形或椭圆形，二层结构，内层孔小，外层孔大，孔的接点发育瘤状构造。2根主刺三棱形，中等粗壮，长度不相同，1根略长，长刺长度大于壳径，短刺等于或小于壳径；1根刺平直，另外1根刺呈轻微右旋扭曲。3根纵脊和3条纵沟相间。

比较：这一新种与 *P. nazarovi* （Kozur et Mostler）的区别为后者的2根主刺均呈轻微左旋扭曲；与 *P. compacta* （Nakaseko et Nishimura）的区别为后者的2根主刺平直而不扭曲。

层位与产地：中三叠统拉丁阶中上部；中国四川理塘地区。

日本假桩球虫 *Pseudostylosphaera japonica* (Nakaseko et Nishimura)

(Pl. 12, fig. 25; pl. 18, figs. 1—9; pl. 23, fig. 31; pl. 24, figs. 10, 11)

Archaeospongoprunum japonica Nakaseko et Nishimura, 1979, pl. 1, figs. 2, 4, 9; Mizutani and Koike, 1982, pl. 3, fig. 3; Sato et al., 1982, pl. 2, figs. 1, 2; Yao, 1982, pl. 1, fig. 21; Adachi and Kojima, 1983, pl. 13, fig. 10; Yoshida, 1986, pl. 9, figs. 1, 2; Feng and Liu, 1993a, pl. 1, fig. 3.

Pseudostylosphaera japonica (Nakaseko et Nishimura); Kozur and Mostler, 1981, p. 31; Lahm, 1984, pl. 4, figs. 9, 10; Blome et al., 1986, pl. 8, 3, figs. 1, 2; Kishida and Hisada, 1986, Figs. 3, 4; Kojima and Mizutani, 1987, Fig. 2 (1); Yeh, 1989, pl. 1, fig. 4, 1990, pl. 4, figs. 5—7; Cheng, 1989, pl. 6, fig. 1, pl. 7, fig. 7; Martini, 1989, pl. 3, fig. 12; Jurkovsek, 1990, pl. 5, fig. 2; Gorican and Buser, 1990, pl. 5, fig. 2; Zhang, 1990, pl. 1, fig. 9; Sashida et al., 1993, Fig. 7 (9, 15); Chiari et al., 1995, pl. 2, fig. 14; Sugiyama, 1997, Fig. 48 (15); Yang et al., 2000, pl. 1, figs. 1, 2; Feng et al., 2001, pl. 5, figs. 1—6; Kamata et al., 2002, Fig. 5 (F).

Pseudostylosphaera coccostyla (Rüst); Jurkovsek, 1990, pl. 5, figs. 4a, 4b, 5a, 5b, 1991, pl. 5, figs. 4, 5.

"*Stylosphaera*" *japonica* (Nakaseko et Nishimura); Bragin, 1991, pl. 1, figs. 11, 13, pl. 9, figs. 13, 14.

描述：这个种的特点是壳球形或椭圆形，壳上发育瘤状构造。2根三棱形主刺平直，长度大致相等。刺的宽度在中部略有膨大，至末端又变锐尖。Jurkovsek（1990）鉴定的 *P. coccostyla* （Rüst）由于2根主刺中部都有轻微膨大而我们把它作为该种的同义名称。

层位与产地：中上三叠统；世界各地。

长刺假桩球虫 *Pseudostylosphaera longispinosa* Kozur et Mostler

(Pl. 19, figs. 28, 49; pl. 23, fig. 32; pl. 24, figs. 4, 5)

Pseudostylosphaera longispinosa Kozur et Mostler, 1981, pl. 1, fig. 6; Gorican and Jurkovsek, 1984, pl. 6, fig. 4; Lahm, 1984, pl. 4, figs. 11, 12; Yeh, 1990, pl. 4, fig. 2; Jurkovsek, 1990, pl. 5, figs. 6a, b; Gorican and Buser, 1990, pl. 5, figs. 3—5; Sugiyama, 1997, Fig. 48 (16); Tekin, 1999, pl. 25, fig. 14; Feng et al., 2001, pl. 5, figs. 22—24; Wang et al., 2002, pl. 2, figs. 35, 36.

Pseudostylosphaera longobardica Kozur et Mostler, 1981, pl. 49, fig. 3.

Pseudostylosphaera cf. *longispinosa* Kozur et Mostler; Kamata et al., 2002, Fig. 5 (E).

描述：这个种的特点是壳小，近球形，壳表发育瘤状构造。2根三棱形主刺细长，形状和大小大致相似，长度约为壳径的2倍。

层位与产地：中上三叠统拉丁阶—卡尼阶；意大利，斯洛文尼亚，土耳其，泰国，日本，中国云南、青海可可西里地区和四川理塘地区。

大刺假桩球虫 *Pseudostylosphaera magnispinosa* Yeh

(Pl. 18, fig. 14; pl. 19, figs. 23, 32; pl. 23, figs. 1, 3)

Pseudostylosphaera magnispinosa Yeh, 1989, pl. 1, figs. 1, 2, 11, 18; Sashida et al., 1993, Fig. 7 (18); Sugiyama,

1997，Fig. 48（23）.

Pseudostylosphaera sp. A Kojima et Mizutani, 1987, Fig. 2（2）.

描述：这个种的特点是壳小，近球形，壳表发育瘤状构造。2根三棱形主刺特别粗壮，长度不相等，1根主刺比另外1根略短。无次生沟。

层位与产地：中三叠统拉丁阶；美国俄勒冈，日本，中国青海可可西里地区。

纳扎罗夫假桩球虫 *Pseudostylosphaera nazarovi*（Kozur et Mostler）
（Pl. 18, figs. 3a, 10; pl. 19, figs. 29—31; pl. 21, fig. 15; pl. 24, figs. 20, 32）

Stylosphaera nazarovi Kozur et Mostler, 1979, pl. 1, fig. 15, pl. 14, figs. 4, 6, pl. 18, fig. 1.

Pseudostylosphaera nazarovi（Kozur et Mostler）; Kozur and Mostler, 1981, p. 31; Cordey et al., 1988, pl. 2, fig. 6; Sugiyama, 1997, Fig. 48（17）; Tekin, 1999, pl. 25, fig. 15; Yang et al., 2000, pl. 1, figs. 3, 4; Wang et al., 2002a, pl. 1, figs. 24, 25, 2005, pl. 1, figs. 1—4.

Pseudostylosphaera sp. B Yeh, 1992, pl. 8, fig. 11.

Archaeospongoprunum nazarovi（Kozur et Mostler）; Takashima and Koike, 1982, pl. 2, figs. 5, 6.

描述：这个种的特点是壳球形或椭圆形，壳上发育瘤状构造。2根三棱形主刺呈轻微左旋扭曲。

层位与产地：中上三叠统拉丁阶上部—卡尼阶下部；奥地利，土耳其，日本，菲律宾，中国西藏双湖地区、青海可可西里地区和四川理塘地区。

拟纤细假桩球虫（新种） *Pseudostylosphaera paragracilis* Wang sp. nov.
（Pl. 24, figs. 14, 15, 19）

词源：Para，希腊词，拟；gracil，拉丁词，纤细的。

描述：外壳近球形，二层结构，内层孔小，外层孔大，孔的接点发育瘤状构造。2根三棱形主刺比较粗壮，3根纵脊和3条纵沟相间，长度大致相似，在刺的末端膨大。1根刺平直，另外1根刺呈轻微左旋扭曲。

比较：这个新种与 *P. gracilis* Kozur et Mock 的区别为后者的2根主刺都呈左旋扭曲。

层位与产地：中三叠统拉丁阶中上部；中国四川理塘地区。

青海假桩球虫（新种） *Pseudostylosphaera qinghaiensis* Wang sp. nov.
（Pl. 21, fig. 19）

Pseudostylosphaera sp. Sashida et al., 1993, Fig. 7（19）.

词源：qinghai，标本产地所属省。

描述：壳小，椭圆形，二层结构，孔的接点发育瘤状构造。2根三棱形主刺较粗壮，相向而生，3根纵脊和3条纵沟相间，在纵脊上具次生沟。2根刺长度不相同，但均大于壳径，刺呈轻微右旋扭曲。

比较：这个新种与 *P. goestlingensis*（Kozur et Mostler）的区别为后者壳体较大，主刺较细，呈较强的右旋扭曲。

层位与产地：中三叠统拉丁阶上部；日本，中国青海可可西里地区。

假桩球虫（未定种A） *Pseudostylosphaera* sp. A
（Pl. 21, fig. 8）

描述：外壳椭圆形，二层结构，内层孔小，外层孔大，孔的接点发育瘤状构造。有2根粗壮的三棱形主刺，3根纵脊和3条纵沟相间，纵脊上发育次生沟。主刺长度略大于壳的长径。

比较：这个未定种 A 与 *P. compacta*（Nakaseko et Nishimura）的区别是后者的纵脊上不发育次生沟。

层位与产地：中三叠统拉丁阶上部；中国青海可可西里地区。

假桩球虫（未定种 C） *Pseudostylosphaera* sp. C

(Pl. 21, figs. 12, 13)

描述：壳中等大小，近球形，二层结构，壳表发育瘤状构造。有2根相对的三棱形主刺，刺平直，宽度匀称，至末端变锐，脊上无次生沟。

比较：这个未定种 C 与 *P. compacta*（Nakaseko et Nishimura）的区别为后者的壳体较小，主刺基部最宽，向远端逐渐变窄，至末端锐尖。

层位与产地：中三叠统拉丁阶上部；中国青海可可西里地区。

假桩球虫（未定种 D） *Pseudostylosphaera* sp. D

(Pl. 23, figs. 8)

描述：壳椭圆形，二层结构，壳表发育瘤状构造。2根三棱形主刺较粗壮，刺的基部宽，向远端逐渐变窄，至末端锐尖，2根刺不位于同一直线上。脊上无次生沟。

层位与产地：中三叠统拉丁阶上部；中国青海可可西里地区。

多弓形脊虫科 Multiarcusellidae Kozur et Mostler, 1979
多弓形脊虫亚科 Multiarcusellinae Kozur et Mostler, 1979
贝图尔虫属 *Beturiella* Dumitrica, Kozur et Mostler, 1980

模式种：*Beturiella robusta* Dumitrica, Kozur et Mostler, 1980

鉴定要点：外壳球形，6根三棱形主刺粗壮，弓形较弱。

时代与分布：中三叠世；欧洲特提斯地区，中国云南、青海和四川。

强壮贝图尔虫 *Beturiella robusta* Dumitrica, Kozur et Mostler

(Pl. 22, figs. 24, 31)

Beturiella robusta Dumitrica, Kozur et Mostler, 1980, pl. 3, fig. 5, pl. 12, figs. 1—3; Lehm, 1984, pl. 3, fig. 9; Gorican and Buser, 1990, pl. 1, fig. 3; Feng et al., 1996, pl. 2, fig. 8.

描述：这个种的特点是外壳球形或亚球形，壳孔中等大小，壳表发育瘤状构造。6根主刺粗壮，三棱形，平直，在末端变尖，3根纵脊和3条纵沟相间，脊上无次生沟。

层位与产地：中三叠统安尼阶上部—拉丁阶；意大利，斯洛文尼亚，中国四川理塘地区和青海可可西里地区。

贝图尔虫（未定种 A） *Beturiella* sp. A

(Pl. 22, figs. 28, 30)

描述：外壳近球形，壳孔较小，有时被微粒硅所覆盖。6根三棱形主刺细长，长度大于壳径，其大小和形状相似。

比较：这个未定种 A 与 *B. robusta* Dumitrica, Kozur et Mostler 的区别为后者主刺粗壮。

层位与产地：中三叠统拉丁阶上部；中国青海可可西里地区。

变刺贝图尔虫（新种） *Beturiella variospinosa* Wang sp. nov.

(Pl. 22, figs. 26, 27, 29, 32)

词源：vario，拉丁词，变化；spinosus，拉丁词，多刺的。

描述：壳近球形，壳孔小，具6根三棱形主刺，主刺长度小于壳径，其中1根主刺特别粗壮，其余5根主刺大小和形状相似，较小，最宽处约为大刺的1/2。3根纵脊和3条纵沟相间，脊窄沟宽。

比较：这个新种与 *B. robusta* Dumitrica, Kozur et Mostler 的区别为后者6根主刺都十分粗壮。

层位与产地：中三叠统拉丁阶上部；中国青海可可西里地区。

奥地利农神亚科 Austrisaturnalinae Kozur et Mostler, 1983
蒂博尔虫属 *Tiborella* Dumitrica, Kozur et Mostler, 1980

模式种：*Tiborella magnidentata* Dumitrica, Kozur et Mostler, 1980

鉴定要点：壳近方形，壳壁厚，壳孔大。4根三棱形主刺位于同一平面中，2根极主刺与其余2根刺呈直角相间。主刺平直或扭曲。

时代与分布：中晚三叠世；欧洲特提斯地区，日本，俄罗斯远东地区，中国华南和青海。

华丽蒂博尔虫 *Tiborella florida* (Nakaseko et Nishimura)

(Pl. 12, figs. 1—6; pl. 19, figs. 40—45)

Cecrops floridus Nakaseko et Nishimura, 1979, pl. 2, figs. 5, 8; Feng and Liu, 1993, pl. 1, fig. 20.

Staurolonche floridus (Nakaseko et Nishimura); Bragin, 1991, pl. 10, fig. 4.

Tiborella florida (Nakaseko et Nishimura); Kozur and Mostler, 1994, p. 52; Ramovs and Gorican 1995, pl. 4, figs. 1—4 (non 5); Feng et al., 2001, pl. 5, fig. 17, pl. 8, figs. 8—12.

Tiborella floridus floridus (Nakaseko et Nishimura); Kozur et al., 1996, pl. 3, figs. 5, 13—16.

描述：这个种的特点是壳近方形，壳孔大。孔的接点发育瘤状构造。4根三棱形主刺大小和形状相似，刺平直，3根窄的纵脊和3条较宽纵沟相间，脊上无次生沟。

层位与产地：中三叠统安尼阶上部；奥地利，斯洛文尼亚，俄罗斯远东地区，日本，中国云南、四川理塘地区和青海可可西里地区。

罩笼虫目 Nassellaria Ehrenberg, 1875
章鱼虫科 Poulpidae De Wever, 1981
章鱼虫属 *Poulpus* De Wever, 1981

模式种：*Poulpus piabyx* De Wever, 1981

鉴定要点：头很大，半球形，壳孔也很大，3根足，没有顶角。

时代与分布：中三叠世安尼期晚期—拉丁期；匈牙利，斯洛文尼亚，希腊，奥地利，土耳其，菲律宾，日本，中国青海。

章鱼虫（未定种A） *Poulpus* sp. A

(Pl. 23, fig. 35)

描述：头大，半球形，孔较大，多边形至卵形。3根足纤细，短，长度与头径相近。

比较：这个未定种A与 *P. curvispinus* Dumitrica, Kozur et Mostler 的区别为后者的3根足更长，更粗壮。

层位与产地：中三叠统拉丁阶上部；中国青海可可西里地区。

扎莫依达虫属 *Hozmadia* Dumitrica, Kozur et Mostler, 1980

模式种：*Hozmadia reticulata* Dumitrica, Kozur et Mostler, 1980

鉴定要点：头大，半球形，具肋或蜂巢构造。具1个顶角和3根分散足。

时代与分布：中三叠世；欧洲特提斯地区，中国青海。

拟圆形扎莫依达虫（新种） *Hozmadia pararotunda* Wang sp. nov.
(Pl. 23, figs. 19, 27)

词源：para，希腊词，拟，相似；rotund，拉丁词，圆形。

描述：头小，亚球形，内层孔小，外层孔较大，孔的接点发育瘤状构造。顶刺粗壮，三棱形，3根三棱形足比较粗壮，平直而不弯曲。顶刺和足的长度都大于头径。

比较：这个新种与 *H. rotunda*（Nakaseko et Nishimura）的区别为后者的足比较短，并且向内弯曲，顶刺和足的长度分别等于和小于头径。

层位与产地：中三叠统拉丁阶上部；中国青海可可西里地区。

棘束虫超科 Acanthodesmiacea Hertwig, 1879
鲁斯特笼虫科 Ruestcyrtiidae Kozur et Mostler, 1979
环三叠塔虫属 *Annulotriassocampe* Kozur, 1994

模式种：*Annulotriassocampe baldii* Kozur, 1994

鉴定要点：壳多节，锥形，头胸节往往部分或全部被微粒硅覆盖，头胸节、腹节和后腹节收缩部分无孔。后腹节发育桶箍状围脊，围脊上只发育1列环孔。

时代与分布：三叠纪—侏罗纪；世界各地。

多节环三叠塔虫 *Annulotriassocampe multisegmantatus* Tekin
(Pl. 22, figs. 1—11; pl. 25, figs. 4—10)

Annulotriassocampe multisegmantatus Tekin, 1999, pl. 41, figs. 3—6.
Triassocampe sp. A Kamata et al., 2002, Fig. 7 (A, B).

描述：这个种的特点是壳长，亚锥形至亚圆柱形，具13—14个后腹节。头胸节穹隆形，光滑，无孔。腹节亚梯形至桶箍形，大部分为微粒硅覆盖。后腹节宽度逐渐增大到第9节，然后，宽度又变小，呈圆柱形至反亚梯形，所有后腹节只有1列围脊孔。

比较：这个种与 *A. sulovensis*（Kozur et Mock）的区别为前者壳体较长，壳节多，所有节的下部边缘环退化。

层位与产地：中上三叠统拉丁阶上部—卡尼阶下部；土耳其，泰国，中国青海可可西里地区和四川理塘地区。

新环三叠塔虫 *Annulotriassocampe nova*（Yao）
(Pl. 20, figs. 26—28, 31)

Dictyomitrella sp. B Yao et al., 1980, pl. 3, figs. 1—3.
Triassocampe nova Yao; Yao et al., 1980, pl. 1, fig. 14, 1982, pl. 2, figs. 1—4; Nishizono and Murata, 1983, pl. 2,

fig. 18; Bragin, 1986, pl. 3, fig. 4, 1991, pl. 5, figs. 12, 16.

描述：这个种的特点是外壳圆锥形或塔形，具有发育很好的围栅脊，其下有1列孔环。节间缢较浅。壳的宽度向远端逐渐增大。

层位与产地：中上三叠统拉丁阶上部—诺利阶中部；日本，俄罗斯远东地区，中国青海可可西里地区。

特别环三叠塔虫 *Annulotriassocampe proprium* (Blome)
(Pl. 22, fig. 12)

Triassocampe proprium Blome, 1984, pl. 16, figs. 4, 11, 14.

Annulotriassocampe proprium (Blome); Kozur, 1994, p. 249; Tekin, 1999, pl. 41, fig. 7.

描述：这个种的特点是头穹隆形，具有1根三棱形顶刺，胸节和腹节亚梯形，高度和宽度逐渐增大，围栅脊明显，在基部发育1列大的圆至亚圆形的孔环。

层位与产地：中上三叠统拉丁阶上部—诺利阶；美国，土耳其，中国青海可可西里地区。

萨洛夫环三叠塔虫 *Annulotriassocampe sulovensis* (Kozur et Mock)
(Pl. 19, figs. 5—7; pl. 20, figs. 30, 36, 38, 39; pl. 22, fig. 13; pl. 25, figs. 16, 17, 20, 23)

Triassocampe sulovensis Kozur et Mock, 1981, pl. 13, fig. 3; Yeh, 1989, pl. 2, fig. 13, 1990, pl. 7, fig. 8; Gorican and Buser, 1990, pl. 12, figs. 4, 5; Kellici and De Wever, 1995, pl. 6, figs. 13, 14.

Triassocampe cf. *sulovensis* Kozur et Mock; Kamata et al., 2002, Fig. 7 (D).

Annulotriassocampe sulovensis (Kozur et Mock); Kozur, 1994, p. 249; Tekin, 1999, pl. 41, fig. 8; Wang et al., 2002a, pl. 2, figs. 39—42, 2002b, pl. 2, figs. 5—7.

描述：这个种的特点是头胸节锥形，其余各节近圆柱形，腹节和后腹节具围栅脊，其下发育1列孔环。围栅脊间无孔。节间缢较深。

比较：这个种与 *A. nova* (Yao) 的区别为壳形和节间缢的深浅不一样。

层位与产地：中上三叠统拉丁阶上部—卡尼阶下部；西喀尔巴阡山和北阿尔卑斯山，美国俄勒冈，菲律宾，土耳其，中国西藏双湖地区、青海可可西里地区和四川理塘地区。

拟三叠塔虫属 *Paratriassocampe* Kozur et Mostler, 1994

模式种：*Paratriassocampe gaetanii* Kozur et Mostler, 1994

鉴定要点：壳多节，圆锥形。头圆锥形至穹隆形，光滑，无孔。腹节和首2节后腹节桶箍形，这些节通常被微粒硅覆盖。其余后腹节较窄，具2列孔环。

时代与分布：中三叠世；欧亚特提斯地区，中国四川和青海。

短壳拟三叠塔虫 *Paratriassocampe brevis* Kozur et Mostler
(Pl. 22, figs. 14, 15)

Paratriassocampe brevis Kozur et Mostler, 1994, pl. 42, figs. 14, 15.

描述：这个种的特点是壳体短，圆锥形。头穹隆形，光滑，在远端具少量孔。胸节和腹节桶箍形，具1列不规则的孔环。后腹节由桶箍形至亚圆柱形再至倒梯形，具2列不规则的孔环。所有后头节之间缢较深、较窄，光滑，无孔。

比较：这个种以其短小的壳体与这个属的其他种相区别。

层位与产地：中三叠统拉丁阶上部；匈牙利，中国青海可可西里地区。

盖坦尼拟三叠塔虫 *Paratriassocampe gaetanii* Kozur et Mostler

(Pl. 25，figs. 14，15，18)

Paratriassocampe gaetanii Kozur et Mostler，1994，pl. 42，figs. 7—11.

描述：这个种的特点是壳体锥形，头穹隆形，光滑，无孔。胸节和腹节桶箍形，后腹节的宽度逐渐增大。每节上具有2列孔环。节间缢较深。

比较：这个种与 *P. brevis* Kozur et Mostler 的区别为后者壳体的壳节少，呈短锥形，胸节和腹节桶箍形，具有1列孔环。

层位与产地：中三叠统拉丁阶上部；意大利，中国四川理塘地区。

条纹三叠塔虫属 *Striatotriassocampe* Kozur et Mostler，1994

模式种：*Striatotriassocampe nodosoannulata* Kozur et Mostler，1994

鉴定要点：壳很长，细亚圆形，每个后头节环状，具有1列光滑或瘤状环。缢往往被垂直肋覆盖。

时代与分布：中三叠世拉丁期中晚期；意大利，中国四川和云南。

光瘤条纹三叠塔虫 *Striatotriassocampe laeviannulata* Kozur et Mostler

(Pl. 25，figs. 1—3)

Striatotriassocampe laeviannulata Kozur et Mostler，1994，pl. 43，figs. 3，7，8.

比较：这个种与 *S. nodosoannulata* Kozur et Mostler 的区别为整个壳上发育光滑的瘤环，垂直肋决没有越过这些环。

层位与产地：中三叠统拉丁阶中上部；意大利，中国四川理塘地区。

多节瘤状条纹三叠塔虫 *Striatotriassocampe nodosoannulata* Kozur et Mostler

(Pl. 25，figs. 11—13，19)

Striatotriassocampe nodosoannulata Kozur et Mostler，1994，pl. 43，figs. 4，10；Feng et al.，2001，pl. 1，figs. 6，7.

描述：这个种的特点是壳细长，节数多，亚圆柱形。头穹隆形，光滑，无孔。壳的宽度在前6个后腹节缓慢增大，之后各节宽度保持稳定。胸节、腹节和前6—7个后腹节围栅脊和垂直肋发育，并出现半列孔环。

比较：这个种与 *S. laeviannulata* Kozur et Mostler 的区别为每个节上在壳的近端部分发育1列瘤状环。垂直肋较强壮，在壳的顶部越过环。

层位与产地：中三叠统拉丁阶中上部；意大利，中国云南和四川理塘地区。

三叠塔虫属 *Triassocampe* Dumitrica, Kozur et Mostler，1980

模式种：*Triassocampe scalaris* Dumitrica, Kozur et Mostler，1980

鉴定要点：壳体长，多节，圆锥形至圆柱形，没有顶刺。各节的上部宽度最大，并具有1列或几列瘤状孔环。

时代与分布：中晚三叠世；世界各地。

冠状三叠塔虫 *Triassocampe coronata* Bragin

(Pl. 12, figs. 14, 15)

Triassocampe coronata Bragin, 1991, pl. 1, fig. 15; Sugiyama, 1992, Fig. 11 (5, 6), 1997, Fig. 27 (6, 7); Feng and Ye, 1996, pl. 2.1, fig. 6; Feng et al., 1996, pl. 1, figs. 1, 2; Spiller, 2002, pl. 5, figs. a, b; Kamata et al., 2002, Fig. 7 (C).

Triassocampe sp. A Mizutani et Koike, 1982, pl. 4, fig. 3; Ishida, 1984, pl. 1, figs. 5—9.

Triassocampe deweveri (Nakaseko et Nishimura); Hattori and Yoshimura, 1982, pl. 1, fig. 4; Kishida and Sugano, 1982, pl. 4, figs. 9, 10; Nishizono et al., 1982, pl. 3, fig. 13.

Triassocampe coronata coronata Bragin; Feng et al., 2001, pl. 1, fig. 15.

Dictyomitrella sp. A De Wever et al., 1979, pl. 5, figs. 12, 16.

描述：这个种的特点是每个后腹节的上部具有明显的围栅脊，在每个围栅脊之下发育1列孔环。

比较：这个种与 *T. deweveri* (Nakaseko et Nishimura) 的区别为有1个发育很好的无孔穹隆形头节和在每个围栅脊之下发育的1列孔环。

层位与产地：中三叠统安尼阶上部—拉丁阶下部；土耳其，马来西亚，泰国，日本，俄罗斯远东地区，中国云南、青海可可西里地区和四川理塘地区。

德韦弗三叠塔虫 *Triassocampe deweveri* (Nakaseko et Nishimura)

(Pl. 20, figs. 29, 32, 34, 35, 37; pl. 25, figs. 22, 24)

Dictyomitrella deweveri Nakaseko et Nishimura, 1979, pl. 10, figs. 8, 9.

Dictyomitrella sp. A De Wever et al., 1979, pl. 5, figs. 12, 16; Yao et al., 1980, pl. 1, figs. 9, 11.

Triassocampe deweveri (Nakaseko et Nishimura); Yao, 1982, pl. 1, fig. 3 (non 1, 2), 1983, Fig. 2 (1); Mizutani et al., 1982, pl. 4, fig. 1; Ishida, 1984, pl. 1, figs. 10—12, 1989, Fig. 2 (4); Kozur and Reti, 1986, Fig. 5 (E); Kishida and Hisada, 1986, Fig. 3 (2); Kojima and Mizutani, 1987, Fig. 3 (2, 3); Cheng, 1989, pl. 6, figs. 13, 14, pl. 7, figs. 10, 11; Zhang Q Y, 1990, pl. 3, fig. 4; Yeh, 1990, pl. 7, figs. 7, 18, 20, pl. 11, figs. 2, 3, 7, 8, 13, 14; Kojima, 1989, pl. 1, fig. 3; Yao, 1990, pl. 1, fig. 1; Sashida et al., 1993a, pl. 2, figs. 8—12, 1993d, Fig. 5 (3—7); Feng and Liu, 1993, pl. 3, figs. 1—4; Kozur and Mostler, 1994, pl. 42, fig. 1, pl. 44, fig. 14, pl. 45, fig. 6; Jasin, 1994, pl. 1, fig. 10; Ramovs and Gorican, 1995, pl. 7, figs. 13, 14; Feng and Ye, 1996, pl. 2, figs. 1, 2; Feng et al., 2001, pl. 3, figs. 1—6; Spiller, 2002, pl. 5, figs. c—h.

描述：这个种的特点是壳圆锥形，多节，无顶刺。围栅脊发育，在每一个腹节和后腹节的围栅脊之下发育2—3列孔环。

层位与产地：中上三叠统安尼阶上部—诺利阶；世界各地。

南畔三叠塔虫 *Triassocampe nanpanensis* (Feng)

(Pl. 12, fig. 13)

Shengia nanpanensis Feng, 1992, pl. 2, figs. 7, 8; Feng and Liu, 1993, pl. 3, figs. 10—13; Fang and Feng, 1996, pl. 2.1, figs. 11, 12.

Triassocampe nanpanensis (Feng); Feng et al., 2001, pl. 1, figs. 21—24.

描述：这个种的特点是壳节少，一般为7节或更多，亚圆柱形，向远端宽度和高度逐渐增大。头穹隆形，光滑，无孔。胸节、腹节和第1后腹节缺乏明显的缢和围栅脊。其余后腹节具有窄的围栅脊和宽的、浅的缢，宽度是高度的2倍，在缢中发育3列孔环。

比较：这个种与 *T. goricani* Feng, Zhang et Ye 的区别为后者每个围栅脊之下只发育1列孔环。

层位与产地：中三叠统安尼阶下部；中国云南和西藏双湖地区。

梯形三叠塔虫 *Triassocampe scalaris* Dumitrica, Kozur et Mostler

(Pl. 25, figs. 21, 25)

Triassocampe scalaris Dumitrica, Kozur et Mostler, 1980, pl. 9, figs. 5, 6, 11, pl. 14, fig. 2; Mizutani and Koike, 1982, pl. 4, fig. 4; Gorican and Buser, 1990, pl. 12, figs. 2, 3; Yeh, 1990, pl. 7, figs. 9, 19, pl. 11, fig. 6; Tekin, 1999, pl. 41, fig. 10; Sashida et al., 2000, Fig. 8 (6, 9, 14—18, 29, 30); Feng et al., 2001, pl. 3, figs. 14—16.

Triassocampe deweveri (Nakaseko et Nishimura); Yao, 1982, pl. 1, figs. 1, 2 (non 3); Yao et al., 1982, pl. 1, fig. 1.

Triassocampe sp. Martini et al., 1989, pl. 1, fig. 1.

Triassocampe scalaris scalaris Dumitrica, Kozur et Mostler; Kozur and Mostler, 1994, pl. 44, figs. 1—6, 10—12, pl. 45, figs. 1, 2, pl. 47, figs. 2, 3.

描述：这个种的特点是壳节9—15节，缺顶刺，长梯形。头节和胸节无孔，腹节和后腹节具有2—4列，一般为3列瘤状孔环。

比较：这个种与 *T. deweveri* (Nakaseko et Nishimura) 的区别为后者的壳节较少，壳体形状不一样，其胸节为桶箍形，腹节圆柱形，所有后腹节为倒转梯形。

层位与产地：中三叠统拉丁阶中上部；世界各地。

三叠塔虫（多未定种） *Triassocampe* spp. Gorican et Buser

(Pl. 20, figs. 40, 41)

Triassocampe spp. Gorican et Buser, 1990, pl. 12, fig. 7 (non 8).

描述：这个多未定种的特点是壳圆锥形，至远端宽度逐渐增大。头节和胸节无孔。腹节和后腹节具小孔，通常被大小不同的瘤所覆盖。瘤纵向相连较明显，各节间缢弱。

比较：这个多未定种以其瘤纵向排列较明显与这个属的其他种相区别，因此，它很可能是一个新种。

层位与产地：中三叠统拉丁阶上部；斯洛文尼亚，中国青海可可西里地区。

莫尼卡星虫科 Monicastericidae Kozur et Mostler, 1994, emend. Dumitrica, 2001

管状三叠笼虫属 *Tubotriassocyrtis* Kozur et Mostler, 1994

模式种：*Tubotriassocyrtis angustituba* Kozur et Mostler, 1994

鉴定要点：壳小，长纺锤形至亚圆柱形。头大，圆锥形至亚半球形，具有1根三棱形顶刺。其余各节圆柱形，最后一节倒锥形，其后为1个长的，窄到宽，无孔管。胸节和腹节往往被不规则粗孔的外层所覆盖。后腹节被具微粒硅的外层完全或部分封闭，不封闭的部分可见1—2列大孔环。

时代与分布：中三叠世拉丁期晚期；意大利，日本，土耳其，俄罗斯远东地区，中国。

纺锤形管状三叠笼虫（新种） *Tubotriassocyrtis fusiformis* Wang sp. nov.

(Pl. 25, fig. 26)

词源：fusiform，拉丁词，纺锤形。

描述：壳小，长纺锤形，约12节。头节较小，圆锥形，无孔，具有1根顶刺。胸节和腹节较短。后腹节8节，从头节至第4后腹节，宽度逐渐增大，从第5后腹节开始至末端宽度又逐渐减小，呈倒梯形。从第4后腹节开始，围栅脊明显，其上发育1列孔环，围栅脊间缢较浅，光滑，无孔。末端具1根长而窄的管子，光滑无孔。

比较：这个新种的形状和构造十分相似于 Triassocampe fusiformis Bragin，区别在于后者无顶刺，各节间缢较深，特别是后腹节具有3列孔环。

层位与产地：中三叠统拉丁阶中上部；中国四川理塘地区。

管状三叠笼虫（未定种A） *Tubotriassocyrtis* sp. A
(Pl. 19, fig. 16)

比较：这个未定种A只保存了部分壳体，其构造与 T. angustituba Kozur et Mostler 的上半部壳体十分相似。

层位与产地：中三叠统拉丁阶；中国青海可可西里地区。

古网冠虫超科 Archaeodictyomitracea Pessagno, 1976
古网冠虫科 Archaeodictyomitridae Pessagno, 1976
古网冠虫属 *Archaeodictyomitra* Pessagno, 1976, emend. Pessagno, 1977

模式种：*Archaeodictyomitra squinaboli* Pessagno, 1976

鉴定要点：壳圆锥形，头节、胸节、腹节和后腹节具有线状排列的连续纵肋，肋间有1列孔。

时代与分布：中晚三叠世—早白垩世；世界各地。

古网冠虫（未定种A） *Archaeodictyomitra* sp. A
(Pl. 19, fig. 8)

描述：壳圆锥形，头节和胸节光滑，腹节和后腹节每半圈被10根线状排列的纵肋分开，2肋间发育1列小孔。节间缢不明显。壳宽度向远端逐渐增大，至最后1—2后腹节宽度略有收缩。

比较：这个未定种A与 A. chalilovi (Aliev) 外形有点相似，区别在于后者各节和缢比较明显。

层位与产地：中三叠统拉丁阶上部；中国青海可可西里地区。

古网冠虫（未定种B） *Archaeodictyomitra* sp. B
(Pl. 20, fig. 33)

描述：壳圆锥形，壳节较少，壳节宽度向远端逐渐增大，至最后2后腹节宽度略有收缩。头节和胸节无孔，腹节和后腹节为纵肋分开，缢和肋间孔不明显。

层位与产地：中三叠统拉丁阶上部；中国青海可可西里地区。

第7章 龙木错-双湖缝合带的古地理意义

羌塘龙木错-双湖缝合带内保存了多套晚古生代—早中生代蛇绿岩和放射虫硅质岩建造，并且具有丰富的深水相放射虫化石。这些记录不仅为我国特提斯演化，还为古特提斯洋的形成与闭合和新特提斯洋打开的时代以及南、北大陆的分界，提供了新的佐证。

7.1 蛇绿岩和放射虫动物群

根据李才等（2016）的专著《羌塘地质》所论，羌塘龙木错-双湖缝合带内有寒武纪—三叠纪各纪蛇绿岩记录。

目前已在这一缝合带内的蛇绿岩套的基质（硅质岩）和/或放射虫硅质岩中发现了晚泥盆世弗拉期、早二叠世隆林期早期、中二叠世栖霞期晚期、晚二叠世长兴期和中三叠世安尼期早期（如：朱同兴等，2010；李曰俊等，1997）放射虫动物群，甚至还可能存有晚三叠世放射虫动物群（李曰俊等，1997）。

按照地区划分，龙木错-双湖缝合带晚古生代—早中生代蛇绿岩分为红脊山蛇绿混杂岩、角木日蛇绿混杂岩、双湖蛇绿混杂岩（表1.1）。

以上三地蛇绿混杂岩，特别是双湖-角木日地区的蛇绿岩基质和/或放射虫硅质岩产有5个放射虫动物群。由老至新，它们分别为 *Helenifore robustum* 动物群、*Pseudoalbaillella sakmarensis-P. lomentaria* 动物群、*Pseudoalbaillella ishigai* 动物群、*Neoalbaillella ornithoformis* 动物群和 *Eptingium nakasekoi* 动物群（表7.1）。

第一个放射虫动物群——*Helenifore robustum* 动物群的主要分子为 *Trilonche pittmani*（26-1）（图版26，图1，后同）、*T. davidi*（26-2）、*T. echinata*（26-3）、*T. elegans*（26-4）、*T. tretactinia*（26-5）、*Traenosphaera sicarius*（26-6）和 *Tetrentactinia spongacea*（26-7）。除羌塘地区外，它们主要发现于澳大利亚东部、美国西海岸、俄罗斯南乌拉尔地区和新西伯利亚鲁德内阿尔泰地区、哈萨克斯坦、泰国、马来西亚以及中国云南、广西、贵州和新疆中泥盆世吉维特期（Givetian）—晚泥盆世法门期（Famennian）地层中，其中以晚泥盆世弗拉期（Frasnian）最为普遍。虽然在当前的放射虫动物群中没有发现这一时期的化石带指引属 *Helenifore*，我们仍把这个放射虫动物群称为 *Helenifore robustum* 带，时代为晚泥盆世弗拉期。它可以同澳大利亚东部（Aitchison，1988；Ishiga，1988）、美国西海岸（Boundy-Sanders and Murchey，1999）、泰国（Sashida et al.，1993，1998）、马来西亚（Spiller，2002）以及中国云南（Wang et al.，2000，2003）、贵州（Wang and Luo，2006）、广西（Wang et al.，2003，2012）和新疆（Wang et al.，2013）的同名放射虫带对比。

第二个放射虫动物群——*Pseudoalbaillella sakmarensis-P. lomentaria* 动物群以丰富的假阿尔拜虫类为特征，其中 *Pseudoalbaillella sakmarensis*（26-18—20）和 *P. lomentaria*（26-12—14）为优势种。除了羌塘地区外，它们主要见于俄罗斯南乌拉尔地区和远东地区、美国、日本、马来西亚、泰国以及中国云南、广西和青海可可西里地区早二叠世狼营期（Wolfcampian）中晚期地层。我们把这个放射虫动物群称为 *Pseudoalbaillella sakmarensis-P. lomentaria* 带，时代为早二叠世隆林期（Longlinian）早期，与美国

表 7.1 双湖-角木日蛇绿混杂岩带晚古生代—早中生代放射虫带与邻区放射虫带对比表

Table 7.1 Correlation chart of Late Palaeozoic—Early Mesozoic radiolarian zones between Shuanghu-Jiaomuri ophiolite mélange belts and adjacent area

系	统	阶	本文	昌宁-孟连地区（王玉净等，2000 冯庆来等，1997，2001）	可可西里地区（王玉净，1995 王玉净等，2005 冯庆来等，2007，2009 李红生等，1993）	华南地区（王玉净等，2007，2012 冯庆来、刘本培，1993 冯庆来等，1997，2001）	日本（Ishiga, 1986, 1990 Sugiyama, 1997）
三叠系	中统	拉丁阶		Muellentortis cochleata	Spongoserrulcla rarauana Annulotriassocampe multisegmantatus	M. cochleata	Spongoserrula dehil M. cochleata Yeharaia elegans
		安尼阶	Eptingium nakasekoi	Triassocampe deweveri T. coronata coronata T. coronata inflata T. dumitricai	Tiborella frorida Pseudostylosphaera jmaponica Eptingium nakasekoi	T. deweveri T. coronata coronata T. coronata inflata T. dumitricai	T. deweveri T. coronata E. nakasekoi
	下统	奥伦尼克阶					Parentactinia nakatsugawaensis
		印度阶					
二叠系	上统	长兴阶	Neoalbaillella ornithoformis	N. optima N. grithoformis A. levis-A. excelsa A. protolevis	A. protolevis	N. optima N. ornithoformis A. levis-A. excelsa A. protolevis	N. optima N. ornithoformis
		吴家坪阶		Follicucullus charveti		Foremanhelena triangula F. bipartitus-F. charveti	F. charveti A. yamejitai
	中统	茅口阶		F. scholasticus-F. ventricosus F. monacanthus P. fusiformis	P. globosa	F. scholasticus-F. ventricosus P. monacanthus P. globosa	F. scholasticus-F. ventricosus P. monacanthus P. globosa
		栖霞阶	Pseudoalbaillella ishigai		A. xiaodongensis	P. ishigai A. sinuata A. xiaodongensis	P. longtanensis A. sinuata
	下统	隆林阶	P. sakmarensis-P. lomentaria	P. rhombothoracata P. sakmarensis	P. rhombothoracata P. elongata P. sakmarensis-P. lomentaria		
		紫松阶			P. chilensis	P. u-forma P. elegans P. bulbosa	P. u-forma m. II P. u-forma m. I P. bulbosa
石炭系	上统					P. nodosa	P. nodosa
	下统	维宪阶		A. cartala Eostylodictia rota	A. cartala Polyentactinia sp. A	A. cartala E. rota	
		杜内阶		A. indensis A. deflandrei A. paradoxa	Cyrtisphaeractenia aff. crassum-Astroentactinia Multispinosa	A. indensis A. deflandrei A. paradoxa	
泥盆系	上统	法门阶		Holoeciscus foremanae	Holoeciscus foremanae	H. foremanae	
		弗拉阶	Helenifore robustum	H. robustum		H. robustum H. laticlavium	

狼营期中晚期和俄罗斯萨克马尔期（Sakmarian）相当。它可以同日本（Ishiga，1986，1990）、泰国（Sashida et al.，1998，2000）、马来西亚（Spiller，2002）、美国西海岸（Blome and Reed，1992，1995）的 *Pseudoalbaillella lomentaria* 带，俄罗斯远东地区（Rudenko and Panasenko，1997）的 *Pseudoalbaillella sakmarensis* 带，中国华南（Wang et al.，1994）和青海可可西里地区的同名放射虫带相对比。

第三个放射虫动物群——*Pseudoalbaillella ishigai* 动物群以 *Pseudoalbaillella ishigai*（26-10，11）的出现为特征，与其共存的放射虫有 *Latentifistula texana*（26-17）、*L. patagilaterala*（26-9）、*Albaillella sinuata*（26-8）和 *Pseudotormentus kamigoriensis*（26-15，16）。*P. ishigai* 与 Ishiga 等（1982）鉴定的 *P.* sp. C 为同种。此种在我国的华南主要发现于中二叠世栖霞期（Chihsian）晚期地层中，并为这一时代的一个标准化石带。因此当前研究的放射虫动物群归为 *P. ishigai* 化石带。此带可与日本锅山（Nabeyama）阶的 *P.* sp. C 带（Ishiga，1986）和后来的 *P. longtanensis* 带（Ishiga，1990）、美国俄勒冈（Oregon）地区伦纳德阶（Leonardian）的 *P.* sp. C 带（Blome and Reed，1992）及俄罗斯远东地区阿丁斯克阶（Artinskian）的 *P. corniculata* 带（Rudenko et al.，1997）相对比。

第四个放射虫动物群——*Neoalbaillella ornithoformis* 动物群以丰富的新阿尔拜虫类为特征。在这个放射虫动物群中，*Neoalbaillella ornithoformis*（26-24，28，29）是一个优势种，因此这个放射虫动物群被命名为 *N. ornithoformis* 带。此带是我国华南晚二叠世长兴期（Changhsingian）第3个化石带（Wang et al.，2006）。它的伴存放射虫异常丰富，包括 *Trilonche pseudocimelia*（26-21，36，37）、*Archaeospongoprunum chiangdaoensis*（26-22）、*Raciditor gracilis*（26-23）、*Ishigaum trifistis*（26-25，40）、*I. craticula*（26-26）、*I. obesum*（26-34）、*Triplanospongos musashiensis*（26-27）、*Neoalbaillella optima*（26-30）、*N. pseudogrypa*（26-31）、*N. gracilis*（26-32，33）、*Ormestonella robusta*（26-35）、*Albaillella levis*（26-38）、*A. lauta*（26-39）、*A. triangularis*（26-44）和 *Copicyntra akikawaensis*（26-41）。这个带完全可以同日本（Ishiga，1990；Kuwahara et al.，1998）、马来西亚（Sashida et al.，1995；Spiller，2002）、菲律宾（Cheng，1989；Tumanda et al.，1990）、泰国（Sashida et al.，2000）、俄罗斯远东地区（Rudenko et al.，1997）、美国西海岸（Blome and Reed，1992，1995）以及中国云南、贵州和广西的同名带相对比（Wang et al.，2006）。

第五个放射虫动物群——*Eptingium nakasekoi* 动物群以中生代特有的笼罩虫类为特征。在这一动物群中，我们共发现6属8种，即 *Eptingium nakasekoi*（26-42，52）、*Parasepsagon praetetracanthus*（26-43）、*Pseudostylosphaera japonica*（26-45）、*P. compacta*（26-46）、*Tiborella florida*（26-50）和 *Triassospongosphaera triassica*（26-51）。它们在世界上主要发现于中三叠世安尼期（Anisian）早期的地层中。这个放射虫动物群的特征种是 *Eptingium nakasekoi*。它在匈牙利（Kozur and Mostler，1994）、斯洛文尼亚（Ramovs and Gorica，1995）、泰国（Kamata et al.，2002）、马来西亚（Spiller，2002）、菲律宾（Cheng，1989）、日本（Nakaseko and Nishimura，1979；Sugiyama，1997）和中国云南（Feng et al.，2001）主要见于中三叠世安尼期早期的地层。Sugiyama（1997）把此种作为安尼期早期的带化石种，并建立了 *Eptingium nakasekoi* 带。这个动物群可以同日本（Sugiyama，1997）和中国青海可可西里地区（Feng et al.，2007）的同名带、中国云南的 *Triassocampe dumitricai* 带（Feng et al.，1996）、欧洲的 *Parasepsagon robustum* 带（Kozur and Mostler，1994）、俄罗斯远东地区的 *Hozmadia* 带（Bragin，1991）和马来西亚的 *Triassocampe coronata* 带（Spiller，2002）相对比。

7.2 古地理意义

以上蛇绿岩和放射虫硅质岩建造，特别是其中放射虫化石的发现和化石带的建立及对比不仅证明了晚古生代—早中生代期间，龙木错-双湖沿线一带确有深海洋盆或裂谷的存在，还特别清楚地展示了我国北特提斯演化的主要历程。

（1）羌塘龙木错-双湖和可可西里龙木错-玉树缝合带内的最低放射虫化石带的时代均为晚泥盆世，早至晚泥盆世早期弗拉期（表7.1），证明那里的古特提斯洋开始于晚泥盆世甚至更早。

（2）龙木错-双湖和龙木错-玉树缝合带内均缺乏晚二叠世早期吴家坪期（Wuchiapingian）和早三叠世蛇绿岩和放射虫硅质岩建造和放射虫化石记录（表7.1），揭示我国北特提斯龙木错-双湖和龙木错-玉树古特提斯洋或裂谷可能于中二叠世末—晚二叠世早期吴家坪期就开始关闭，晚二叠世或二叠纪末完全消失，变为开阔浅海。但是，这一浅海海底于中三叠世安尼期早期再度拉张或沉降，造就了一个新洋盆——新特提斯洋。

可可西里龙木错-玉树缝合带西金乌兰湖-移山湖一线以北汉台山蛇形沟汉台山群碳酸盐岩地层灰岩段产有典型且丰富的晚二叠世早期吴家坪期和早三叠世晚期奥伦尼克期（Olenekian）浅海相特提斯生物群化石（沙金庚等，1992；张以弗，郑健康，1994；张以弗，沙金庚，1995；沙金庚，1998；Sha et al.，2004）。吴家坪期动物群由 *Gymnocodium* cf. *exile*、*Permocalculus plumosus*、*Mizzia velebitana*、*Pseudovermiporella elliotti* 等钙藻（鲍惠铭，1995），*Colaniella-Baisalina pulchra reitlingerae* 有孔虫组合（罗辉，1995），*Codonofusiella lui?* 带（包括 *C*. cf. *pseudolui*）（张遴信，1995）以及水螅、海绵、苔藓虫、海百合茎等（沙金庚，1995a）等组成，产于汉台山群灰岩段底部。奥伦尼克期生物群包括 *Neospathodus waageni-N. timorensis* 牙形刺组合（徐珊红，1995），*Bakevellia costata-Leptochondria virgalensis-Entolium microtis* 双壳类组合（沙金庚，1995b；Sha and Grant-Macie，1996），*Neritaria intexa*、*N. comensis*、*Naticopsis eyerichi* 等腹足类（朱祥根，1995），海百合茎碎片（沙金庚，1995a）等，产于汉台山群灰岩段下部。

以上龙木错-玉树缝合带内晚二叠世早期吴家坪期和早三叠世蛇绿岩-放射虫岩岩系和放射虫化石消失，但却出现了碳酸盐岩和浅海相生物群的层序记录，有力支持了以上关于古特提斯洋于中二叠世末—晚二叠世初开始闭合、晚二叠世或二叠纪末消失和新特提斯洋于中三叠世初打开的结论。

（3）尽管昌宁-双江-孟连缝合带[即澜沧江缝合带（刘本培等，1993）]的放射虫化石带序列较齐全，但其中最老的放射虫化石带的时代也为晚泥盆世早期弗拉期，并同样缺乏二叠纪末—早三叠世早期印度期（Induan）甚至可能的早三叠世晚期奥伦尼克期放射虫和含放射虫的蛇绿岩（表7.1）。这些地层和化石记录说明昌宁-双江-孟连特提斯洋盆的主要演化过程与上述羌塘龙木错-双湖和可可西里龙木错-玉树洋盆相同：古特提斯洋打开于晚泥盆世早期，甚至更早，闭合于晚二叠世或二叠纪末；新特提斯洋开始于中三叠世初或早三叠世末。

（4）喀喇昆仑山口-龙木错-玉树-金沙江-昌宁-双江-孟连缝合带为北主缝合带或古特提斯缝合带（黄汲清，陈炳蔚，1987）。类同或相同的放射虫动物群（表7.1）表明了龙木错-双湖、龙木错-玉树和昌宁-双江-孟连（澜沧江）缝合带记录的3条晚古生代—早中生代的深水洋盆或裂谷相通，并与浅海特提斯构成了浩瀚的古特提斯和新特提斯大洋。龙木错-双湖缝合带西端截止并与龙木错-玉树缝合带相连于龙木错（图1），这一缝合带地理分布格架也无疑显示了龙木错-双湖缝合带西端确实与北主缝合带相连。这一缝合带与澜沧江缝合带[即昌宁-双江-孟连缝合带（刘本培等，1993）]相连成为冈瓦纳与欧亚大陆的界线（如：Metacalfe，2002，2013；李才等，2009），但目前在连接这两条缝合带的北澜沧江地带尚未发现晚古生代—早中生代放射虫化石证据记录。

第8章 结 束 语

含有放射虫化石的蛇绿岩以及放射虫硅质岩和放射虫化石是判别洋盆或裂谷的最可靠的标志和必不可少的依据。我国青藏高原北部羌塘地区龙木错-双湖和可可西里龙木错-玉树缝合带内不仅记录着晚古生代—早中生代蛇绿岩-放射虫硅质岩岩系，而且其中保存着丰富多彩并具国际对比意义的放射虫化石，那里是记录包括古特提斯洋的形成和关闭与新特提斯洋打开时代在内的我国北特提斯演化的罕见档案馆。

目前已从那里识别出晚泥盆世—中三叠世放射虫25科53属155种（包括25新种）、放射虫化石带14个。那里的蛇绿岩和放射虫硅质岩建造，特别是放射虫化石带层序显示：位于我国古特提斯北部的古特提斯洋形成于晚古生代晚泥盆世甚至更早。但随着古特提斯洋洋壳的俯冲和羌塘地块的向北漂移，古特提斯洋于中二叠世末—晚二叠世初（如：张以弗，1991；沙金庚等，1992；边千韬等，1993，1996，1997；青海可可西里综合科学考察队，1994；张以弗，郑健康，1994；王玉净，1995；张俊明，1995；沙金庚，1998，2009，2018；Sha and Fürsich，1999；Sha et al.，2004）或晚二叠世（如：张以弗，1993）就开始闭合，晚二叠世或二叠纪末完全消失，进入了特提斯演化的重要转折点（黄汲清，陈炳蔚，1987），变成了以碳酸盐沉积为主的开阔浅海，并与其以南的浅海特提斯相连为浩瀚无垠但波光粼粼、生机勃勃的蓝色大海——古特提斯浅海（沙金庚等，1992；Sha and Fürsich，1999；Sha et al.，2004；沙金庚，1998，2009，2018）。中三叠世早期安尼期早期，北特提斯海底拉张裂解，导致了那里新特提斯洋的诞生。

尽管很多特提斯研究者认为新特提斯洋于晚古生代中甚至早二叠世就已随着阿拉伯-澳大利亚的冈瓦纳东缘裂解，以及基墨里/滇缅泰马大陆/地体与冈瓦纳分解而形成（如：Muttoni et al.，2009；Zanchi and Gaetani，2011；Angiolini et al.，2013，2015；Metcalfe，2013；Berra and Angiolini，2014），甚至拉萨与喜马拉雅地块之间的新特提斯洋（即印度斯-雅鲁新特提斯洋）在中二叠世前就已打开（张以春，王玥，2019），但迄今尚未找到支持上述结论的放射虫化石证据。

北主缝合带或喀喇昆仑山口-龙木错-玉树-金沙江-昌宁-双江-孟连缝合带（包括龙木错-双湖缝合带）以南还有2条记录我国南特提斯演化的缝合带，即南主缝合带印度斯/河-雅鲁藏布缝合带及其分支班公湖-丁青-怒江缝合带（黄汲清，陈炳蔚，1987）。最近，在藏南泽当的印度斯-雅鲁藏布缝合带蛇绿混杂岩中的硅质岩内，陈迪舒等（2019）发现了中三叠世安尼期晚期的 *Oertlispongus inaequispinosus* 放射虫动物群，在班公湖-丁青-怒江缝合带蛇绿混杂岩内的硅质岩中，李亚等（2018）在改则地区、尼玛次仁等（2005）在那曲地区都发现了这个放射虫动物群，在北主缝合带龙木错-玉树缝合带蛇绿混杂岩中，冯庆来等（2009）在巴颜喀拉盆地硬水泉组硅质岩中也发现了这个动物群，在南土耳其Merlin混杂岩带3个地层剖面中都发现了这类化石（Tekin et al.，2016）。此外，本书在龙木错-双湖缝合带蛇绿岩混杂岩的硅质岩中在角木日地区硅质岩中找到了早安尼期放射虫 *Eptingium nakasekoi* 动物群。这些资料证明，在北主缝合带以南的我国南特提斯迄今未见早于中三叠世的放射虫化石记录，那里新特提斯洋初始于中三叠世安尼期而不是二叠纪。我国南、北新特提斯洋与其他地区洋盆一样于早三叠世末或中三叠世初同时打开，但闭合的时间不一致（王玉净等，2021）。

龙木错-双湖、龙木错-玉树和昌宁-双江-孟连缝合带晚古生代—早中生代放射虫动物群的相似或共同性（表7.1），以及龙木错-双湖缝合带西端截止并与龙木错-玉树缝合带相连于龙木错（图1）的缝合带地理分布格架，均显示了龙木错-双湖缝合带与北主缝合带或古特提斯缝合带（喀喇昆仑山口-龙木错-玉树-金沙江-昌宁-双江-孟连缝合带）（黄汲清，陈炳蔚，1987）相通。但它何时向东南方向延伸至澜沧江缝合带［即昌宁-双江-孟连缝合带（刘本培等，1993）］（Metcalfe，2002，2013；李才等，2009），目前尚缺乏连接这2条缝合带的过渡带——北澜沧江带的晚古生代—早中生代放射虫化石证据。但是我们相信，在不久的将来，一定能找到这些化石证据。

参 考 文 献
(REFERENCES)

王玉净, 1995. 放射虫 [M] // 沙金庚. 青海可可西里地区古生物. 北京: 科学出版社: 58-69, 146-148.

王玉净, 1997. 新疆北部碳酸盐岩中一个晚泥盆世法门期放射虫动物群 [J]. 微体古生物学报, 14 (2): 149-160.

王玉净, 2005. 日喀则晚白垩世放射虫 [M] // 沙金庚, 王启飞, 卢辉楠. 羌塘盆地微体古生物. 北京: 科学出版社: 139-151, 266-267.

王玉净, 王建平, 裴放, 2002b. 西藏丁青蛇绿岩中一个晚三叠世动物群 [J]. 微体古生物学报, 19 (4): 323-336.

王玉净, 方宗杰, 杨群, 等, 2000. 云南西部中—晚泥盆世硅质岩相地层及其放射虫动物群 [J]. 微体古生物学报, 17 (3): 235-254.

王玉净, 邝国敦, 1993. 广西东南部钦州地区早石炭世放射虫化石 [J]. 微体古生物学报, 10 (3): 275-287.

王玉净, 齐敦伦, 1995. 苏皖南部孤峰组放射虫动物群 [J]. 微体古生物学报, 12 (4): 374-387.

王玉净, 杨群, 2007. 中国石炭—二叠纪放射虫化石带及古生物地理学意义 [J]. 微体古生物学报, 24 (4): 337-345.

王玉净, 杨群, 松冈笃, 等, 2002a. 藏南泽当雅鲁藏布缝合带中的三叠纪放射虫 [J]. 微体古生物学报, 19 (3): 215-227.

王玉净, 杨群, 郭通珍, 2005. 青海可可西里地区中三叠世晚期放射虫 *Spongoserrula rarauana* 动物群 [J]. 微体古生物学报, 22 (1): 1-9.

王玉净, 沙金庚, 徐波, 等, 2021. 西藏那曲地区中三叠统嘎加组放射虫动物群 [J]. 微体古生物学报, 38 (1): 25-47.

王永胜, 曲永贵, 等, 2012. 中华人民共和国区域地质调查报告: 帕度错 (西藏) 幅 (I45C004003) (1:250000) [M]. 武汉: 中国地质大学出版社: 1-188.

尹集祥, 1997. 青藏高原及邻区冈瓦纳相地层地质学 [M]. 北京: 地质出版社: 1-175.

尼玛次仁, 谢尧武, 2005. 藏北那曲地区中三叠叠层的新发现及其地质意义 [J]. 地质通报, 24 (12): 1141-1149.

冯心涛, 朱同兴, 李才, 等, 2005. 藏北双湖地区上三叠统肖茶卡群的重新厘定及其地质意义 [J]. 地质通报, 24 (12): 135-1140.

边千韬, 沙金庚, 郑祥身, 1993. 西金乌兰晚二叠—早三叠世石英砂岩及其大地构造意义 [J]. 地质科学, 23 (4): 327-335.

边千韬, 郑祥身, 1991. 西金乌兰和冈齐曲蛇绿岩的发现 [J]. 地质科学 (3): 304.

边千韬, 郑祥身, 叶建青, 等, 1996. 构造 [M] // 张以弗, 郑祥身. 青海可可西里地区地质演化. 北京: 科学出版社: 132-148.

边千韬, 郑祥身, 李红生, 等, 1997. 青海可可西里地区蛇绿岩的时代及形成环境 [J]. 地质论评, 43 (4): 347-355.

朱同兴, 冯心涛, 等, 2012. 中华人民共和国区域地质调查报告: 黑虎岭 (西藏) 幅 (I45C002003) (1:250000) [M]. 武汉: 中国地质大学出版社: 1-221.

朱同兴, 李宗亮, 等, 2010. 中华人民共和国区域地质调查报告: 江爱达日那 (西藏) 幅 (I45C003003) (1:250000) [M]. 武汉: 中国地质大学出版社: 1-284.

朱同兴, 李宗亮, 李才, 等, 2005. 藏北双湖地区三叠纪地层新资料 [J]. 地质通报, 24 (12): 1127-1134.

朱同兴, 张启跃, 董瀚, 等, 2006. 藏北双湖地区才多茶卡一带构造混杂岩中发现晚泥盆世和晚二叠世放射虫硅质岩 [J]. 地质通报, 25 (12): 1413-1418.

朱同兴, 董瀚, 李才, 等, 2005. 青藏高原北羌塘地区晚三叠世地层展布和沉积型式 [J]. 沉积与特提斯地质, 25 (3): 18-23.

朱祥根, 1995. 腹足类 [M] // 沙金庚. 青海可可西里地区古生物. 北京: 科学出版社: 69-81, 148-149.

刘本培, 冯庆来, 方念乔, 等, 1993. 滇西南昌宁-孟连带和澜沧江带古特提斯多岛洋构造演化 [J]. 地球科学 (中国地质大学), 18 (5): 529-539, 671.

刘本培, 崔新省, 1983. 西藏阿里日土县宽铰蛤 (Eurydesma) 动物群的发现及其生物地理区系意义 [J]. 地球科学 (武汉地质学院学报）（1）：79-92.

李才，谢尧武，董永胜，等，2009. 北澜沧江带的性质：是冈瓦纳板块与扬子板块的界线吗？［J］. 地质通报，28：1711-1719.

李才，解超明，王明，等，2016. 羌塘地质［M］. 北京：地质出版社：1-681.

李曰俊，吴浩若，李红生，等，1997. 藏北阿木岗群、查桑群和鲁谷组放射虫的发现及有关问题讨论［J］. 地质论评，43（3）：250-256.

李亚，纪占胜，武桂春，等，2018. 西藏南羌塘盆地中三叠世座倾错组放射虫动物群及其地层学意义［J］. 地球科学，43（11）：3932-3946.

李红生，边千韬，1993. 可可西里西金乌兰湖-冈齐曲蛇绿混杂岩中晚古生代放射虫［J］. 现代地质，7（4）：410-420.

沙金庚，1995a. 前言［M］// 沙金庚. 青海可可西里地区古生物. 北京：科学出版社：141-142.

沙金庚，1995b. 双壳类［M］// 沙金庚. 青海可可西里地区古生物. 北京：科学出版社：82-116，149-151.

沙金庚，1998. 青海可可西里地区的古生物特征及其古地理学意义［J］. 古生物学报，37（1）：85-96.

沙金庚，2009. 沧海桑田长江源：古生物化石讲述的故事［M］// 沙金庚. 世纪飞跃：辉煌的中国古生物学. 北京：科学出版社：376-385.

沙金庚. 2018. 古生物化石见证青藏高原的隆升［M］// 戎嘉余. 生物演化与环境. 合肥：中国科学技术大学出版社：298-329.

沙金庚，张遴信，罗辉，等，1992. 论可可西里晚古生代裂谷的消亡时代［J］. 微体古生物学报，9（2）：177-182.

张以春，王玥，2019. 西藏普兰县叶玛地区中二叠纪西兰塔组中的有孔虫及其地质意义［J］. 古生物学报，53（3）：311-323.

张以弗，1991. 可可西里-巴颜喀拉及邻区特提斯特征［J］. 西藏地质，6（2）：62-72.

张以弗，1993. 青藏高原北部地质构造演化初论［J］. 青海地质（2）：1-7.

张以弗，沙金庚，1995. 地层概述［M］// 沙金庚. 青海可可西里地区古生物. 北京：科学出版社：1-9，142.

张以弗，郑健康，1994. 青海可可西里及领区地质概论［M］. 北京：地震出版社：1-177.

张俊明，1995. 生物相、沉积相特征和古地理变迁［M］// 沙金庚. 青海可可西里地区古生物. 北京：科学出版社：133-140，153-154.

张遴信，1995. —［M］// 沙金庚. 青海可可西里地区古生物. 北京：科学出版社：54-58，146-147.

青海可可西里综合科学考察队，1994. 青藏高原腹地：可可西里综合科学考察 地质演化话沧桑［M］. 上海：上海科学技术出版社：10-29.

青海省区域地质测量队，1970. 区域地质调查报告书：一分册. 温泉（青海）幅（I-46）（1∶1 000 000）（内部报告）［M］. 西宁：青海省地质局：1-267.

赵政璋，李永铁，叶和飞，等，2001. 青藏高原地层［M］. 北京：科学出版社：1-542.

罗辉，1995. 第二章 二叠纪有孔虫［M］//沙金庚. 青海可可西里地区古生物. 北京：科学出版社：23-48，144-145.

范影年，1985. 中国西藏石炭—二叠纪皱纹珊瑚的地理区系［J］. 西藏高原地质文（16）：87-106.

苟金，邢国忠，1990. 青海可可西里巴音查乌马地区的上三叠统［J］. 西北地质（2）：1-5.

黄汲清，陈炳蔚，1987. 中国及邻区特提斯海的演化［M］. 北京：地质出版社：1-109（英文），1-78（中文）.

夏代祥，刘世坤，1997. 西藏自治区岩石地层［M］. 武汉：中国地质大学出版社：1-302.

徐珊红，1995. 牙形刺［M］// 沙金庚. 青海可可西里地区古生物. 北京：科学出版社：118-123，151.

梁定益，聂泽同，郭铁鹰，等，1983. 西藏阿里喀喇昆仑南部的冈瓦纳—特提斯相石炭二叠系［J］. 地球科学（武汉地质学院学报）（1）：9-27.

曾庆高，毛国政，王保弟，等，2011. 中华人民共和国区域地质调查报告：日干配错（西藏）幅（I45C004002）（1∶250000）［M］. 武汉：中国地质大学出版社：1-206.

谢义木，1983. 改则北就下石炭统的发现. 中国区域地质，4辑：107-108.

翟庆国，李才，黄小鹏，2006. 西藏羌塘中部角木日地区二叠纪玄武岩的地球化学特征及其构造意义［J］. 地质通报，25（12）：1419-1427.

鲍惠铭，1995. 钙藻［M］// 沙金庚. 青海可可西里地区古生物. 北京：科学出版社：10-15，142-143.

Afanasieva M S, Amon E Q, 2006. Radiolaria［M］. Moscow：PIN RAS, 2006：1-320（in Russian）.

Aitchison J C, 1988. Late Paleozoic radiolarian ages from the Gwydir terrane, New England orogen, eastern Australia [J]. Geology, 16: 793-795.

Aitchison J C, 1990. Significance of Devonian-Carboniferous radiolarians from accretionary terranes of the New England Orogen, eastern Australia [J]. Marine Micropaleontology, 15: 365-378.

Aitchison J C, 1993a. Devonian (Frasnian) radiolarians from the Gogo Formation, Canning Basin, western Australia [J]. Palaeontographica Abteilung A, 228 (4-6): 105-128.

Aitchison J C, 1993b. Albaillellaria from the New England Orogen, NSW, Australia [J]. Marine Micropaleontology, 21: 353-367.

Aitchison J C, Stratford J M, 1997. Middle Devonian (Givetian) Radiolaria from eastern New South Wales, Australia: A reassessment of the Hinde (1899) fauna [J]. Neues Jahrbuch für Geologie und Pal-ontologie, Abhandlungen, 180 (1): 1-19.

Aitchison J C, Davis A M, Stratford J M C, 1999. Lower and Middle Devonian radiolarian biozonation of the Gamilaroi terrane New England Orogen, eastern Australia [J]. Micropaleontology, 45 (2): 138-162.

Angiolini L, Zanchi A, Zanchetta S, et al., 2013. The Cimmerian geopuzzle: new data from South Pamir [J]. Terra Nova, 25: 352-360.

Angiolini L, Zanchi A, Zanchetta S, et al., 2015. From rift to dift in south Pamir (Tajikistan): Permian evolution of a Cimmerian terrane [J]. Journal of Asian Earth Sciences, 102: 146-169.

Bao H M, 1995. Calcareous Algae [M] // Sha J G. Palaeontology of the Hoh Xil region, Qinghai. Beijing: Science Press: 16-15, 142-143 (in Chinese with English abstract).

Berra F, Angiolini L, 2014. The evolution of the Tethys region throughout the Phanerozoic: A brief tectonic reconstruction [M] // Marlow L, Kendall C, Yose L. Petroleum systems of the Tethyan region: American Association of Petroleum Geologists Memoir, 106: 1-27.

Bian Q T, Sha J G, Zheng X S, 1993. The Late Permian-Early Triassic beach-subfacies quartzose sandstone in the Xijir Ulan area and its tectonic significance [J]. Scientia Geologica Sinica, 28 (4): 327-335 (in Chinese with English abstract).

Bian Q T, Zheng X S, 1991. Ophiolite is found in the areas of Xijir Ulan and Gangqiqu [J]. Scientia Geologica Sinica (3): 304 (in Chinese).

Bian Q T, Zheng X S, Ye J Q, et al., 1996. Structure [M] // Zhang Y F, Zheng X S. Geological evolution of the Hoh Xil region. Beijing: Science Press: 80-148 (in Chinese with English abstract).

Bian Q T, Zheng X S, Li H S, et al., 1997. Age and tectonic setting of ophiolite in the Hoh Xil region, Qinghai province [J]. Geological Review, 43 (4): 347-355 (in Chinese with English abstract).

Blome C D, Reed K M, Tailleur I L, 1989. Radiolarian biostratigraphy of the Otuk Formation in and near the national Potroleum reserve in Alaska [J]. Professional Paper of United States Geological Survey, 1399: 725-776.

Blome C, Jones D, Murchey B L, et al., 1986. Geologic implication of radiolarian-bearing Paleozoic and Mesozoic rocks from the Blue Mountains Province Eastern, Oregon [J]. Professional Paper of United States Geological Survey, 1435, 79-93.

Blome C D, Reed K M, 1992. Permian and Early (—) Triassic Radiolarian Faunas from the Grindstone Terrane, central Oregon [J]. Journal of Palaeogeography, 66 (3): 351-383.

Blome C D, Reed K M, 1995. Radiolarian Biostratigraphy of the Quinn River Formation. Black Rock Terrane, North-Central Nevada: Correlations with Eastern Klamath Terrane Geology [J]. Micropaleontology, 41 (1): 49-68.

Boundy-Sanders S Q, Sandberg C A, Murchey B L, et al., 1999. A late Frasnian (Late Devonian) radiolarian, sponge spicule, and Conodont fauna from the Slaven Chert, northern Shoshone Range, Roberts Mountains allochthon, Nevada [J]. Micropaleontology, 45 (1): 62-68.

Bragin N J, 1986. Biostratigraphy of Triassic deposits of southern Sakhalin [J]. Bulletin of Academy of Sciences of the USSR, Geological Series, 4: 61-67 (in Russian).

Bragin N J, 1991. Radiolaria and Lower Mesozoic units of the USSR east regions [J]. Transactions of Academy of Sciences of the USSR, 469: 1-125 (in Russian with English Summary).

Braun A, 1990. Radiolarien aus dem Unter-Karbon Deutschlands CourierForsch-Inst [J]. Senckenberg, 133: 1-43.

Braun A, Schmidt-Effing R, 1993. Biozonation, diagenesis and evolution of radiolarians in the Lower Carboniferous of Germany [J]. Marine Micropaleontology, 21: 369-383.

Caridroit M, De Wever P, 1984. Description de quelques nouvelles especes de Follicucullidae et d'Entactinidae (Radiolaires Polycystines) du Permien du Japon [J]. Geobios, 17: 639-644.

Caridroit M, De Wever P, 1986. Some late Permian radiolarians from pelitic rocks of the Tatsuno Formation (Hyogo Prefecture), southwest Japan [J]. Marine Micropaleontology, 11: 55-90.

Caridroit M, 1993. Permian radiolarian from NW Thailand [M] //Thanasuthipitak T. International Symposium on Biostratigraphy of Mainland Southeast Asia: Facies and Paleontology: Volume I. Chiang Mai, Thailand: 83-96.

Catalano R, Stefano P D, Kozur H, 1991. Permian circum-Pacific deep-water fauna from the western Tethys (Sicily, ltaly): new evidences for the position of the Permian Tethys [J]. Palaeogeography Palaeoclimatology Palaeoecology, 87: 75-108.

Chen D S, Luo H, Wang X H, et al., 2019. Late Anisian radiolarian assemblages from the Yarlung-Tsangpo Suture Zone in the Jinlu area, Zedong, southern Tibet: Implications for the evolution of Neotethys [J]. Island Arc, 29: 1-10.

Cheng Y N, 1986. Taxonomic studies in Upper Paleozoic Radiolaria [J]. National Museum Natural Science, Taiwan, Special Publication, 1: 1-310.

Cheng Y N, 1989. Upper Paleozoic and Lower Mesozoic Radiolarian Assemblages from the Busuanga Islands, North Palawan Block, Philippines [J]. Bulletin of the National Museum of Natural Science, 1: 129-175.

Corde F, De Wever P, Dumitrica P, et al., 1988. Description of some new Middle Triassic radiolarians from the Camp Cove Formation, southern British Columbia, Canada [J]. Revue de Micropaléontologie, 31 (1): 30-37.

Deflandre G, 1952. Albaillella nov. gen., Radiolaire fossile du Carbonifére inférieur, type dúne ligée aberrante éteinte [J]. Comptes-rendus de l'Academie des Sciences de Paris, 234: 872-874.

Deflandre G, 1972. Le systeme trabeculaire interne chez les Pylentomendes et les Popofskyellildes, Radiolaries du Paleozoique, phylogenese des Nassellaires [J]. Comptes Rendus de I' Academie des Sciences, Paris, 274 D: 3535-3540.

De Wever P, 1982. Nassellaria (Radiolaires Polycystines) du Lias de Turquie [J]. Revue de Micropaléontologie, 24 (4): 189-232.

De Wever O, 1984. Triassic Radiolarians from the Darnó area (Hungary) [J]. Acta Geologica Hungarica, 27 (3-4): 295-306.

De wever P, Caridroit M, 1984. Descrption de quelques nouveaux Latentifistulidae (Radiolarires polycystines) paleozoiques du Japon [J]. Revue de Micropaléontologie: Paris, 27 (2): 98-106.

De Wever P, Dumitrica P, Caulet J P, et al., 2001. Radiolarians in the Sedimentary Record [M]. Amsterdam: Gordon and Breach Science Publishers: 1-533.

De Wever P, Sanfilippo A, Riedel W R, et al., 1979. Triassic radiolarians from Greece, Sicily and Turkey [J]. Micropaleontology, 25 (1): 75-110.

Dosztaly L, 1991. Triassic radiolarians from the Balaton upland. A Magyar Allami Foldtani Intezet Evi Jelentese, 1989: 333-355.

Dosztaly L, 1993. The Anisian/Ladinian and Ladinian/Carnian boundaries in the Balaton Highland based on radiolarians [J]. Acta Geologica Hungarica, 36, 1, 59-72.

Dumitrica P, 1982. Triassic Oertlisponginae (Radiolaria) from eastern Carpathians and Southern Alps. D. S. S. [J]. Institutul de Geologie si Geofizica, 67 (3): 57-74.

Dumitrica P, 1999. The Oertlispongidae (Radiolaria) from the Middle Triassic of Masirah Island (Oman) [J]. Revue de Micropaléontologie, 42 (1): 33-42.

Dumitrica P, Kozur H, Mostler H, 1980. Contribution to the radiolarian fauna of the Middle Triassic of the Southern Alps [J]. Geologisch-Pal-ontologische Mitteilungen Innsbruck, 10 (1): 1-46.

Dumitrica P, Tekin U K, Bedi Y, 2010. Eptingiacea and Saturnaliacea (Radiolaria) from the middle Carnian of Turkey and some late Ladinian to early Norian samples from Oman and Alaska [J]. Pal-ontologische Zeitschrift, 84: 259-292.

Fan Y N, 1985. A division of zoogeographical provonces by Permo-Carbonoferous corals in Xizang (Tibet), China [J]. Contribution to the Geology of the Qinghai-Xizang (Tibet), (16): 87-106 (in Chinese with English abstract).

Fang N Q, Feng Q L, 1996. Devonian to Triassic Tethys in Western Yunnan, China [M]. Wuhan: China University of Geosciences Press: 1-135.

Feng Q L, 1992. Permian and Triassic Radiolarian Biostratigraphy in South and Southwest China [J]. Journal of China University of Geosciences, 3 (1): 51-62.

Feng Q, Fang N Q, Zhang Z F, et al., 1998. Uppermost Permian Radiolaria from Southwestern China [J]. Journal of China University of Geosciences, 9 (3): 238-245.

Feng Q L, Helmcke D, Chonglakmani C, et al., 2004. Early Carboniferous radiolarians from North-West Thailand: Palaeogeographical implications [J]. Palaeontology, 47 (2): 377-393.

Feng Q L, Liu B P, 1993a. Radiolaria from Late Permian and Early-Middle Triassic in Southwest Yunnan [J]. Earth Science: Journal of China University of Geosciences, 18 (5): 540-552 (in Chinese).

Feng Q L, Liu B P, 1993b. Permian radiolarian on southwest Yunnan [J]. Earth Science: Journal of China University of Geosciences, 18: 553-563.

Feng Q L, Ye M, Zhang Z F, 1996. Triassic Radiolarian fauna from Southwest China [J]. Scientia Geologica Sinica, 5 (3): 380-394.

Feng Q L, Ye M, Zhang Z J, 1997. Early Carboniferous radiolarians from Western Yunnan [J]. Acta Micropalaeontologica Sinica, 14 (1): 79-92 (in Chinese with English abstract).

Feng Q L, Yang Z, Crasquin S, et al., 2007. Permian and Triassic radiolarians from northern Tibet Correlation between radiolarian and Conodont biozones [J]. Bulletin de la Societe Geologique de France, 178 (6): 485-495.

Feng Q L, Yang Z, Li X, et al., 2009. Middle and Late Triassic radiolarians from northern Tibet: Implications for the Bayan. Har Basin evolution [J]. Geobios, 42 (5): 581-601.

Feng Q L, Zhang Z F, Ye M, 2001. Middle Triassic radiolarian fauna from Southwest Yunnan, China [J]. Micropaleontology, 47 (3): 173-204.

Feng X T, Zhu T X, Li C, et al., 2005. Redefinition of the Upper Triassic Xiaochaka Group in the Shuanghu area, northern Tibet, China, and its geological implications [J]. Geological Belletin of China, 24 (12): 1135-1140 (in Chinese with English abstract).

Foreman H P, 1963. Upper Devonian Radiolaria from the Huron Member of the Ohio Shale [J]. Micropaleontology, 9 (3): 267-304.

Glongka J, 2004. Plate tectonic evolution of the southern margin of Eurasia in the Mesozoic and Zenozoic [J]. Tectonophysics, 381: 235-273.

Gorican S, Buser S, 1990. Middle Triassic radiolarians from Slovenia (Yugoslavia) [J]. Geologija, 31/32: 133-197.

Gorican S, Halamic J, Grgasovic T, et al., 2005. Stratigraphic evolution of Triassic arc-backare system in northwestern Croatia [J]. Bulletin de la Socite geologique de France, 176 (1): 3-22.

Gou J, Xing G Z, 1990. Upper Triassic of Bayinchawuma of Hoh Xil, Qinghai [J]. Xibei Dizhi (Northwestern Geology of China) (2): 1-5 (in Chinese).

Gourmelon F, 1987. Les Radiolaria Tournaisiens des Nodules phosphates de la Montagne Noire et Des Pyrenees Centrales [J]. Biostratigraphie du Paleozoique, 6: 1-172.

Hinde G J, 1899. On the Radiolaria in the Devonian Rocks of New South Wales [J]. Quarterly Journal of the Geological Society of London, 55: 38-64.

Holdsworth B K, Jones D L, 1980. Preliminary radiolarian zonation for Late Devonian through Permain time [J]. Geology, 8: 281-285.

Holdsworth B K, Murchey B L, 1989. Paleozoic Radiolarian Biostratigraphy of the northern Brooks Range, Alaska [J]. Professional Papers of United States Geological Survey (USA), 1399: 777-791.

Huang J Q, Chen B W, 1987. The evolution of the Tethys in China and adjacent region [M]. Beijing: Geological Publication, 1-109. (English version); 1-78 (Chinese version).

Ishiga H, 1982. Late Carboniferous and Early Permian radiolarians from the Tamba Belt, Southwest Japan [J]. Earth Science, 36 (6): 333-339.

Ishiga H, 1983. Morphological change in the Permian Radiolaria, Pseudoalbaillella scalprata in Japan [J]. Transactions and Proceedings, Palaeontological Society of Japan, New Series, 129: 1-8.

Ishiga H, 1986a. Late Carboniferous-Permian radiolarian biostratigraphy in Japan with special reference to distribution and phyletic lineage of Follicucullus in Late Permian time [J]. New of Osaka Micropaleontologists: Special Volume, 7: 1-8.

Ishiga H, 1986b. Late Carboniferous and Permian radiolarian biostratigraphy of Southwest Japan [J]. Journal of Geosciences, Osaka City University, 29: 89-100.

Ishiga H, 1988. Paleontological study of radiolarians from the Southern New England Fold Belt, Eastern Australia [M] //Preliminary Report on the Geology of the New England Fold Belt, Australia, 1: 77-93.

Ishiga H, 1990. Palaeozoic Radiolarian [M] // Ichikawa K, et al. Pre-Cretaceous Terranes of Japan. Publ. Project 224, Pre-Jurassic Evolution of Eastern Asia, IGCP project 224, Osaka Japan: 285-295.

Ishiga H, 1991. Description of a new Follicucullus species from Southwest Japan [J]. Memoirs of the Faculty of Science, Shimane University, 25: 107-118.

Ishiga H, Imoto N, 1980. Some Permian radiolarians in the Tamba district, Southwest Japan [J]. Earth Science, 34 (6): 333-345.

Ishiga H, Kito T, Imoto N, 1982a. Late Permian radiolarian assemblages in the Tamba district and an adjacent area, Southwest Japan [J]. Earth Science, 36: 10-22.

Ishiga H, Kito T, Imoto N, 1982b. Permian radiolarian biostratigraphy [J]. News of Osaka Micropaleontologist: Special Volume, 5: 17-26.

Ishiga H, Kito T, Imoto N, 1982c. Middle Permian radiolarian assemblages in the Tamba district and an adjacent area, Southwest Japan [J]. Earth Science, 36 (5): 272-281.

Ishiga H, Leitch E C, Watanake T, et al., 1988. Radiolarian and Conodont biostratigraphy of siliceous rocks from the New England Fold Belt. Australian Jour [J]. Earth Science, 35: 73-80.

Jolivet M, 2017. Mesozoic tectonic and topographic evolution of Central Asia and Tibet: a preliminary synthesis [M] // Brunet M-F, McCann T, Sobel E R. Geologicalmevolution of Central Asian Basins and the western Tien Shan Range. Geological Society, London. Special Publication, 427: 19-35.

Kamata Y, Sashida K, Ueno K, et al., 2002. Triassic radiolarian faunas from the Mae Sariang area, northern Thailand and their paleogeographic significance [J]. Journal of Asian Earth Sciences, 20: 491-506.

Kellici I, De Wever P, 1995. Radiolaries Triasiques de Massif de la Marmolada, Italie du Nord [J]. Revue de Micropaléontologie, 38 (2): 139-167.

Kiessling W, Tragelehn H, 1994. Devonian Radiolarian Faunas of conodont-dated localities in the Frankenwald (Northern Bavaria, Germany) [J]. Abhandlungen der geologisches Bundesanstali in Wien, 50: 219-255.

Kolar-Iurkock T, 1989. New Radiolaria from the Ladinian substage (Middle Triassic) of Slovenia (NW Yugoslavia) [J]. Neues Jahrbuch für Geologie und Pal-ontologie, Monatshefte, 3: 155-165.

Kozur H, 1980. Ruzhencevispongidae, eine neus Spumellaria Familie aus dem oberen Kungurian (Leonardian) und Sakmarian des Vorurals [J]. Geologisch-Paläontologische Mitteilungen Innsbruck, 10 (6): 235-242.

Kozur H, 1981. Albaillellidae (Radiolaria) aus dem Unter Perm des Vorurals [J]. Geologisch-Paläontologische Mitteilungen Innsbruck, 10: 263-274.

Kozur H, 1988. Muelleritortiidae n. fam., eine characteristiche Longobardische (Oberladinische) Radiolarienfamilie: Teil. 1 [M] // Freiberger Forschungshefte. C, Geowissenschaften Geologie, 419 (7): 51-61, Leipzig.

Kozur H, 1993. Upper Permian radiolarians from the Sosio Valley Area, Western Sicily (Italy) and from the Uppermost Lamar Limestone of West Texas [J]. Jahrbuch der Geologischen Bundesanstalt, Wien, 136: 99-123.

Kozur H, Mostler H, 1979. Beitr—zur Erforschung der mesozoischen radiolarian, Teil Ⅲ: Die Oterfamilien Actinommacea Haeckel,

1862, emend. , Artiscacea Haeckel, 1882, Multiarcusellacea nov. der Spumellaria und triassische Nassellaria [J]. Geologisch-Pal-ontologische Mitteilungen Innsbruck, 9 (1/2): 1-132.

Kozur H, Mostler H, 1981. Beiträzur Erforschung der mesozoischen radiolarian, Teil IV: Thalassosphaeracea Haeckel, 1862, Hexasty-lacea Haeckel 1882, emend. Petrushevskaja, 1979, Sponguracea Haeckel, 1862, emend und Weitere triassische Lithocycliacea, Trematodiscacea, Actiommacea und Nassellaria [J]. Geologisch-Paläontologische Mitteilungen Innsbruck, Sonderband: 1-208.

Kozur H, Krahl, 1984. Erster Nachweis triassischer Radiolaria in der Phyllit-Grupe auf der Insel Kreta [J]. Neues Jahrbuch für Geologie und Paläontologie, Monatshefte, 1984/7, 400-404.

Kozur H, Reti Z, 1986. The first paleontological evidence of Triassic Ophiolites in Hungary [J]. Neues Jahrbuch für Geologie und Paläontologie, Monatshefte, 5: 284-292.

Kozur H, Krahl H, 1987. Erster Nachweis von Radiolarien im tethyan Perm Europas [J]. Neues Jahrbuch für Geologie und Paläontologie, Abhandlungen, 175: 375-372.

Kozur H, Krainer H, Mostler H, 1996. Radiolarians and facies of the Middle Triassic Loibl Formation South Alpine Karawanken Mountains (Carinthia Austria) [J]. Geologisch- Paläontologische Mitteilungen Innsbruck,, 4: 195-269, Innsbruck.

Kozur H, Mostler H, 1994. Anisian to Middle Carnian Radiolarian zonation and description of some stratigraphically important Radiolarians [J]. Geologisch-Paläontologische Mitteilungen Innsbruck, Sonderband, 3: 39-255.

Kozur H, Mostler H, 1996a. Longobardian (Late Ladinian) Muelleritortidae (Radiolaria) from the Republic of Bosnia-Herzegovina [J]. Geologisch-Paläontologische Mitteilungen Innsbruck, Sonderband, 4: 83-103.

Kozur H, Mostler H, 1996b. Longobardian (Late Ladinian) Oertlispongidae (Radiolaria) from the Republic of Bosnia-Herzegovina and the Stratigraphic value of advanced Oertlispongidae [J]. Geologisch-Paläontologische Mitteilungen Innsbruck, Sonderband, 4: 105-193.

Kuwahara K, 1999. Middle-Late Permian radiolarian assemblages from China and Japan [M] //Yao Akira, Ezaki Yoichi, Hao Weicheng, et al. Biotic and Geological Development of the Paleo-Tethys in China. Beijing: Peking University Press: 43-54.

Kuwahara K, Yao A, An T, 1997. Paleozoic and Mesozoic complex in the Yunnan area, China, Pt. 1, Preliminary report of Middle-Late Permian radiolarian assemblages [J]. Journal of Geoscience, Osaka City University, 40 (3): 37-49.

Kuwahara K, Yao A, Yamakita S, 1998. Reexamination of Upper Permian radiolarian biostratigraphy [J]. Earth Science (Chikyu Kagaku), 52 (5): 391-404.

Lahm B, 1984. Spumellarienfaunen (Radiolaria) Aus den mitteltriassischen Buchensteiner-Schichten von Recoaro (Norditalien) und den Obertriassischen Reiflingerkalken von Grossreifling (Österreich) - Systematik-Stratigraphie. Munchner Geowiss [J]. Geologisch Paläontologische, Abhandlung, Reihe A, 1: 1-161.

Li C, Xie Y W, Dong Y S, et al. , 2009. The north Lancangjiang suture: the boundary between Gondwana and Yangtze? [J]. Geological Bulletin of China, 28 (12): 1711-1719 (in Chinese with English abstract).

Li C, Xie C M, Wang M, et al. , 2016. Geology in the Qiangtang region [M]. Beijing: Geological Press: 1-681 (in Chinese with English abstract).

Li H S, Bian Q T, 1993. Upper Paleozoic Radiolaria of the Xijin Ulan-Gangqiqu ophilite complex, Kekexili [J]. Geoscience (Journal of Graduate School, China University of Geosciences), 7 (4): 410-420 (in Chinese with English abstract).

Li Y, Ji Z S, Wu G C, et al. , 2018. Radiolarian fauna from Middle of Zuoqingcuo Formation, Southern Qiangtang Basin, Xizang, China, and its stratigraphic significance [J]. Earth Science, 43 (11): 3932-3946 (in Chinese with English abstract).

Li Y J, Wu H R, Li H S, 1997. Discovery of radiolarians in the Amugang and Chasang groups and Lugu Formation in northern Tibet and some related geological problems [J]. Geological Review, 43 (3): 250-256 (in Chinese with English abstract).

Li Y X, Wang Y J, 1991. Upper Devonian (Frasnian) radiolarian fauna from the Liukiang Formation, eastern and southeastern Guangxi [J]. Acta Micropalaeontologica Sinica, 8 (4): 395-404 (in Chinese with English Abstract).

Liang D Y, Nie Z T, Guo T Y, et al. , 1983. Permo-Carboniferous Gondwana? Tethys facies in southern Karakoran, Ali, Xiznag

(Tibet) [J]. Earth Science (Journal of Wuhan College of Geology) (1): 9-27 (in Chinese with English abstract).

Liu B P, Cui X S, 1983. Discovery of Eurydesma-Fauna from Rutog, northwest Xizang (Tibet), and its biogeographic significance [J]. Earth Science (Journal of Wuhan College of Geology) (1): 79-92 (in Chinese with English abstract).

Liu B P, Feng Q L, Fang N Q, et al., 1993. Tectonic evolution of Palaeo-Tethys poly-island-ocean in Changning-Menglian and Lancnagjiang belts, southwestern Yunnan, Chinna [J]. Earth Science (Journal of China University of Geosciences), 18 (5): 529-539, 671 (in Chinese with English abstract).

Luo H, 1995. Permian foramiifers [M] // Sha J G. Palaeontology of the Hoh Xil region, Qinghai. Beijing: Science Press: 23-48, 144-145 (in Chinese with English abstract).

Metcalf I, 2002. Permian Tectonic framework and palaeogeography of SE Asia [J]. Journal of Asian Earth Sciences, 22: 551-556.

Metcalfe I, 2013. Gondwana dispersion and Asian accretion: Tetonic and palaeogeographic evolution of eastern Tethys [J]. Journal of Asian Earth Sciences, 66: 1-33.

Mikiko S, Yao A, 2006. Lower-Middle Permian radiolarian biostratigraphy in the Qinzhou area, South China [J]. Jourmal of Geoscience, Osaka City University, 49 (3): 31-47.

Mizutani S, Koike T, 1982. Radiolarians in the Jurassic siliceous shale and in the Triassic bedded chert of Unuma, Kagamigahere City, Gifu Prefecture, central Japan [J]. News of Osaka Micropaleontologists, Special Volume, 5: 117-134 (in Japanese with English abstract).

Murchey B L, 1990. Age and depositional setting of siliceous in the Upper Paleozoic Havallah sequence near Battle Mountain Nevada, Implications for the Paleogeography and structural evolution of the western margin of North America [J]. Geological Society of America: Special Paper, 255: 137-155.

Murchey B L, Jones D L, 1992. A mid-Permian Chert event: Widespread deposition of biogenic siliceous sediments in coastal island arc and oceanic basins [J]. Palaeogeography, Palaeoclimatology, Palaeoecology, 96: 161-174.

Muttoni G, Gaetani M, Kent D V, et al., 2009. Opening of the Neo-Tethys Ocean and the Pangea B to Pangea A transformation during the Permian [J]. GeoArabia, 14 (4): 17-48.

Nakaseko K, Nishimura A, 1979. Upper Triassic Radiolaria from Southwest Japan [J]. Science Reports, College of General Education, Osaka University, 28 (2): 61-109.

Nazarov B B, Ormiston A R, 1985. Radiolaria from the Late Paleozoic of the South Urals, USSR and West Texas, USA [J]. Micropaleontolopy, 31 (1): 1-54.

Nimaciren, Xie R W, 2005. Discovery of Middle Triassic strata in Nagqu area, northern Tibet, China, and its geological implications [J]. Geological Bulletin of China, 24 (12): 1141-1149 (in Chinese with English abstract).

Pullen A, Kapp P, Gehrels G E, et al., 2008. Triassic continental subduction in central Tibet and Mediterranean-style closure of the Paleo-Tethys Ocean [J]. Geology, 36: 351-354.

Ramovs A, Gorian, 1995. Late Anisian-Early Ladinian radiolarians and conodonts from —marna Goranear Ljubljana, Slovenia [J]. Razprave Ⅳ Razreda Sazu, 36 (9): 179-221.

Regional Geological Survey Team of Qinghai Province, 1970. Regional Geological survey report (1st Volume). Wenquan (Qinghai) (I-46) (1: 1 000 000) (Internal Report) [M]. Xining: Qinghai Geological Bureau: 1-267 (in Chinese).

Riedel W R, 1967a. Some new family of Radiolaria [J]. Proc. Geol. Soc., London, 1640: 148-149.

Riedel W R, 1967b. Subclass Radiolaria [M] // Harland, W. R, et al. The Fossil Record. Geological Society of London, 291-298.

Rudenko V S, Panasenko E S, 1990. Permian Albaillellaria (Radiolaria) of the Pantovyi Creek sequence in Primorye [M] // Zakharov Y D, et al. New Data on Paleozoic and Mesozoic Biostratigraphy of the South Far East. 181-193 (in Russian).

Rudenko V S, Panasenko E S, Rubalka S V, 1997. Radiolaria from Permian-Triassic boundary beds in cherty deposits of Primarye (Sikhote-Alin) [M] // Dickins J M, Yang Z Y, et al. Late Paleozoic and Early Mesozoic Circum-Pacific Events and Their Global Correlation. World and Regional Geology, 10: 147-151.

Sashida K, Igo H, Hisada K, et al., 1993. Occurrence of Paleozoic and Early Mesozoic Radiolaria in Thailand (preliminary report)

[J]. Micropaleontology, 48 (1): 129-143.

Saesaengseerung D, Sashida K, Sardsud A, 2007. Late Devonian to Carboniferous radiolarian fauna from the Pak Chom area, Loei Province, northeastern Thailand [J]. Paleontological Research, 11 (2): 109-121.

Sashida K, Adachi S, Igo H, et al., 1995. Middle and Late Permian radiolarians from the Semanggol Formation, Northwest Peninsular Malaysia [J]. Transactions and Proceedings, Palaeontological Society of Japan, New series, 177: 43-58.

Sashida K, Adachi S, Igo H, et al., 1997. Middle to Upper Permian and Middle Triassic radiolarians from Eastern Thailand [J]. Science Reports of the University of Tsukuba, Institute of Geoscience, Section B, 18: 1-17.

Sashida K, Igo H, Adachi S, et al., 2000. Late Permian to Middle Triassic radiolarians faunas from Northern Thailand [J]. Journal of Palaeogeography, 74 (5): 789-811.

Sashida K, Igo H, Adachi S, et al., 1998. Late Paleozoic radiolarian faunas from northern and northeastern Thailand [J]. Science Reports of the University of Tsukuba, Institute of Geoscience, Section B, 19: 1-27.

Sashida K, Tonishi K, 1985. Permian Radiolarians from the Kanto Mountains, Central Japan: Some Upper Permian Spumellaria from Itsukaichi Western Part of Tokyo Prefecture [J]. Science Reports of the Institute of Geoscience, University of Tsukuba, 6: 1-19.

Sashida K, Salyapongse S, Charusiri P, 2002. Lower Carboniferous radiolarian fauna from the Safa Yoi-Kabang Area, southernmost part of Peninsular Thailand [J]. Micropaleontology, 48 (1): 129-143.

Sashida K, Kamata Y, Adachi S, et al., 1999. Middle Triassic radiolarians from West Timor, Indonesia [J]. Journal of Palaeogeography, 73 (5): 765-786.

Sashida K, Nishimura H, Igo H, et al., 1993. Triassic radiolarian fauna from Kiso-fukushima, Kiso Mountains, central Japan [J]. Science Reports of the University of Tsukuba, Institute of Geoscience, Section B, 14: 77-97.

Schmidt-Effing R, 1988. Eine Radiolarien-Fauna des Famenne (Ober Devon) aus dem Frankenwale (Bayern) Geologica et Paleontologica [J]. 22: 33-41.

Schwartzapfel J A, Holdsworth B K, 1996. Upper Devonian and Mississippian radiolarian zonation and biostratigraphy of the woodford, Sycamore, Cany and Goodard Formations, Oklahoma [J]. Cushman Found. Journal of Foraminiferal Research: Special Publication, 33: 1-275.

Sha J G, 1995a. Introduction [M] // Sha J G. Palaeontology of the Hoh Xil region, Qinghai. Beijing: Science Press: 141-142 (in Chinese with English abstract).

Sha J G, 1995b. Bivalves [M] // Sha J G. Palaeontology of the Hoh Xil region, Qinghai. Beijing: Science Press: 82-116, 149-151 (in Chinese with English abstract).

Sha J G, 1998. Characteristics of stratigraphy and palaeontology of Hoh Xil, Qinghai: Geographic significance [J]. Acta Palaeontologica Sinica, 37 (1): 85-96 (in Chinese with English abstract).

Sha J G, 2009. From seas into mulberry fields and from mulberry fields into seas of the source area of the Yangtze River (Tibet Plateau): A story narrated by fossils [M] // Sha J G. Commemoration of the 80th anniversary of the founding of Palaeontological Society of China. Century leap: Splendid achievement of palaeontology in China. Beijing: Science Press: 376-385 (in Chinese).

Sha J G, 2018. Fossils witnessed the uplift of Tibetan Plateau [M] // Rong J Y, Yuan X, Zhan R B, et al., Biological evolution and environment. Heifei: University of Sciences and Technology of China Press: 298-329 (in Chinese).

Sha J G, Fürsich F T, 1999. Palaeotethys Ocean closed before Capitanian times in Hoh Xil area (W China): New data on the temporal extend of the Palaeotethys [J]. Neues Jahrbuch für Geologie und Pal-ontologie, Abhandlung, (7): 440-448.

Sha J G, Grant-Mackie J A, 1996. Late Permian to Miocene bivalve assemblages from Hoh Xil, Qinghai-Xizang Plateau, China [J]. Journal of the Royal Society of New Zealand, 26 (4): 429-455.

Sha J G, Johnson A L A, Fürsich F T, 2004. From deep-sea to high mountain ranges: Palaeogeographic and biotic changes in Hoh Xil, the source area of the Yangtze River (Tibet Plateau) since the Late Palaeozoic [J]. Neues Jahrbuch für Geologie und Pal-ontologie, 233 (2): 169-195.

Sha J G, Zhang L X, Luo H, et al., 1992. On the geological age of the closure of the Late Palaeozoic rift in Hoh Xil, Qinghai [J].

Acta Micropalaeontologica Sinica, 9 (2): 177-182 (in Chinese with English abstract).

Sha J G, 1995. Palaeontology of the Hoh Xil Region, Qinghai [M]. Beijing: Science Press: 1-140 (in Chinese with English abstract).

Shang Q, Caridroit M, Wang Y, 2001. Radiolarians from the Uppermost Permian Changhsingian of Southern Guangxi [J]. Acta Micropalaeontologica Sinica, 18 (3): 229-240.

Sheng J Z, Wang Y J, 1985. Fossil Radiolaria from Kufeng Formation at Longtan, Nanjing [J]. Acta Palaeontologica Sinica, 24 (2): 171-180.

Spiller F C P, 1996. Late Paleozoic radiolarians from the Bentong-Raub suture zone, Peninsular Malaysia [J]. Island Arc, 5: 91-103.

Spiller F C P, 2002. Radiolarian Biostratigraphy of Peninsular Malaysia and Implications for Regional Palaeotectonics and Palaeogeog [J]. Palaeontographica Abteilung A Stuttgart, 266 (1/2/3): 1-91.

Spiller F C P, Metcalfe I, 1995. Late Paleozoic radiolarians from the Bentong-Raub suture zone, and the Semanggol Formation of Peninsular Malaysia-initial results [J]. Journal of Southeast Asian Earth Sciences, 11 (3): 217-224.

Straford J M C, Aitchison J C, 1997. Lower to Middle Devonian radiolarian assemblages from the Gamilaroi terrane, Glenrock station, NSW, Australia [J]. Marine Micropaleontology, 30: 225-250.

Sun Donggying, Xia Wenchen, Liu Dongjie, 2002. Reexamination of Radiolarian Biostratigraphy in Permian in Pelagic Chert Sequences at Dachonling Section [J]. South China. Journal of China University of Geosciences, 13: 207-214.

Sugiyama K, 1992. Lower and Middle Triassic radiolarians from M. T. Kinkazan, Gifu Prefecture, Central Japan [J]. Transactions and Proceedings of the Palaeontological Society of Japan, New Series, 167: 1180-1223.

Sugiyama K, 1997. Triassic and Lower Jurassic radiolarian biostratigraphy in the siliceous claystone and bedded chent units of the southeastern Mino Terrane, Central Japan [J]. Bulletin of the Mizunami Fossil Museum, 24: 79-193.

Takemura A, Nakaseko K, 1981. A new Radiolarian Genus from the Tamba Belt, Southwest Japan [J]. Transactions and Proceedings of the Palaeontological Society of Japan, New Series, 124: 208-214.

Tekin U K, 1999. Biostratigraphy and systematics of Late Middle to Late Triassic radiolarians from the Taurus Mountains and Ankara region, Turkey [J]. Geologisch Pal-ontologische Mitteilungen Innsbruck, Sonderband, 5: 1-296.

Tekin U K., Mostler H, 2005. Longobardian (Middle Triassic) Entactinarian and Nassellarian Radiolaria from the Dinarides of Bosnia and Herzegovina [J]. Journal of Palaeogeography, 79 (1): 1-20.

Tekin U K, Bedi Y, Okuyucu C, et al., 2016. Radiolarian biochronology of upper Anisian to upper Ladinian (Middle Triassic) blocks and tectonic slices of volcano-sedimentary successions in the Mersin Melange, southern Turkey: New insights for the evolution of Neotethys [J]. Journal of African Earth Sciences, 124: 409-426.

The Integrated Scientific Survey Team to Hoh Xil, 1994. The hinterland of the Qinghai-Xizang Plateau: A integrated scientific survey to Hoh Xil. Geological evolution [M]. Shanghai: Shanghai Scientific and Technical Publishers: 10-29 (in Chinese and English).

Tumanda F P, Sato T, Sashida K, 1990. Preliminary Late Permian radiolarian biostratigraphy of Busuanga Island, Palawan, Philippines [J]. Annual Report—University of Tsukuba, Institute of Geoscience, 16: 39-45.

Wang R J, 1993a. Fossil Radiolaria from the Kufeng Formation of Chaohu, Anhui [J]. Acta Micropalaeontologica Sinica, 32: 442-457.

Wang R J, 1993b. Fossil radiolarians from the Kufeng Formation, Hushan area, Nanjing [J]. Acta Micropalaeontologica Sinica, 10 (4): 459-468 (in Chinese with English abstract).

Wang Y J, 1995. Radiolarians [M] // Sha J G. Palaeontology of Hoh Xil region, Qinghai. Beijing: Science Press, 58-69, 146-148 (in Chinese with English abstract).

Wang Y J, 1997. An Upper Devonian (Famennian) radiolarian fauna from carbonate rocks, northern Xinjiang [J]. Acta Micropalaeontologica Sinica, 14 (2): 149-160 (in English with Chinese abstract).

Wang Y J, 2005. Late Cretaceous radiolariansfrom Xigazê [M] // Sha J G, Wang Q F, Lu H N. Micropaleontology of the Qiangtang Basin. Beijing: Science Press: 139-151, 266-267.

Wang Y J, Aitchison J C, Luo Hui, 2003. Devonian raduilarian faunas from South China [J]. Micropaleontology, 49 (2): 127-145.

Wang Y J, Cheng Y N, Yang Q, 1994. Biostratigraphy and systematics of Permian radiolarians in China [J]. Palaeoworld, 4: 172-202.

Wang Y J, Duan B X, Luo H, et al., 2013. *Helenifore robustum* radiolarian fauna from the Hanjica Formaion, Middle Tianshan, Xinjiang, China and the Chronostratigraphic significance [J]. Acta Micropalaeontologica Sinica, 30 (3): 217-227.

Wang Y J, Fan Z X, 1997. Discovery of Permian radiolarians in ophiolite belt on northern side of Xar Moron River, Nei Monggol and its geological significance [J]. Acta Micropalaeontologica Sinica, 36 (1): 58-69.

Wang Y J, Fang Z J, Yang Q, et al., 2000. Middle-Late Devonian strata of cherty facies and radiolarian faunas from West Yunnan [J]. Acta Micropalaeontologica Sinica, 17 (3): 235-254 (in Chinese with English abstract).

Wang Y J, Kuang G D, 1993. Early Carboniferous radialarians from Qinzhou, southeastern Guangxi [J]. Acta Micropalaeontologica Sinica, 10 (3): 275-287 (in Chinese with English abstract).

Wang Y J, Luo H, 2009. Upper Devonian (Frasnian) *Helenifore robustum* radiolarian fauna from the Bazhai Village in Ziyun County, Guizhou province [J]. Acta Micropalaeontologica Sinica, 26 (2): 129-138.

Wang Y J, Luo H, Aitchison C, 2006. Influence of the Frasnian - Famennian event on radiolarian faunas, Eclogae geol [J]. Eclogae Geologicae Helvetiae, 99 (1): 127-132.

Wang Y J, Luo H, Kuang G D, et al., 1998. Late Devonian-Late Permian strata of cherty facies at Xiaodong and Bancheng counties of the Qinzhou area, SE Guangxi [J]. Acta Micropalaeontologica Sinica, 15 (4): 351-356 (in Chinese with English abstract).

Wang Y J, Luo H, Yang Q, 2012. Late Paleozoic radiolarians in the Qinfang area, Southeast Guangxi [M]. Hefei: University of Science and Technology of China Press: 1-127 (in Chinese with English abstract).

Wang Y J, Qi D L, 1995. Radiolarian fauna of Kuhfeng Formation in southern Part of Jiangsu and Anhui provinces [J]. Acta Micropalaeontologica Sinica, 12 (4): 374-387 (in English with Chinese abstract).

Wang Y J, Qu Y G, et al., 2012. Regional geological survey report of People's Republic of China. Paduco (Tibet) (I45C004003) (1 : 250000) [M]. Wuhan: China University of Geosciences Press: 1-188 (in Chinese).

Wang Y, J, Shang Q H, 2001. Discovery of the Neoalbaillella radiolarian fauna in the Shaiwa Group of Ziyun district, Guizhou [J]. Acta Micropalaeontologica Sinica, 18 (2): 111-121.

Wang Y J, Wang J P, Pei F, 2002b. A Late Triassic radiolarian fauna in the Dingqing Ophiolite Belt, Zizang (Tibet) [J]. Micropalaeontologica Sinica, 19 (4): 323-336 (in Chinese with English abstract).

Wang Y J, Yang Q, 2007. Carboniferous-Permian radialarian biozones of China and their palaeobiogeographic implications [J]. Acta Micropalaeontologica Sinica, 224 (4): 337-345 (in Chinese with English abstract).

Wang Y J, Yang Q, 2011. Biostratigraphy, Phylogeny and Paleobiogeography of Carboniferous - Permian radiolarians in South China [J]. Palaeoworld, 20 (2/3): 134-145.

Wang Y J, Yang Q, Cheng Y N, et al., 2006. Lopingian (Upper Permian) radiolarian biostratigraphy of South China [J]. Palaoworld 15: 31-53.

Wang Y J, Yang Q, Guo T Z, 2005. The Late Middle Triassic radiolarians Spongoserrula rarauana fauna from the Hoh Xil region, Qinghai [J]. Acta Micropalaeontologica Sinica, 22 (1): 1-9.

Wang Y J, Yang Q, Matsukawa A, et al., 2002a. Triassic radiolarians from the Yalung Zangbo suture zone in the Jinlu area, Zetang county, southern Tibet [J]. Micropalaeontologica Sinica, 19 (3): 215-227 (in Chinese with English abstract).

Wang Y J, Li J X, 1994. Discovery of the Follicucullus bipartitus-F. chrveti radiolarian assemblage zone and its geological significance [J]. Acta Micropalaeontologica Sinica, 11 (2): 201-212.

Won M-Z, 1983. Radiolarien aus dem Unter-Karbon des Rheinischen Schiefergbirges (Deutschland) [J]. Palaeontographica Abteilung A, 182 (4-6): 116-175.

Won M-Z, 1991. Phylogenetic study of some species of genus Albaillella Deflandre 1952 and Radiolarian zonation in the Rheinische Schiefergebirge, West Germany [J]. Journal of the Paleontological Society of Korea, 7 (1): 13-25.

Won M-Z, Blodgett R, Clautice K H, et al., 1999. Late Devonian (Late Famennian) radiolarians from the Chulitna terrane, south-central Alaska [J]. Short Notes on Alaska Geology, 145-152.

Wonganan N, Caridroit M, 2005. Middle and Upper Devonian radiolarian fauna from Chiang Dao area, Chiang Mai Province, northern Thailand [J]. Micropaleontology, 51 (1): 39-57.

Wu H R, Xian X Y, Kuang G D, 1994. Late Paleozoic radiolarian assemblages of southern Guangxi and its geological significance [J]. Scientia Geologica Sinica, 29 (4): 339-345.

Xia D X, Liu S K, 1997. Stratigraphy (Lithostratic) of Xizang Autonomous Region [M]. Wuhan: China University of Geosciences Press: 1-302 (in Chinese).

Xu S H, 1995. Conodonts [M] // Sha J G. Palaeontology of Hoh Xil region, Qinghai. Beijing: Science Press: 118-123, 151 (in Chinese with English abstract).

Yang Q, Matsuoka A, Wang Y J, et al., 2000. A Middle Triassic radiolarian assemblage from Quxia, Lhaze County, Southern Tibet [J]. Science Reports of Niigata University, Series E (Geology), 15: 59-65.

Yao A, 1982. Middle Triassic to Early Jurassic Radiolarians from the Inuyama Area, Central Japan [J]. Journal of Geosciences, Osaka City University, 25 (4): 53-70.

Yao A, Kuwahara K, 1999. Paleozoic and Mesozoic radiolarian from the Changning-Menglian Terrane, Western Yunnan, China. [M] // Yao Biotic and Geological Development of the Paleo-Tethys in China. Beijing: Beijing University Press: 17-42.

Yao A, Kuwahara K, 2000. Permian and Triassic radiolarian from the southern Guizhou Province, China [J]. Journal of Geosciences, Osaka City University, 43 (1): 1-19.

Yeh K Y, 1989. Studies of Radiolaria from the Fields Creek Formation, East-Central Oregon, U.S.A [J]. Taibei: Bulletin of the "National" Museum of Natural Science: 1: 43-109.

Yeh K Y, 1990. Taxonomic studies of Triassic Radiolaria from Busuanga Island, Philippines [J]. Taibei: Bulletin of the "National" Museum of Natural Science: 2: 1-63.

Yeh K Y, 1992. Triassic Radiolaria from Uson Island, Philippines [J]. Taibei: Bulletin of the "National" Museum of Natural Science: 3: 51-91.

Yin J X, 1997. Stratigraphic geology of Gondwana facies of Qinghai-Xizang Plateau and adjacent areas [M]. Beijing: Geological Publishing House: 1-175 (in Chinese).

Zanchi A, Gaetani M, 2011. The geology of the Karakoram range, Pakistan: The new 1 : 100000 geological map of Cenral-Western Karakoram [J]. Italian Journal of Geosciences (Bollettin delia Societa geologica italiana e del Servizio geologico d'Italia), 130 (2): 161-246.

Zeng Q G, Mao G Z, Wang B D, et al., 2011. Regional geological survey report of People's Republic of China. Rigain Pünco (Tibet) (I45C004002) (1 : 250000) [M]. Wuhan: China University of Geosciences Press: 1-206 (in Chinese).

Zhai Q G, Li C, Huang X P, 2006. Geochemistry of Permian basalt in the Jiaomuri area, central Qiangtang, Tibet, China, and its tectonic significance [J]. Geological Bulletin of China, 25 (12): 1419-1427 (in Chinese with English abstract).

Zhang J M, 1995. Biofacies, sedimentary chracteristics and palaeogeographical changes [M] // Sha J G. Palaeontology of Hoh Xil region, Qinghai. Beijing: Science Press: 133-140, 153-154 (in Chinese with English abstract).

Zhang L X, 1995. Fusulinids [M] // Sha J G. Palaeontology of Hoh Xil region, Qinghai. Beijing: Science Press: 54-58, 146-147 (in Chinese with English abstract).

Zhang Y C, Wang Y, 2019. Middle Permian Forminifers from the Xilanta Formation in the Gyanyina area, Burang county, Tibet and their geological implications [J]. Acta Palaeotologica Sinica, 58 (3): 311-323 (in Chinese with English abstract).

Zhang Y F, 1991. Characteristics of Kekexili-Bayanahr and neighbouring Tethys [J]. Tibet Geology, 6 (2): 62-72 (in Chinese with English abstract).

Zhang Y F, 1993. Primary discussion on geological tectonic evolution of the northern Qinghai-Xizang Plateau [J]. Tibet Geology (2): 1-7 (in Chinese with English abstract).

Zhang Y F, Sha J G, 1995. General stratigraphy situation [M] // Sha J G. Palaeontology of Hoh Xil region, Qinghai. Beijing: Science

Press: 1-9, 142 (in Chinese with English abstract).

Zhang Y F, Zheng J K, 1994. Geological overview in Kekeshili, Qinghai and adjacent areas [M]. Beijing: Seismological Publishing House: 1-177 (in Chinese with English abstract).

Zhao Z Z, Li Y T, Ye H F, et al., 2001. Strata of Qinghai-Tibet Plateau [M]. Beijing: Science Press: 1-542 (in Chinese with English abstract).

Zhu T X, Dong H, Li C, et al., 2005. Distribution and sedimentary model of the Late Triassic strata in northern Qiangtang on the Qinghai-Xizang Plateau [J]. Sedimentary Geology and Tethyan Geology, 25 (3): 18-23.

Zhu T. X., Feng X T, et al., 2012. Regional geological survey report of People's Republic of China. Heihuling (Tibet) (I45C002003) (1∶250000) [M]. Wuhan: China University of Geosciences Press: 1-221 (in Chines).

Hu T X, Li Z L, et al., 2010. Regional geological survey report of People's Republic of China. Jiangaidarina (Tibe) (I45C003003) (1∶250000) [M]. Wuhan: China University of Geosciences Press: 1-284 (in Chinese).

Zhu T X, Li Z L, Li C, et al., 2005. New data of Triasic strata in the Shuanghu area, northern Tibet, China [J]. Geological Bulletin of China, 24 (12): 1127-1134 (in Chinese with English abstract).

Zhu T X, Zhang Q Y, Dong H, et al., 2006. Discovery of the Late Devonian and Late Permian radiolarian cherts in tectonic mélanges in the Cêdo Caka area, Shuanghu, northern Tibet, China [J]. Geological Bulletin of China, 25 (12): 1413-1418 (in Chinese with English abstract).

Zhu X G, 1995. Gastropods [M] // Sha J G. Palaeontology of Hoh Xil region, Qinghai. Beijing: Science Press: 69-81, 148-149 (in Chinese with English abstract).

索 引
(INDEX OF GENERA AND SPECISE)

1. 拉—汉属种名称

Albaillella indensis 英德阿尔拜虫 17（1，2）

A. lauta 优美阿尔拜虫 14（58）

A. levis 光壳阿尔拜虫 14（50，52—54，59，60）

A. sinuata 曲状阿尔拜虫 7（15—18），10（9—12）

A. triangularis 三角形阿尔拜虫 14（42—49，51，55—57，61—65）

A. undulata 波状阿尔拜虫 17（3）

A. xiadongensis 小董阿尔拜虫 7（3，11—14）

Annulotriassocampe multisegmantatus 多节环三叠塔虫 22（1—11），25（4—10）

A. nova 新环三叠塔虫 20（26—28，31）

A. proprium 特别环三叠塔虫 22（12）

A. sulovensis 萨洛夫环三叠塔虫 19（5—7），20（30，36，38，39），22（13），25（16，17，20，23）

Archaeodictyomitra sp. A 古网冠虫（未定种 A）19（8）

A. sp. B 古网冠虫（未定种 B）20（33）

Archaeospongoprunum chiangdaoensis 清道古海绵梅虫 12（38）

A. globosum 球形古海绵梅虫 23（33，34）

A. sinisterispinosum 左旋刺古海绵梅虫 6（28）

Archocyrtium castuligerum 卡斯塔利格原笼虫 2（2，3）

A. diductum 双管原笼虫 2（4，6），3（2）

A. ferreum 铁色原笼虫 2（8，9）

A. lagabriellei 拉加布赖尔原笼虫 17（5）

A. ludicrum 荒谬原笼虫 2（7）

A. strictum 窄原笼虫 2（5）

A. wonae 旺氏原笼虫 3（1）

A. sp. A 原笼虫（未定种 A）3（3，4）

Astroentactinia biaciculata 双尖星内射虫 1（11，15，16，20，22，24，25），2（37，39—42，44），17（18—20）

A. mirousi 米劳斯星内射虫 2（38，45）

A. multispinosa 多刺星内射虫 3（26，28—30）

A. paronae 帕罗纳星内射虫 1（21，23，26—28，36—38），2（11，13—15，25）

A. stellata 星状星内射虫 1（12—14，17—19），2（16）

Beturiella robusta 强壮贝图尔虫 22（24，31）

B. variospinosa 变刺贝图尔虫 22（25—27，29，32）

B. sp. A 贝图尔虫（未定种 A）22（28，30）

Cauletella manica 袖状考勒特虫 13（37）

Copicyntra akikawaensis 秋川多内圆虫 6（33），13（38—43）

C. cuspidata 长刺多内圆虫 6（40，45—47）

Copiellintra diploacantha 双刺多内椭圆虫 6（29—31）

Cryptostephanidium cornigerum 带角隐王冠虫 18（22，23，25），21（33—36），23（13），25（41）

C. longispinosum 长刺隐王冠虫 23（16，28）

C. sp. A 隐王冠虫（未定种 A）21（11，37）

Eptingium manfredi 曼弗雷德埃普廷格虫 19（34—39），23（15，20—26）

E. nakasekoi 中世古幸次郎埃普廷格虫 12（16—23）

Hegleria mammilla 乳头状赫格勒虫 6（39，44）

Hindeosphaera bella 华丽欣德球虫 24（25）

H. bispinosa 双刺欣德球虫 21（20—32），24（27，28，31）

H. spinulosa 小刺欣德球虫 19（27），23（10，12），24（26，29，30）

Holoeciscus elongatus 长全角虫 2（1）

Hozmadia pararotunda 拟圆形扎莫依达虫 23（19，27）

Ishigaum craticula 格状石贺裕明虫 14（12，16，21，22，27—29，34，36，38）

I. obesum 肥胖石贺裕明虫 13（36），14（15，18，23，41）

I. trifistis 三棍石贺裕明虫 12（40，42），13（34），14（4，7—11，19，26，35）

I. sp. A 石贺裕明虫（未定种 A）14（30）

Latentibifistula asperspongiosa 粗海绵双隐管虫 8（41，42，45）

L. triacanthophora 三刺双隐管虫 4（38—45）

Latentifistula conica 圆锥形隐管虫 11（1—4），16（14，18）

L. patagilaterala 侧翼隐管虫 7（19，20，22，25—28），10（2，3）

L. sp. A 隐管虫（未定种 A）7（21）

L. sp. B 隐管虫（未定种 B）10（8）

L. texana 德克萨斯隐管虫 7（23，24），10（1，4，5），16（13）

Muelleritortis cochleata 匙形缪勒旋扭虫 25（38，40）

M. expansa 膨大缪勒旋扭虫 24（33，34）

M. koeveskalensis 凯维斯卡尔缪勒旋扭虫 24（35），25（37，39）

Neoalbaillella gracilis 纤细新阿尔拜虫 13（15—19）

N. optima 最优新阿尔拜虫 13（10，24）

N. ornithoformis 鸟形新阿尔拜虫 12（31—34），13（1—5，7，8，11—13，20—22，24—27）

N. pseudogrypa 假鹰钩新阿尔拜虫 13（14）

Ormistonella robusta 强壮奥米斯顿虫 14（17，24，25，37，40）

Palaeoscenidium cladophorum 具枝古蓬虫 2（24）

Pantanellium multiporum 多孔潘坦内尔虫 23（29，30）

P. sp. A 潘坦内尔虫（未定种 A）24（22，24）

Parasepsagon firmum 坚固拟彼萨格诺虫 18（30，31，33，35—39，42—45），21（1—7，14）

P. hohxiliensis 可可西里拟彼萨格诺虫 18（32，34，40，46）

P. praetetracanthus 前四刺拟彼萨格诺虫 12（7）

P. sp. A 拟彼萨格诺虫（未定种 A）19（9）

P. sp. B 拟彼萨格诺虫（未定种 B）21（9，10）

P. variabilis 变异拟彼萨格诺虫 19（33），23（14）

Paratriassocampe brevis 短壳拟三叠塔虫 22（14，15）

P. gaetanii 盖坦尼拟三叠塔虫 25（14，15，18）

Paroertlispongus longispinosus 长刺拟奥特尔海绵虫 19（1，2，21，22），22（16—20）

P. opiparus 美丽拟奥特尔海绵虫 22（22，23）

P. sp A 拟奥特尔海绵虫（未定种A）22（21）

Paurinella aequispinosa 等刺保林虫 19（13，14），20（23—25），25（31）

P. latispinosa 宽刺保林虫 25（29，30）

P. sp. A 保林虫（未定种A）19（15）

Polyentactinia aranea 蜘蛛多内射虫 2（17）

P. sp. A 多内射虫（未定种A）17（16）

Poulpus sp. A 章鱼虫（未定种A）23（35）

Pseudoalbaillella elongata 长形假阿尔拜虫 8（1—5），16（1，2）

P. fusiformis 纺锤形假阿尔拜虫 4（11—16）

P. globosa 球形假阿尔拜虫 3（42—44，47—49），4（17—30）

P. ishigai 石贺裕明假阿尔拜虫 10（15，16，18，20—22，24，27，30，31，41）

P. litangensis 理塘假阿尔拜虫 3（31—40）

P. lomentaria 豆荚状假阿尔拜虫 11（7—18），16（3—9）

P. longtanensis 龙潭假阿尔拜虫 4（1—10）

P. monopteryla 单翼假阿尔拜虫 10（17，19，23，25，26，33—36，38，40，42）

P. nonpteryla 无翼假阿尔拜虫 10（28，29，32，37，39，43—48）

P. postscalprata 后锐边假阿尔拜虫 15（13—17，19，24）

P. rhombothoracata 菱形假阿尔拜虫 7（4—10），8（6—17，35）

P. sakmarensis 萨克马尔假阿尔拜虫 8（18—32），11（5，6，19—39，41），15（1—9）

P. scalprata 锐边假阿尔拜虫 3（41，45，46，50），4（31—37），15（10—12，18，20—23）

P. simplex 简单假阿尔拜虫 8（33，34，36—40）

Pseudostylosphaera coccostyla 颗石柱假桩球虫 18（19）

P. compacta 致密假桩球虫 12（24，26—30），18（20，21），19（12，24，25，47，48），21（16），23（2，4—7）

P. goestlingensis 戈斯特林假桩球虫 23（11），24（21，23）

P. gracilis 纤细假桩球虫 18（11—13，15—18），21（17，18），23（17，18），24（16—18）．

P. helicata 螺旋状假桩球虫 23（9）

P. imperspicua 不明显假桩球虫 24（7—9）

P. inaequispinosa 不等刺假桩球虫 24（1—3，6，12，13）

P. japonica 日本假桩球虫 12（25），18（1—9），23（31），24（10，11）

P. longispinosa 长刺假桩球虫 19（28，49），23（32），24（4，5）

P. magnispinosa 大刺假桩球虫 18（14），19（23，32），23（1，3）

P. nazarovi 纳扎罗夫假桩球虫 18（3a，10），19（29—31），21（15），24（20，32）

P. paragracilis 拟纤细假桩球虫 24（14，15，19）

P. qinghaiensis 青海假桩球虫 21（19）

P. sp. A 假桩球虫（未定种A）21（8）

P. sp. C 假桩球虫（未定种C）21（12，13）

P. sp. D 假桩球虫（未定种D）23（8）

Pseudotormentus kamigoriensis 上郡假石片虫 5（44，45，53，54），7（39—43），8（43，44），10（13，14）

Pylentonema mira 奇异门孔虫 17（6，7）

P. sp. A 门孔虫（未种A）2（10）

Quadricaulis femoris 股状四茎虫 5（46—52），16（12）

Quadriremis minima 微小四浆虫 7（46）

Quinqueremis arundinea 芦苇状五浆虫 8（46—50）

Q. robusta 强壮五浆虫 6（20，21），16（19）

Q. sp. A 五浆虫（未定种A）19（20）

Raciditor gracilis 纤细卡里特罗伊特虫 6（1，3，5—10），10（7），12（41），14（31—33）

R. inflata 膨胀卡里特罗伊特虫 10（6），14（20，39）

R. oblatum 扁球形卡里特罗伊特虫 5（43），6（11—19），7（44），16（11，15—17，20）

R. phlogidea 火焰状卡里特罗伊特虫 5（32—42），6（2，4）

R. sp. A 卡里特罗伊特虫（未定种A）16（10）

Spongentactinia exilispina 弱刺海绵内射虫 2（26，32—36）

S. indisserta 欠发育海绵内射虫 3（16，25）

S. spongites 海绵状海绵内射虫 3（18，23）

S. sp. A 海绵内射虫（未定种A）3（6—11）

S. sp. B 海绵内射虫（未定种B）1（2，5，6）

S. sp. C 海绵内射虫（未定种C）1（7，8），3（12）

Spongoserrula rarauana 拉劳海绵锯齿虫 19（10，11），20（1—10）

Spongostephanidium spongiosum 海绵质海绵王冠虫 18（27）

Stigmosphaerostylus cruciformis 十字形斑点球桩虫 5（12，20—22），7（30—34）

S. diversitus 分异斑点球桩虫 1（3，4，9，29—31，33—35）

S. gracilentus 细弱斑点球桩虫 5（23，24，29，30）

S. ichikawai 市川浩一郎斑点球桩虫 6（22）

S. itsukaichiensis 五日市斑点球桩虫 5（2—11，14，17）

S. micula 碎屑状斑点球桩虫 2（27—31，43，45），3（27）

S. modestus 中等斑点球桩虫 5（1，13，15，16，18，19）

S. pantotolma 大无畏斑点球桩虫 3（24）

S. proceraspina 长刺斑点球桩虫 1（32）

S. spiciosus 华丽斑点球桩虫 1（39—41）

S. sp. A 斑点球桩虫（未定种A）6（25，32），7（35—38）

S. sp. B 斑点球桩虫（未定种B）6（23，24，26，27）

S. sp. C 斑点球桩虫（未定种C）6（34，35，46，47）

S. sp. D 斑点球桩虫（未定种D）16（21）

S. variospina 变刺斑点球桩虫 1（1，10），3（5，13—15，19，22），9（26—28），17（4）

S. vetulus 古老斑点球桩虫 5（25—28，31），6（36—38）

S. vulgaris 普通斑点球桩虫 17（8—15，17）

Striatotriassocampe laeviannulata 光瘤条纹三叠塔虫 25（1—3）

S. nodosoannulata 多节瘤状条纹三叠塔虫 25（11—13，19）

Tetrentactinia gigantia 大壳四内射虫 3（20）

T. spongacea 海绵四内射虫 2（12，18，23），9（18）

Tiborella florida 华丽蒂博尔虫 12（1—6），19（40—45）

Triaenosphaera hebes 年轻三叉球虫 2（21），3（17，21）

T. robustispina 粗刺三叉球虫 9（22，23）

T. sicarius 剑三叉球虫 2（19，20，22），9（19—21，25，30，41，42）

Triassistephanidium laticornis 宽角三叠王冠虫 18（24，26，41）

Triassocampe coronata 冠状三叠塔虫 12（14，15）

T. deweveri 德韦弗三叠塔虫 20（29，32，34，35，37），25（22，24）

T. nanpanensis 南畔三叠塔虫 12（13）

T. scalaris 梯形三叠塔虫 25（21，25）

T. spp. 三叠塔虫（多未定种）20（40，41）

Triassospongosphaera brevispinosa 短刺三叠海绵球虫 22（33）

T. latispinosa 宽刺三叠海绵球虫 22（37）

T. qinghaiensis 青海三叠海绵球虫 22（34—36）

T. sp. A 三叠海绵球虫（未定种A）12（8—11）

T. triassica 三叠三叠海绵球虫 12（12），19（17—19），20（11—22），22（38—40），25（27，28）

Trilonche echinatum 刺三矛虫 9（8，10—16）

T. davidi 达维德三矛虫 9（6，7，9，24，29，32，33）

T. echinatum 刺三矛虫 9（8，10—16）

T. elegans 华美三矛虫 9（17，35）

T. minax 小三矛虫 9（5）

T. pittmani 皮特曼三矛虫 9（2—4）

T. pseudocimelia 假美丽三矛虫 12（37），14（13，14）

T. sp. B 三矛虫（未定种B）7（45）

T. tretactinia 三射三矛虫 9（31，34，36—40，43）

Triplanospongos musashiensis 武藏三面海绵虫 12（35，36，39），13（35，45—48），14（1—3，5，6）

Tritortis dispiralis 双旋三旋扭虫 18（28，29），25（32，33）

T. kretaensis 克兰特三旋扭虫 25（35，36，42）

T. robustispinosa 粗刺三旋扭虫 25（34）

Tubotriassocyrtis fusiformis 纺锤形管状三叠笼虫 25（26）

T. sp. A 管状三叠笼虫（未定种A）19（16）

2. 汉—拉属种名称

十字形斑点球桩虫 *Stigmosphaerostylus cruciformis* 5（12，20—22），7（30—34）

三矛虫（未定种B）*T. sp. B* 7（45）

三角形阿尔拜虫 *Albaillella triangularis* 14（42—49，51，55—57，61—65）

三刺双隐管虫 *Latentibifistula triacanthophora* 4（38—45）

三射三矛虫 *Trilonche tretactinia* 9（31，34，36—40，43）

三棍石贺裕明虫 *Ishigaum trifistis* 12（40，42），13（34），14（4，7—11，19，26，35）

三叠三叠海绵球虫 *Triassospongosphaera triassica* 12（12），19（17—19），20（11—22），22（38—40），25（27，28）

三叠海绵球虫（未定种 A）*T.* sp. A 12（8—11）

三叠塔虫（多未定种）*Triassocampe* spp. 20（40，41）

大壳四内射虫 *Tetrentactinia gigantia* 3（20）

大无畏斑点球桩虫 *Stigmosphaerostylus pantotolma* 3（24）

大刺假桩球虫 *Pseudostylosphaera magnispinosa* 18（14），19（23，32），23（1，3）

门孔虫（未定种 A）*Pylentonema?* sp. A 2（10）

小三矛虫 *Trilonche minax* 9（5）

贝图尔虫（未定种 A）*Beturiella* sp. A 22（28，30）

小刺欣德球虫 *Hindeosphaera spinulosa* 19（27），23（10，12），24（26，29，30）

保林虫（未定种 A）*Paurinella* sp. A 19（15）

小董阿尔拜虫 *Albaillella xiadongensis* 7（3，11—14）

上郡假石片虫 *Pseudotormentus kamigoriensis* 5（44，45，53，54），7（39—43），8（43，44），10（13，14）

不明显假桩球虫 *Pseudostylosphaera imperspicua* 24（7—9）

不等刺假桩球虫 *P. inaequispinosa* 24（1—3，6，12，13）

中世古幸次郎埃普廷虫 *Eptingium nakasekoi* 12（16—23）

中等斑点球桩虫 *Stigmosphaerostylus modestus* 5（1，13，15，16，18，19）

分异斑点球桩虫 *S. diversita* 1（3，4，9，29—31，33—35）

五日市斑点球桩虫 *S. itsukaichiensis* 5（2—11，14，17）

双尖星内射虫 *Astroentactinia biaciculata* 1（11，15，16，20，22，24，25），2（37，39—42，44），17（18—20）

五浆虫（未定种 A）*Quinqueremis* sp. A 19（20）

双刺多内椭圆虫 *Copiellintra diploacantha* 6（29—31）

双刺欣德球虫 *Hindeosphaera bispinosa* 21（20—32），24（27，28，31）

双旋三旋扭虫 *Tritortis dispiralis* 18（28，29），25（32，33）

双管原笼虫 *Archocyrtium diductum* 2（4，6），3（2）

戈斯特林假桩球虫 *Pseudostylosphaera goestlingensis* 23（11），24（21，23）

无翼假阿尔拜虫 *Pseudoalbaillella nonpteryla* 10（28，29，32，37，39，43—48）

日本假桩球虫 *Pseudostylosphaera japonica* 12（25），18（1—9），23（31），24（10，11）．

欠发育海绵内射虫 *Spongentactinia indisserta* 3（16，25）

火焰状卡里特罗伊特虫 *Raciditor phlogidea* 5（32—42），6（2，4）

长全角虫 *Holoeciscus elongatus* 2（1）

长形假阿尔拜虫 *Pseudoalbaillella elongata* 8（1—5），16（1，2）

长刺多内圆虫 *Copicyntra cuspidata* 6（40，45—47）

长刺拟奥特尔海绵虫 *Paroertlispongus longispinosus* 19（1，2，21，22），22（16—20）

长刺假桩球虫 *Pseudostylosphaera longispinosa* 19（28，49），23（32），24（4，5）．

长刺隐王冠虫 *Cryptostephanidium longispinosum* 23（16，28）

长刺斑点球桩虫 *Stigmosphaerostylus proceraspina* 1（32）

卡里特罗伊特虫（未定种 A）*Raciditor* sp. A 16（10）

卡斯塔利格原笼虫 *Archocyrtium castuligerum* 2（2，3）

古网冠虫（未定种 A）*Archaeodictyomitra* sp. A 19（8）

古网冠虫（未定种 B）*A.* sp. B 20（33）

古老斑点球桩虫 *Stigmosphaerostylus vetulus* 5（25—28，31），6（36—38）

可可西里拟彼萨格诺虫 Parasepsagon hohxiliensis 18（32，34，40，46）
左旋刺古海绵梅虫 Archaeospongoprunum sinisterispinosum 6（28）
市川浩一郎斑点球桩虫 Stigmosphaerostylus ichikawai 6（22）
皮特曼三矛虫 Trilonche pittmani 9（2—4）
石贺裕明假阿尔拜虫 Pseudoalbaillella ishigai 10（15，16，18，20—22，24，27，30，31，41）
石贺裕明虫（未定种 A）Ishigaum sp. A 14（30）
鸟形新阿尔拜虫 Neoalbaillella ornithoformis 12（31—34），13（1—5，7，8，11—13，20—22，24—27）
龙潭假阿尔拜虫 Pseudoalbaillella longtanensis 4（1—10）
达维德三矛虫 Trilonche davidi 9（6，7，9，24，29，32，33，33A）
优美阿尔拜虫 Albaillella lauta 14（58）
光壳阿尔拜虫 A. levis 14（50，52—54，59，60）
光瘤条纹三叠塔虫 Striatotriassocampe laeviannulata 25（1—3）
华丽蒂博尔虫 Tiborella florida 12（1—6），19（40—45）
华丽欣德球虫 Hindeosphaera bella 24（25）
华丽斑点球桩虫 Stigmosphaerostylus spiciosus 1（39—41）
华美三矛虫 Trilonche elegans 9（17，35）
后锐边假阿尔拜虫 Pseudoalbaillella postscalprata 15（13—17，19，24）
多孔潘坦内尔虫 Pantanellium multiporum 23（29，30）
多节环三叠塔虫 Annulotriassocampe multisegmantatus 22（1—11），25（4—10）
多节瘤状条纹三叠塔虫 Striatotriassocampe nodosoannulata 25（11—13，19）
多刺星内射虫 Astroentactinia multispinosa 3（26，28—30）
年青三叉球虫 Triaenosphaera hebes 2（21），3（17，21）
曲状阿尔拜虫 Albaillella sinuata 7（15—18），10（9—12）
米劳斯星内射虫 Astroentactinia mirousi 2（38，45）
纤细卡里特罗伊特虫 Raciditor gracilis 6（1，3，5—10），10（7），12（41），14（31—33）
纤细假桩球虫 Pseudostylosphaera gracilis 18（11—13，15—18），21（17，18），23（17，18），24（16—18）
纤细新阿尔拜虫 Neoalbaillella gracilis 13（15—19）
克兰特三旋扭虫 Tritortis kretaensis 25（35，36，42）
坚固拟彼萨格诺虫 Parasepsagon firmum 18（30，31，33，35—39，42—45），21（1—7，14）
拟纤细假桩球虫 Pseudostylosphaera paragracilis 24（14，15，19）.
拟彼萨格诺虫（未定种 A）Parasepsagon sp. A. 19（9）
拟彼萨格诺虫（未定种 B）P. sp. B. 21（9，10）
拟圆形扎莫依达虫 Hozmadia pararotunda 23（19，27）
拟奥特尔海绵虫（未定种 A）Paroertlispongus sp. A 22（21）
纳扎罗夫假桩球虫 Pseudostylosphaera nazarovi 18（3a，10），19（29—31），21（15），24（20，32）
纺锤形假阿尔拜虫 Pseudoalbaillella fusiformis 4（11—16）
纺锤形管状三叠笼虫 Tubotriassocyrtis fusiformis 25（26）
芦苇状五桨虫 Quinqueremis arundinea 8（46—50）
豆荚状假阿尔拜虫 Pseudoalbaillella lomentaria 11（7—18），16（3—9）
乳头状赫格勒虫 Hegleria mammilla 6（39，44）
侧翼隐管虫 Latentifistula patagilaterala 7（19，20，22，25—28），10（2，3）

具枝古蓬虫 *Palaeoscenidium cladophorum* 2（24）

凯维斯卡尔缪勒旋扭虫 *Muelleritortis koeveskalensis* 24（35），25（37，39）

刺三矛虫 *Trilonche echinatum* 9（8，10—16）

单翼假阿尔拜虫 *Pseudoalbaillella monopteryla* 10（17，19，23，25，26，33—36，38，42）

帕罗纳氏星内射虫 *Astroentactinia paronae* 1（21，23，26—28，36—38），2（11，13—15，25）

拉加布赖尔原笼虫 *Archocyrtium lagabriellei* 17（5）

拉劳海绵锯齿虫 *Spongoserrula rarauana* 19（10，11），20（1—10）

空球虫（未定种 A）*Cenosphaera* sp. A 19（3，4）

旺氏原笼虫 *Archocyrtium wonae* 3（1）

武藏三面海绵虫 *Triplanospongos musashiensis* 12（35，36，39），13（35，45—48），14（1—3，5，6）

波状阿尔拜虫 *Albaillella undulata* 17（3）

放射虫（属种不能鉴定）*Radiolariads* gen. et sp. indet. 11（40，42）

细弱斑点球桩虫 *Stigmosphaerostylus gracilentus* 5（23，24，29，30）

股状四茎虫 *Quadricaulis femoris* 5（46—52），16（12）

肥胖石贺裕明虫 *Ishigaum obesum* 14（15，18，23，41）

英德阿尔拜虫 *Albaillella indensis* 17（1，2）

青海三叠海绵球虫 *Triassospongosphaera qinghaiensis* 22（25，34—36）

青海假桩球虫 *Pseudostylosphaera qinghaiensis* 21（19）

奇异门孔虫 *Pylentonema mira* 17（6，7）

冠状三叠塔虫 *Triassocampe coronata* 12（14，15）

前四刺拟彼萨格诺虫 *Parasepsagon praetetracanthus* 12（7）

剑三叉球虫 *Triaenosphaera sicarius* 2（19，20，22），9（19—21，25，30，41，42）

南畔三叠塔虫 *Triassocampe nanpanensis* 12（13）

变异拟彼萨格诺虫 *Parasepsagon variabilis* 19（33），23（14）

变刺贝图尔虫 *Beturiella variospinosa* 22（26，27，29，32）

变刺斑点球桩虫 *Stigmosphaerostylus variospina* 1（1，10），3（5，13—15，19，22），9（26—28），17（4）

带角隐王冠虫 *Cryptostephanidium cornigerum* 18（22，23，25），21（33—36），23（13），25（41）

扁圆形卡里特罗伊特虫 *Raciditor oblatum* 5（43），6（11—19），7（44），16（11，15—17，20）

星状星内射虫 *Astroentactinia stellata* 1（12—14，17—19），2（16）

秋川多内圆虫 *Copicyntra akikawaensis* 6（33），13（38—43）

美丽拟奥特尔海绵虫 *Paroertlispongus opiparus* 22（22，23）

荒谬原笼虫 *Archocyrtium ludicrum* 2（7）

原笼虫（未定种 A）*Archocyrtium* sp. A 3（3，4）

圆锥形隐管虫 *Latentifistula conica* 11（1—4），16（14，18）

宽角三叠王冠虫 *Triassistephanidium laticornis* 18（24，26，41）

宽刺三叠海绵球虫 *Triassospongosphaera latispinosa* 22（37）

宽刺保林虫 *Paurinella latispinosa* 25（29，30）

弱刺海绵内射虫 *Spongentactinia exilispina* 2（26，32—36）

格状石贺裕明虫 *Ishigaum craticula* 14（12，16，21，22，27—29，34，36，38）

海绵四内射虫 *Tetrentactinia spongacea* 2（12，18，23），9（18）

海绵内射虫（未定种 A）*Spongentactinia* sp. A 3（6—11）

海绵内射虫（未定种 B）*S.* sp. B 1（2，5，6）
海绵内射虫（未定种 C）*S.* sp. C 1（7，8），3（12）
海绵状海绵内射虫 *S. spongites* 3（18，23）
海绵质海绵王冠虫 *Spongostephanidium spongiosum* 18（27）
特别环三叠塔虫 *Annulotriassocampe proprium* 22（12）
窄原笼虫 *Archocyrtium strictum* 2（5）
致密假桩球虫 *P. compacta* 12（24，26—30），18（20，21），19（12，24，25，47，48），21（16），23（2，4—7）
袖状考勒特虫 *Cauletella manica* 13（37）
铁色原笼虫 *Archocyrtium ferreum* 2（8，9）
假美丽三矛虫 *Trilonche pseudocimelia* 12（37），14（13，14）
假桩球虫（未定种 A）*Pseudostylosphaera* sp. A 21（8）
假桩球虫（未定种 C）*P.* sp. C 21（12，13）
假桩球虫（未定种 D）*P.* sp. D 23（8）
假鹰钩新阿尔拜虫 *Neoalbaillella pseudogrypa* 13（14）
匙形缪勒旋扭虫 *Muelleritortis cochleata* 25（38，40）
曼弗雷德埃普廷虫 *Eptingium manfredi* 19（34—39），23（15，20—26）
梯形三叠塔虫 *Triassocampe scalaris* 25（21，25）
清道古海绵梅虫 *Archaeospongoprunum chiangdaoensis* 12（38）
球形古海绵梅虫 *A. globosum* 23（33，34）
球形假阿尔拜虫 *Pseudoalbaillella globosa* 3（42—44，47—49），4（17—30）
理塘假阿尔拜虫 *P. litangensis* 3（31—40）
盖坦尼拟三叠塔虫 *Paratriassocampe gaetanii* 25（14，15，18）
章鱼虫（未定种 A）*Poulpus* sp. A 23（35）
粗刺三叉球虫 *Triaenosphaera robustispina* 9（22，23）
粗刺三旋扭虫 *Trilortis robustispinosa* 25（34）
粗海绵双隐管虫 *Latentibifistula asperspongiosa* 8（41，42，45）
菱形假阿尔拜虫 *Pseudoalbaillella rhombothoracata* 7（4—10），8（6—17，35）
萨克马尔假阿尔拜虫 *P. sakmarensis* 8（18—32），11（5，6，19—39，41），15（1—9）
萨洛夫环三叠塔虫 *Annulotriassocampe sulovensis* 19（5—7），20（30，36，38，39），22（13），25（16，17，20，23）
隐王冠虫（未定种 A）*Cryptostephanidium* sp. A 21（11，37）
隐管虫（未定种 A）*Latentifistula* sp. A 7（21）
隐管虫（未定种 B）*L.* sp. B 10（8）
强壮贝图里虫 *Beturiella robusta* 22（24，31）
强壮五桨虫 *Quinqueremis robusta* 6（20，21），16（19）
强壮奥米斯顿虫 *Ormistonella robusta* 14（17，24，25，37，40）
普通斑点球桩虫 *Stigmosphaerostylus vulgaris* 17（8—15，17）
最优新阿尔拜虫 *Neoalbaillella optima* 13（10，24）
短壳拟三叠塔虫 *Paratriassocampe brevis* 22（14，15）
短刺三叠海绵球虫 *Triassospongosphaera brevispinosa* 22（33，33A）
等刺保林虫 *Paurinella aequispinosa* 19（13，14），20（23—25），25（31）
斑点球桩虫（未定种 A）*Stigmosphaerostylus* sp. A 6（25，32），7（35—38）
斑点球桩虫（未定种 B）*S.* sp. B 6（23，24，26，27）
斑点球桩虫（未定种 C）*S.* sp. C 6（34，35，46，47）
斑点球桩虫（未定种 D）*S.* sp. D 16（21），21（12，13）

锐边假阿尔拜虫 *Pseudoalbaillella scalprata* 3（41，45，46，50），4（31—37），15（10—12，18，20—23）
微小四桨虫 *Quadriremis minima* 7（46）
新环三叠塔虫 *Annulotriassocampe nova* 20（26—28，31）
碎屑状斑点球桩虫 *Stigmosphaerostylus micula* 2（27—31，43，45），3（27）
简单假阿尔拜虫 *Pseudoalbaillella simplex* 8（33，34，36—40）
管状三叠笼虫（未定种A）*Tubotriassocyrtis* sp. A 19（16）
蜘蛛多内射虫 *Polyentactinia aranea* 2（17）
颗石柱假桩球虫 *Pseudostylosphaera coccostyla* 18（19）
德韦弗三叠塔虫 *Triassocampe deweveri* 20（29，32，34，35，37），25（22，24）
德克萨斯隐管虫 *Latentifistula texana* 7（23，24），10（1，4，5），16（13）
潘坦内尔虫（未定种A）*Pantanellium* sp. A 24（22，24）
膨大缪勒旋扭虫 *Muelleritortis expansa* 24（33，34）
膨胀卡里特罗伊特虫 *Raciditor inflata* 10（6），14（20，39）
螺旋状假桩球虫 *Pseudostylosphaera helicata* 23（9）

英 文 摘 要
(ENGLISH SUMMARY)

Late Palaeozoic and Early Mesozoic Radiolarian Palaeontology and Stratigraphy of Qiangtang Region

Wang Yujing Sha Jingeng Zhu Tongxing
Guo Tongzhen and Xie Chaoming

Introduction

When did the Palaeo-Tethys Ocean close? What age did the Neo-Tethys Ocean open? These two major scientific issues related to the evolution of Tethys have been concerned and debated for a long time by geoscientists. There exist two suture zones in northern Qinghai-Xizang/Tibet Plateau, i. e., the Longmuco (co=lake) -Shuanghu (hu=lake) suture zone in central Qiangtang and Longmuco-Yushu (Jinshajiang) suture zone in north Qiangtang, in which the Late Palaeozoic—Early Mesozoic ophiolites yielding radiolarian fossils and radiolarian bedded cherts are recorded. These two suture zones are therefore the rare archives for studying the evolution of Tethys.

After around a thirty-year field investigation and laboratory analysis, the authors, Prof. Yujing Wang et al., found the Late Devonian—Middle Triassic radiolarian faunas from Kangtejin of the north side of Gangqiqu, Yishanco of Xijir Ulan Hu (Xijinwulanco, co=lake), north side of Xijir Ulan Hu (Xijinwulanco), Shexinggou of the south side of Hantaishan; Northwest side of the Xijir Ulan Hu (Xijinwulanco) tectonic mélange zone, southeast of Yanghu, north shore of Cêdo Caka (Caiduochaka), Yaqu village, north of Timachaka, northeast of Rongma, Jiaomuri, south of Caimaerco, north of Caimaerco (see in Fig. 1 of Introduction 1) in the two suture zones above.

In this monograph, the Late Palaeozoic—Early Mesozoic stratigraphic systems of Qiangtang region (Chapter 1) and the Late Palaeozoic—Early Mesozoic radiolarian formations (Chapter 2) were summarized, described and correlated; the Late Devonian—Middle Triassic radiolarian faunas, from the Longmuco-Shuanghu and Longmuco-Yushu suture zones, composed of 25 families, 53 genera and 155 species (including 25 new species) (Chapters 4 and 6) were systematically studied and elaborately described, and 14 radiolarian fossil zones were established and locally, nationally and globally correlated (Chapters 3 and 5). At last, the authors, Yujing Wang et al., concluded that the Palaeo-Tethys Ocean closed no later than the end-Permian and the Neo-Tethys Ocean did not open until early Middle Triassic Anisian, rather than Permian, and suggested that the Longmuco-Shuanghu-Lancangjiang suture zone is also a branch of the northern main suture zone, after the Longmuco-Yushu-Jinshajiang suture zone (Chapters 7 and 8).

The contents of stratigraphy, systematic taxonomy of radiolarian faunas, radiolarian zonation and correlation, and the palaeogeographic implications of Longmuco-Shuanghu and Longmuco-Yushu (Jinshajiang) suture zones were summarized and the new taxa of radiolarian fossils were described in the following.

1 Late Palaeozoic and Early Mesozoic stratigraphic system in Qiangtang region

According to the boundary of the southern and northern deep faults in the Longmuco-Shuanghu suture zone the Qiangtang stratigraphic realm is divided into three stratigraphic regions from south to north, namely the south Qiangtang stratigraphic region, the central Qiangtang tectonic stratigraphic region and the north Qiangtang stratigraphic region. Their correlation of the stratigraphy is showed in Fig. 1.1 and

Table 1.1 of Chapter 1.

A set of the conglomerate-bearing slate in the iceborne sediment and the cold bivalvids *Eurydesma* fauna of the Gondwana facies are found in the Late Carboniferous Cameng Formation and in the Early Permian Zhanjin Formation in the south Qiangtang stratigraphic region respectively. These data have proved that the relation between the south Qiangtang stratigraphic region and the south Gondwana stratigraphic realm is very close.

The Fusulinid *Palaeofusulina* fauna, including *P. nana*, *P. sinensis*, *P. fusiformis* etc. is discovered in the limestones of the Rejuecaka Formation in the north Qiangtang stratigraphic region. This fusulinid fauna is a most high fusulinid zone of Permian Changhsingian Formation in South China and is widely distributed in marine strata of the Palaeotethys.

The Cathaysian flora, characterized by the representative of *Gigantonoclea guizhouensis* found in the clasitic rocks of this Formation is a typical and distributed westward mostly Tethyan flora developed in a warm climata. These materials have demonstrated that the paleogeographic unit between the north Qiangtang stratigraphic region and South China is the same.

The Late Palaeozoic and Early Mesozoic five radiolarian faunas found in the cherts of the ophiolite-melange zone in the central Qiangtang tectonic stratigraphic region may entirely compare with contemporary radiolarian zones in the cherts of Hoh Xil ophiolite-mélange zone and of the Changni-Menlian one of South China as well.

2 Late Palaeozoic and Early Mesozoic stratigraphic sections containing radiolarian fossils in Shuanghu of north Tibet and Hoh Xil of Qinghai

2.1 Devonian section of Yishanco in Hoh Xil of Qinghai

The section contains *Holoeciscus foremanae* fauna sample ($8P_2W16-2$, $8P_2W22-2$).

2.2 Northwest Xijinwulanco section in Hoh Xil of Qinghai

The section yields *Albaillella xiaodongensis* fauna sample (W5031), *Pseudoalbaillella globosa* fauna sample (W5032), *Annulotriassocampe multisegmantatum* fauna sample (Bb8311-1) and *Spongoserrula rarauana* fauna sample (8PW5418-2).

2.3 Shexingou section in Hoh Xil of Qinghai

It contains *Pseudoalbaillella sakmarensis-P. lomentaria* fauna sample (b90-125, b90-129), and *P. rhombothoracata* fauna sample (8PFS5014).

2.4 Kangtejin section of Gangqiqu in Hoh Xil of Qinghai

The section contains *Pseudoalbaillella rhombothoracata* fauna sample (b90-45) and *P. sakmarensis-P. lomentaria* fauna sample (FKGK1).

2.5 North Xijinwulanco section in Hoh Xil of Qinghai

It contains *Tiborella florida* fauna sample (8PFS6, 8PFS7), *Astroentactinia multispinosa* fauna sample (b90-64), *Albaillella indensis* fauna sample (FKXG1) and *Cyrtisphaeractenium* aff. *crassium* fauna sample (b90-69).

2.6 Cêdo Caka section in Shuanghu of north Tibet

Two faunas are recorded in this section, *Helenifore robustum* fauna sample (CDP14WF1) and

Neoalbaillella ornithoformis fauna sample (D9002WF3-2, D9002WF1).

2.7 Cherts of Yaqu Village in Shuanghu of north Tibet

The radiolarians of this section consist of two faunas, *Pseudoalbaillella ishigai* fauna sample (D7052WF1) and *P. sakmarensis-P. lomentaria* fauna sample (D7052WF3).

2.8 Cherts in tectonic-melange zone in Litang of Sichuan

This section contains *Pseudoalbaillella globosa* fauna sample (ZD305-90b) and *Muelleritortis cochleata* fauna sample (Z144b) come from this section.

2.9 Basalt section of Jiaomuri in central Qiangtang of north Tibet

Only one fauna *Eptingium nakasekoi* fauna sample (D1339WF1) has been recognized from this section.

3 Composition, age and correlation of Late Palaeozoic radiolarian faunas in Shuanghu of north Tibet and Hoh Xil of Qinghai

3.1 Palaeozoic radiolarian faunas and their ages of Shuanghu of north Tibet and Hoh Xil of Qinghai

3.1.1 Radiolarian fauna in sample $8P_2W22-2$ from Hoh Xil of Qinghai

This radiolarian fauna contains 9 genera and 22 species (including 1 new species, *Stigmosphaerostylus spiciosus* Wang), belonging to 4 orders, 6 superfamilies and 8 families. In which 3 genera, *Astroentactinia*, *Stigmosphaerostylus* and *Archocyrtium*, are numerous in species number, about sixty percentage of the fauna. The species of this fauna and their geological ranges are listed in Table 3.1 of Chapter 3.

Although the Famennian index fossil *Holoeciscus foremanae* has not been found in this fauna, the main members, such as *Holoeciscus elongates*, *Spongentactinia exilispina*, *S. spongites*, *Tetrentactinia spongacea*, *Astroentactinia paronae*, *A. stellata*, *Polyentactinia aranea*, *Archocyrtium diductum*, *A. strictum*, *Triaenosphaera sicarius* all are commonly associated fossils of the *Holoeciscus foremanae* zone. This fauna therefore, is considered as the *Holoeciscus foremanae* zone, having a range of Late Devonian Famennian time. This zone could completely correlate to the same zone of Australia, the west coast of America, Malaysia, Thailand, Germany and Guangxi, Yunnan and Xingjiang of China.

3.1.2 Radiolarian fauna in sample $8P_2W16-2$ from Hoh Xil of Qinghai

This radiolarian fauna is composed of 6 genera, 10 species (including 1 new species, *Tetrentactinia gigantia* Wang) (see Table 3.2 of Chapter 3), belonging to 3 orders, 3 superfamilies and 5 families. Except for a new species and *Astroentactinia multispinosa*, the other 8 species are all same as the radiolarian taxa of sample $8P_2W22-2$ (see Table 3.1 of Chapter 3). Therefore, we classified this radiolarian fauna in the *Holoeciscus foremanae* zone, of which age is Late Devonian Famennian.

3.1.3 Radiolarian fauna in sample W5032 from Hoh Xil of Qinghai

This radiolarian fauna composed of 3 radiolarian forms of pseudoalbaillellids, latentifistullids and entactinids is very high in abundance and diversity, including 11 genera, 21 species (including 5 new species, *Archaeospongoprunum sinisterispinosum* Wang, *Stigmosphaerostylus cruciformis* Wang, *S. gracilentus* Wang, *S. vetulus* Wang, and *Raciditor obtatum* Wang). The genera and species of this radiolarian fauna and their geographic ranges are listed in Table 3.3 of Chapter 3. The *Pseudoalbaillella globosa* is a most important fossil in this fauna. It is distributed in all over the world and has already became an index zonal fossil of Middle Permian early Guadalupian time. Thus we named this fauna as the *Pseudoalbaillella globosa* zone, a global radiolarian zone.

3.1.4 Radiolarian fauna in sample W5031 from Hoh Xil of Qinghai

This radiolarian fauna consists of 9 genera, 11 species (including 2 new species, *Stigmosphaerostylus cruciformis* Wang and *Raciditor oblatum* Wang), attributed to 4 orders, 5 superfamilies and 8 families. The taxa and the geographic ranges of genera and species of this fauna are listed in Table 3.4 of Chapter 3. The species *Albaillella xiaodongensis* Wang found in the cherty strata of southeast Guangxi and of southwest Japan is an important fossil in this fauna. Owing to short living range of this species, it has already been erected as the *Albaillella xiaodongensis* zone by Wang (Wang et al., 2006, 2007, 2012), below the *Albaillella sinuata* zone. Therefore, we see this fauna as the *Albaillella xiaodongensis* zone. It is early chihsian stage of Middle Permian in age, comparable to the early Leonardian stage of USA and Artinskia stage of Russia.

3.1.5 Radiolarian fauna in sample 8PFS5014 from Hoh Xil of Qinghai

This radiolarian fauna, including 4 genera, 7 species, represents 2 order, 3 superfamilies and 4 families. The list and range of the genera and species of this fauna see in Table 3.5 of Chapter 3.

The two most important species, *Pseudoalbaillella rhombothoracata* and *P. elongata* in this fauna have also been discoveried in Early Permian latest Wolfcampian strata of Japan, Malaysia, the East Far of Russia, Sisili island of Italy and South China. The species *P. rhombothoracata* has already been became an index zonal fossil of Wolfcampian. We have also regarded this fauna as the *P. rhombothoracata* zone which definitely can be correlated to the same zone of Japan, Malaysia, the East Far of Russia and South China.

3.1.6 Radiolarian fauna in sample CDP14WF1 of Tibetan Cêdo Caka

This radiolarian fauna consisting of 4 genera, 10 species (including 1 new species, *Triaenosphaera robustispina* Wang) has been placed into 2 order, 3 superfamilies and 3 families. The list and range of the genera and species in this fauna see in Table 3.6 chapter 3. The genus *Trilonche* is one most important fossil in this fauna. It is composed of 6 species which 5 (*T. davidi*, *T. echinata*, *T. elegans*, *T. minax* and *T. pittmani*) have been recorded in Late Devonian Farsnian strata of Australia, America, Southern Ural and Rudny-Altay of Russia and Xinjiang of China. Although the *Helenifore laticlavilum* and the *H. robustum* of Frasnian index fossils are not found in this fauna we have temporarily termed this fauna as the *H. robustum* zone, which could be correlate to same zone found in the locations mentioned above.

3.1.7 Radiolarian fauna in sample D7052WF1 of Tibetan Yaqu

This radiolarian fauna contains 5 genera, 9 species (including 2 new species, *Pseudoalbaillella monopteryla* Wang and *P. nonpteryla* Wang). The list and range of the genera and species of this fauna see in Table 3.7 of Chapter 3. The species *Pseudoalbaillella ishigai* is a critical and abundant taxon in this fauna. It has also been found in Middle Permian late Chihsian strata of South China, and these became a zonal fossil of this time (Wang et al., 2006, 2007, 2012). We acknowledged this fauna as the *Pseudoalbaillella ishigai* zone. It can be completely correlated to same zone of South China, and to the zones of *P. longtanensis* of Japan (Ishiga, 1990, Yao and Kuwahara, 2004), *Pseudoalbaillella corniculata* (= *P. longtanensis*) (Rudenko and Panasenko, 1997), and *P.* sp. C of America (Blome and Reed,

1995), as well.

3.1.8 Radiolarian fauna in sample D7052WF3 of Tibetan Yaqu

This radiolarian fauna contains 2 genera, 3 species (including a new species *Latentifistula conica* Wang). They are belonged to 2 orders, 2 superfamilies and 2 fanilies.

The two species of the genus *Pseudoalbaillella*, *P. sakmarensis* and *P. lomentaria* making up ninety-five percent of this fauna are discoveried in Early Wolfcampian strata of Japan, Oregon of America, Malaysia, Thailand, south Ural and the East Far of Russia, South China and Hoh Xil area, Qinghai of China. The species *P. lomentaria* is already considered as an index zonal fossil of Early Permian middle Sakamotozawan time by Japanese students (Ishiga, 1986, 1990, Yao and Kuwahara, 2004, Shimakawa and Yao, 2006). Owing to the coexistence of these two species, we named this fauna as the *Pseudoalbaillella sakmarensis-P. lomentaria* zone representing the Early Permian early Longlinian in time. It could correlated to the *P. lomentaria* zone of Japan (Ishiga, 1990), Malaysia (Spiller, 2002), Thailand (Sashida et al., 1998, 2000, Wonganan and Caridroit, 2007) and west coast of America (Blome and Reed, 1992, 1995), and to the *P. sakmarensis* zone of Russia (Rudenko and Panasendo, 1997), and to the same zone of South China and Hoh Xil region, Qinghai of China.

3.1.9 Radiolarian fauna in sample D9002WF3-2 of Tibetan Cêdo Caka

This radiolarian fauna with high abundance and diversity of genera and species contains 9 genera, 16 species, representing in 4 orders, 4 superfamilies and 5 families. The list and range of the genera and species in this fauna see in Table 3.8 of Chapter 3.

The two genera *Neoalbaillella* and *Albaillella* are the most important fossil in this fauna. The former is composed of 4 species and one unidentified species, while the latter consists of 3 species.

The species *Neoalbaillella ornithoformis* is numerous in individual number in this fauna. It has been found in Late Permian late Changhsingian strata of Japan, the Philippines, Malaysia, Thailand, the East Far of Russia and South China, and has already became an index zonoal fossil of this time. We referred this fauna to the *Neoalbaillella ormithoformis* zone which can be correlated to the same zone of countries mentioned above.

3.1.10 Radiolarian fauna in sample D9002WF1 of Tibetan Cêdo Caka

This radiolarian fauna includes 6 genera, 6 species. They are placed in 4 orders, 5 superfamilies and 5 families. The list and range of the genera and species in this fauna see in Table 3.9 of Chapter 3.

The two species *Neoalbaillella ornithoformis* and *Archaeospongoprunum chiangdaoensis* are dominant fossils in this fauna. Besides Tibet, they occurr in Late Permian late Changhsingian strata of Japan, Thailand, the East Far of Russia and South China. This fauna is similar to the fauna with *N. ornithoformis* above. It is therefor also known as *N. ornithoformis* zone which can be correlated to the same zone of the countries mentioned above.

3.1.11 Radiolarian fauna in sample FKGK1 from Kongtejin in Gangqiqu of Qinghai

This radiolarian fauna containing 5 genera, 10 species (including 2 new species *Latentifistula conica* Wang and *Raciditor oblatum* Wang) has been placed in 2 orders, 2 superfamilies and 3 families. List and range of the genera and species in this fauna see in Table 3.10 of Chapter 3.

The genus *Pseudoalbaillella* consisting of 5 species is a most important genus of the fauna, since there are about fifty percent species of the fauna belong to the genus.

The species *Pseudoalbaillella sakmarensis* and *P. lomentaria* are frequently associated with each other in Early Wolfcampian strata of Japan, Oregon of America, Malaysia, Thailand, south Ural area and the East Far of Russia, South China and Tibetan Shuanghu area of China. We also see this fauna to be *P. lomentaria-P. sakmarensis* zone. Its range and compoaition are same as those of the same zone of Tibetan Yaqu sample D7052WF3.

3.1.12 Radiolarian fauna in sample FKXG1 from Yishanco of Qinghai

It contains 5 genera, 7 species, belonging to 3 orders, 3 superfamilies and 4 families. The species and their ranges see in Table 3.11 of Chapter 3.

The species *Albaillella indensis* is a critical taxon in this fauna. It is an index zonal fossil of the Early Carboniferous late Tournaisian—early Visean, and has been found in Germany, France, Australia, Alaska of America, and South China and Qinghai. We entitled this fauna as the *Albaillella indensis* zone which can be correlated to the same zone of countries mentioned above.

3.1.13 Radiolarian fauna in sample ZD305-90b1 from Litang of Sichuan

This radiolarian fauna is monotonous in genera and species, only composed of one genus and 3 species (including one new species *Pseudoalbaillella litangensis* Wang), representing into one order, one superfamily and one Family. The species *Pseudoalbaillella globosa* Ishiga et Imoto is distributed in all over the world and has already been regarded as an index zonal species of Middle Permian early Maokouan time. We named this fauna as the *P. globosa* zone which can completely correlated to the same radiolarian zone of all over the world.

3.2 Correlation of radiolarian faunas in Shuanghu of north Tibet and Hoh Xil of Qinghai

There are 14 Late Palaeozoic radiolarian zones which were recognized so far in Shuanghu area of N Tibet and Hoh Xil region of Qinghai. Their correlations have been briefly summarized in section 3.1 above, and the global correlation of the Late Devonian—Early Carboniferous radiolarian zones from Qiangtang region see in Table 3.13 of Chapter 3.

4 Systematic classification and description of new species of Late Palaeozoic radiolarians

Subphylum Radiolaria Müller, 1858
  Superorder Polycystina Ehrenberg, 1838, emend. Riedel, 1967
    Order Albaillellaria Deflandre, 1953, emend. Holdsworth, 1969
      Superfamily Palaeoscenidioidea Riedel, 1967, emend. Nazarov et Rudenko, 1981
        Family Palaeosceniidae Riedel, 1967, emend. Nazarov et Rudenko, 1981
          Subfamily Palaeosceniinae Riedel, 1967, emend. Nazarov et Rudenko, 1981
            Genus *Palaeoscenidium* Deflandre, 1953
              Palaeoscenidium cladophorum Deflandre, 1953, 2 (24)
      Superfamily Ceratoikiscoidea Holdsworth, 1969
        Family Lapidopiscidae Deflandre, 1958

Subfamily Lapidopiscinae Deflandre, 1958
  Genus *Holoeciscus* Foreman, 1963
    Holoeciscus elongatus Kiessling et Tragehn, 1994, 2 (1)
Superfamily Albaillellacea Cheng, 1986
  Family Albaillellidae Cheng, 1986
  Subfamily Albaillellinae Cheng, 1986
    Genus *Albaillella* Deflandre, 1953
      Albaillella indensis Won, 1983, 17 (1, 2)
      A. lauta Kuwahara, 1992, 14 (58)
      A. levis Ishiga, Kito et Imoto, 1982, 14 (50, 52—54, 59, 60)
      A. sinuata Ishiga et Watase, 1986, 7 (15—18), 10 (9—12)
      A. triangularis Ishiga, Kito et Imoto, 1982, 14 (42—49, 51, 55—57, 61—65)
      A. undulata Deflandre, 1952, 17 (3)
      A. xiaodongensis Wang, 1994, 7 (3, 11—14)
Superfamily Follicucullacea Cheng, 1986
  Family Pseudoalbaillellidae Cheng, 1986
    Genus *Pseudoalbaillella* Holdswoth et Jones, 1980
      Pseudoalbaillella elongata Ishiga et Imoto, 1980, 8 (1—5), 16 (1, 2)
      P. fusiformis (Holdsworth et Jones), 1980, 4 (11—16)
      P. globosa Ishiga et Imoto, 1982, 3 (42—44, 47—49), 4 (17—30)
      P. ishigai Wang, 1994, 10 (15, 16, 18, 20—22, 24, 27, 30, 31, 41)
      P. litangensis Wang sp. nov., 3 (31—40)
      P. lomentaria Ishiga et Imoto, 1980, 11 (7—18), 16 (3—9)
      P. longtanensis Sheng et Wang, 1985, 4 (1—10)
      P. monopteryla Wang sp. nov., 10 (17, 19, 23, 25, 26, 33—36, 38, 40, 42)
      P. nonpteryla Wang sp. nov., 10 (28, 29, 32, 37, 39, 43—48)
      P. postscalprata Ishiga, 1983, 15 (13—17, 19, 24)
      P. rhombothoracata Ishiga et Imoto, 1980, 7 (4—10), 8 (6—17, 35)
      P. sakmarensis (Kozur), 1981, 8 (18—32), 11 (5, 6, 19—39, 41), 15 (1—9)
      P. scalprata Holdsworth et Jones, 1980, 3 (41, 45, 46, 50), 4 (31—37), 15 (10—12, 18, 20—23)
      P. simplex Ishiga et Imoto, 1980, 8 (33, 34, 36—40)
  Family Neoalbaillellidae Cheng, 1986
    Genus *Neoalbaillella* Takemura et Nakaseko, 1981
      Neoalbaillella gracilis Takemura et Nakaseko, 1981, 13 (15—19)
      N. optima Ishiga, Kito et Imoto, 1982, 13 (10, 24)
      N. ornithoformis Takemura et Nakaseko, 1981, 12 (31—34), 13 (1—5, 7, 8, 11—13, 20—22, 24—27)
      N. pseudogrypa Sashida et Tonishi, 1988, 13 (14)

Order Latentifistularia Caridroit, De Wever et Dumitrica, 1999
　Superfamily Latentifistuloidea Nazarov et Ormiston, 1999
　　Family Latentifistulidae Nazarov et Ormiston, 1983
　　　Subfamily Latentifistulinae Nazarov et Ormiston, 1983
　　　　Genus *Latentifistula* Nazarov et Ormiston, 1983
　　　　　　Latentifistula conica Wang sp. nov., 11 (1—4), 16 (14, 18)
　　　　　　L. patagilaterala Nazarov et Ormiston, 1985, 7 (19, 20, 22, 25—28), 10 (2, 3)
　　　　　　L. sp. A, 7 (21)
　　　　　　L. sp. B, 10 (8)
　　　　　　L. texana Nazarov et Ormiston, 1985, 7 (23, 24), 10 (1, 4, 5), 16 (13)
　　　Subfamily Latentibifistulinae Afanasieva, 2000
　　　　Genus *Latentibifistula* Nazarov et Ormiston, 1983
　　　　　　Latentibifistula asperspongiosa Sashida et Tonishi, 1986, 8 (41, 42, 45)
　　　　　　L. triacanthophora Nazarov et Ormiston, 1983, 4 (38—45)
　　Family Cauletellidae Caridroit, De Wever et Dumitrica, 1999
　　　Genas *Cauletella* Caridroit, De Wever et Dumitrica, 1999
　　　　　　Cauletella manica (De Wever et Caridroit), 1984, 13 (37)
　　　Genus *Triplanospongos* Sashida et Tonishi, 1988
　　　　　　Triplanospongos musashiensis Sashida et Tonishi, 1988, 12 (35, 36, 39), 13 (35, 45—48), 14 (1—3, 5, 6)
　　　Genus *Ishigaum* De Wever et Caridroit, 1984
　　　　　　Ishigaum craticula Shang, Caridroit et Wang, 2001, 14 (12, 16, 21, 22, 27—29, 34, 36, 38)
　　　　　　I. obesum De Wever et Caridroit, 1984, 13 (36), 14 (15, 18, 23, 41)
　　　　　　I. trifistis De Wever et Caridroit, 1984, 12 (40, 42), 13 (34), 14 (4, 7—11, 19, 26, 35)
　　　　　　I. sp. A Wang et Li, 14 (30)
　　Family Quadriremidae Afanasieva, 2000
　　　Subfamily Quadrireminae Afanasieva, 2000
　　　　Genus *Quadriremis* Nazarov et Ormiston, 1985
　　　　　　Quadriremis minima Nazarov et Ormiston, 1985, 7 (46)
　Superfamily Ruzhencevispongoidea Nazarov et Ormiston, 1983
　　Family Tormentidae Nazarov et Ormiston, 1983
　　　Genus *Pseudotormentus* De Wever et Caridroit, 1984
　　　　　　Pseudotormentus kamigoriensis De Wever et Caridroit, 1984, 5 (44, 45, 53, 54), 7 (39—43), 8 (43, 44), 10 (13, 14)
　　Family Ormistonellidae De Wever et Caridroit, 1984, emend. De Wever et al., 2001
　　　Subfamily Ormistonellinae De Wever et Caridroit, 1984, emend. De Wever et al., 2001
　　　　Genus *Ormistonella* De Wever et Caridroit, 1984

      Ormistonella robusta De Wever et Caridroit, 1984, 14 (17, 24, 25, 37, 40)
    Genus *Raciditor* Sugiyama, 2000
      Raciditor gracilis (De Wever et Caridroit), 1984, 6 (1, 3, 5—10), 10 (7), 12 (41), 14 (31—33)
      R. inflata (Sashida et Tonishi), 1986, 10 (6), 14 (20, 39)
      R. oblatum Wang sp. nov., 5 (43), 6 (11—19), 7 (44), 16 (11, 15—17, 20)
      R. phlogidea (Wang), 1994, 5 (32—42), 6 (2, 4)
      R. sp. A, 16 (10)
    Genus *Quadricaulis* Caridroit et De Wever, 1986
      Quadricaulis femoris Caridroit et De Wever, 5 (46-52), 16 (12)
    Genus *Quinqueremis* Nazarov et Ormiston, 1983
      Quinqueremis robusta Nazarov et Ormiston, 1985, 6 (20, 21), 16 (19)
      Q. arundinea Nazarov et Ormiston, 1983, 8 (46, 50)
      Q. sp. A, 19 (20)
Order Entactinaria Kozur et Mostler, 1982
 Superfamily Entactinioidea Riedel, 1967
  Family Entactiniidae Riedel, 1967, emend. Nazarov, 1975
   Subfamily Entactiniinae Riedel, 1967, emend. Nazarov, 1975
    Genus *Stigmosphaerostylus* Rüst, 1892, emend. Foreman, 1963, Aitchison et Stratford, 1997
      Stigmosphaerostylus cruciformis Wang sp. nov., 5 (12, 20, 22), 7 (30, 34)
      S. diversitus (Nazarov), 1973, 1 (3, 4, 9, 29—31, 33—35)
      S. gracilentus Wang sp. nov., 5 (23, 24, 29, 30)
      S. ichikawai (Caridroit et De Wever), 1984, 6 (22)
      S. itsukaichiensis (Sashida et Tonishi), 1985, 5 (2—11, 14, 17)
      S. micula (Foreman), 1963, 2 (27—31, 43, 45), 3 (27)
      S. modestus (Sashida et Tonishi), 1985, 5 (1, 13, 15, 16, 18, 19)
      S. pantotolma (Braun), 1989, 3 (24)
      S. proceraspina (Aitchison), 1993, 1 (32)
      S. spiciosus Wang sp. nov., 1 (39—41)
      S. sp. A, 6 (25, 32), 7 (35—38)
      S. sp. B, 6 (23, 24, 26, 27)
      S. sp. C, 6 (34, 35, 46, 47)
      S. sp. D, 16 (21)
      S. variospina (Won), 1983, 1 (1, 10), 3 (5, 13—15, 19, 22), 9 (26—28), 17 (4)
      S. vetulus Wang sp. nov., 5 (25—28, 31), 6 (36—38)
      S. vulgaris (Won), 1983, 17 (8—15, 17)
    Genus *Trilonche* Hinde, 1899, emend. Foreman, 1963, Aitchison et Stratford, 1997

Trilonche davidi (Hinde), 1899, 9 (6, 7, 9, 24, 29, 32, 33, 33A)

T. echinatum (Hinde), 1899, 9 (8, 10—16)

T. elegans Hinde, 1899, 9 (17, 35)

T. minax (Hinde), 1899, 9 (5)

T. pittmani Hinde, 1899, 9 (2—4)

T. pseudocimelia (Sashida et Tonishi), 1988, 12 (37), 14 (13, 14)

T. sp. B, 7 (45)

T. tretactinia (Foreman), 1963, 9 (31, 34, 36—40, 43)

Family Astroentactiniidae Nazarov et Ormiston, 1985

Subfamily Astroentactiniinae Nazarov et Ormiston, 1985

Genus *Astroentactinia* Nazarov, 1975

Astroentactinia biaciculata Nazarov, 1975, 1 (11, 15, 16, 20, 22, 24, 25), 2 (37, 39—42, 44), 17 (18—20)

A. multispinosa (Won), 1983, 3 (26, 28—30)

A. paronae (Hinde), 1899, 1 (21, 23, 26—28, 36—38), 2 (11, 13—15, 25)

A. stellata Nazarov, 1975, 1 (12—14, 17—19), 2 (16)

Order Sphaerellaria Haeckel, 1887

Superfamily Polyentactinioidea Nazarov, 1975

Family Polyentactiniidae Nazarov, 1975

Subfamily Polyentactiniinae Nazarov, 1975

Genus *Polyentactinia* Foreman, 1963

Polyentactinia aranea Gourmelon, 1986, 2 (17)

Family Haplentactiniidae Nazarov, 1980

Subfamily Haplentactiniinae Nazarov, 1980

Genus *Triaenosphaera* Deflandre, 1963

Triaenosphaera hebes Won, 1983, 2 (21), 3 (17, 21)

T. robustispina Wang sp. nov., 9 (22, 23)

T. sicarius Deflandre, 1973, 2 (19, 20, 22), 9 (19—21, 25, 30, 41, 42)

Superfamily Spongentactinioidea Nazarov, 1975

Family Spongentactiniidae Nazarov, 1975

Subfamily Spongentactiniinae Nazarov, 1975

Genus *Tetrentactinia* Foreman, 1963

Tetrentactinia gigantia Wang sp. nov., 3 (20)

T. spongacea Foreman, 1963, 2 (12, 18, 23), 9 (18)

Genus *Spongentactinia* Nazarov, 1975

Spongentactinia exilispina (Foreman), 1963, 2 (26, 32—36)

S. indisserta Nazarov, 1975, 3 (16, 25)

S. spongites (Foreman), 1963, 3 (18, 23)

S. sp. A, 3 (6—11)

            S. sp. B, 3 (2, 5, 6)

            S. sp. C, 1 (7, 8), 3 (12)

Order Spumellaria Ehrenberg, 1875, emend. De Wever et al., 2001

  Superfamily Sponguracea Haeckel, 1862, emend. De Wever et al., 2001

    Family Oertlispongidae Kozur et Mostler, 1980

      Subfamily Copicyntrinae Kozur et Mostler, 1980

        Genus *Copicyntra* Nazarov et Ormiston, 1985

            Copicyntra akikawaensis Sashida et Tonishi, 1988, 6 (33), 13 (38—43)

            C. cuspidata Nazarov et Ormiston, 1985, 6 (40, 45—47)

        Genus *Copiellintra* Nazarov et Ormiston, 1985

            Copiellintra diploacantha Nazarov et Ormiston, 1985, 6 (29—31)

        Genus *Hegleria* Nazarov et Ormiston, 1985

            Hegleria mammilla (Sheng et Wang), 6 (39, 44)

    Family Archaeospongoprunidae Pessagno, 1973

      Genus *Archaeospongoprunum* Pessagno, 1973

          Archaeospongoprunum chiangdaoensis (Sashida et Tonishi), 2000, 12 (38)

          A. sinisterispinosum Wang sp. nov., 6 (28)

Order Nassellaria Ehrenberg, 1875

  Superfamily Pylentonemoidea Deflandre, 1963

    Family Pylentonemidae Deflandre, 1963

      Subfamily Pylentoneminae Deflandre, 1963

        Genus *Pylentonema* Deflandre, 1963

            Pylentonema mira Cheng, 1986, 17 (6, 7)

            P.? sp. A Schwartzapfel et Holdsworth, 1996, 2 (10)

      Subfamily Archocyrtiinae Kozur et Mostler, 1981

        Genus *Archocyrtium* Deflandre, 1972

            Archocyrtium castuligerum Deflandre, 1972, 2 (2, 3)

            A. diductum Deflandre, 1973, 2 (4, 6), 3 (2)

            A. ferreum Braun, 1989, 2 (8, 9)

            A. lagabriellei Gourmelon, 1987, 17 (5)

            A. ludicrum Deflandre, 1973, 2 (7)

            A. strictum Deflandre, 1973, 2 (5)

            A. wonae Cheng, 1986, 3 (1)

            A. sp. A, 3 (3, 4)

Pseudoalbaillella litangensis Wang sp. nov.

(Pl. 3, figs. 31—40)

Etymology: After the locality of the species, Litang area, Sichuan Province.

Description: Shell body coniform, with three parts: apical cone strong without segmentation; pseudothorax small, weakly inflated with two short wings; pseudoabdomen with three segments, the first segment short, medium sized and barrel-shaped, third one cylindric. Marked contraction happened between the first and middle, and the middle and third segments. No dorsal and ventral flap observed.

Comparison: This new species resembles *P. fusiformis* (Holdsworth et Jones) in the form and structure of the shell, but differs from the former by having a skirt-like third pseudoabdomen.

Horizon and locality: Middle Permian lower Maokouan; Litang area, Sichuan of China.

Pseudoalbaillella monopteryla Wang sp. nov.
(Pl. 10, figs. 17, 19, 23, 25, 26, 33—36, 38, 40, 42)

Etymology: mono, Latin, single; pteryl, Latin, wing.

Description: Shell body consists of three parts, apical cone without segmentation; pseudothorax subglobular with a lateral wing; pseudoabdomen with five segments, the first segment small, other four segments wider. All the segments are similar in height and gradually widened. No dorsal and ventral flap observed.

Comparison: This new species resembles *P. ishigai* Wang in the form and structure of the shell body, but differs from the latter in having one lateral wing. It differs from *Ps. monacantha* (Ishiga et Imoto) by having five pseudoabdomens.

Horizon and locality: Middle Permian upper Chihsian; Shuanghu area, north Tibet of China.

Pseudoalbaillella nonpteryla Wang sp. nov.
(Pl. 10, figs. 28, 29, 32, 37, 39, 43—48)

Etymology: non, Latin, no; pteryl, Latin, wing.

Description: Shell body with three parts, apical cone lacking segmentation; pseudothorax elliptic, weakly inflated, without marked lateral wings; pseudoabdomen composed of five segments, the first segment small, the other ones gradually increased in width but their height unequal. No dorsal and ventral flap observed.

Comparison: This new species is distinguished from other species of this genus in having a weakly inflated and not marked lateral wings on the pseudothorax.

Horizon and locality: Middle Permian upper Chihsian; Shuanghu area, north Tibet of China.

Latentifistula conica Wang sp. nov.
(Pl. 11, figs. 1—4; pl. 16, figs. 14, 18)

Etymology: conic, Greece, conical.

Description: Shell body small, with three same sized and shaped spongy arms, each arm delicate and short, its basal part widest, distal part slender and conical at tip. Pore frames dense and unequal in size.

Comparison: This new species differs from *L. patagilaterala* Nazarov et Ormiston in that the latter possesses spongy arms with narrow basal part, wide middle part and spear like tip. It resembles *L. turgida* (Ormiston et Lane) in the shell form, but the latter has swollen spongy arms and polygonal and irregular pore frames.

Horizon and locality: Early Permian lower Longlinian; Shuanghu area, north Tibet of China.

Raciditor oblatum Wang sp. nov.
(Pl. 5, fig. 43; pl. 6, figs. 11-19; pl. 7, fig. 44; pl. 16, figs. 11, 15—17, 20)

Etymology: oblatus, Latin, oblate.

Latentifistulidae gen. et sp. indet. Spiller, 2002, pl. 8, figs. I, J (non K); Saesaengseerung et al., 2009, fig. 8 (21, 22).

gen. et sp. indet. Kada, 1990, pl. 1, fig. 17.

incertae sedis E Kuwahara et Yao, 1998, pl. 4, fig. 139.

Nazarovella sp. Feng et Liu, 2002, pl. 2, figs. 6, 7.

Nazarovella phlogidea Wang; Wang et al., 2012, pl. 20, fig. 31 (non 32).

Katroma (?) sp. Xia et Zhang, 1998, pl. 4, figs. 11—14.

Ishigaum? sp. Xia et Zhang, 1998, pl. 4, figs. 15, 16; Feng et Liu, 2002, pl. 2, fig. 8 (non 9, 10).

Quadreiremis sp. Li et Bian, 1993, pl. 2, fig. 9.

Description: Shell body with four different tube arms, three of them similar in form and size, dispersed in the same plane and radiating at angles of about 120°. The fourth arm small and perpendicular to other three arms. An oblate spongy body developing on the upper part of every arm with a robust apical spine at tip.

Comparison: This new species differs from *R. gracilis* (De Wever et Caridroit) in that the latter species has a small and sphaerical spongy body in upper part and a weakened apical spine at tip of each arm. It is distinguished from *R. phlogidea* (Wang) in that the latter has a longer and flaming spongy body on the upper part of the arms and shorter apical spine.

Horizon and locality: Permian; Japan, Malaysia, Thailand, South China and Hoh Xil region, Qinghai of China.

Stigmosphaerostylus craciformis Wang sp. nov.
(Pl. 5, figs. 12, 20, 22; pl. 7, figs. 30—34)

Etymology: cruciform, Latin, crossed.

Description: Lattice shell small and sphaerical, with six robust three-bladed main spines interconnected at 90°. Length of the main spines is more twice than the shell diameter. Pore frames circular to oval in outline and medium sized. No by-spines observed.

Comparison: This new species differs from *S. modesta* (Sashida et Tonishi) in that the latter has a larger shell body, with shorter main spines and many by-spines.

Horizon and locality: Middle Permian lower Guadalupian; Hoh Xil region, Qinghai of China.

Stigmosphaerostylus gracilentus Wang sp. nov.
(Pl. 5, figs. 23, 24, 29, 30)

Etymology: gracilentus, Latin, gracile.

Description: Lattice shell small and sphaerical. Six slender main spines with narrow grooves. Their length are twice over the shell diameter. Pore frames large and circular to polygonal in shape.

Comparison: This new species resembles *S.* sp. B Umed et al., 2004 (= *Borisella* cf. *maksimovae* Afanasieva, 2000, Obut et al., 2008) in the shell form, but differs from it in that the latter has rod-shaped main spines.

Horizon and locality: Upper Devonian Famennian; Hol Xil region, Qinghai of China.

Stigmosphaerostylus spiciosus Wang sp. nov.
(Pl. 1, figs. 39—41)

Etymology: spicios, Latin, beautiful.

Description: Lattice shell sphaerical. Six three-bladed main spines unequal in length, one of them longer, but small or equal with the shell diameter in length. The other five spines have similar length, form and size but smaller than the shell diameter. By-spines spicule-shaped.

Comparison: This new species differs from *S. cometes* (Foreman) in that the latter has one very robust main spine and by-spines, and its length is longer than shell diameter.

Horizon and locality: Upper Devonian Famennian; Hoh Xil region, Qinghai of China.

Stigmosphaerostylus vetulus Wang sp. nov.
(Pl. 5, figs. 25—28, 31; pl. 6, figs. 36—38)

Etymology: vetulus, Latin, ancient.

Description: Cortical shell medium in size and subsphaerical in outline. Six straight three-bladed main spines slender, and larger than shell diameter in length. Three longitudinal grooves alternate with three longitudinal ridges. Pored frames are smaller. By-spines few speculately.

Comparison: This new species differs from *S. itsukaichiensis* (Sashida et Tonishi) in that the latter has longer main spines, of which the length is about twice of the shell diameter and more by-spines.

Horizon and locality: Middle Permian lower Maokoun; Hoh Xil region, Qinghai of China.

Triaenosphaera robustispina Wang sp. nov.
(Pl. 9, figs. 22, 23)

Etymology: robustispinus, Latin, robust, spin, robust spine.

Description: Cortical shell medium in size and subsphaerical in shape. Four strongly three-bladed main spines arranged in the form of tetrahedron are similar in the shape and size, but longer than the shell diameter and widest at the base, rapidly narrowed toward distal end and sharp at tip. Pored frames larger and about 7—8 in a half volume.

Comparison: This new species differs from *T. sicarius* Deflandre in that the latter has shorter base main spines, and smaller width at and numerous pored frames.

Horizon and locality: Upper Devonian Frasnian; Shuanghu area, north Tibet of China.

Tetrentactinia gigantia Wang sp. nov.
(Pl. 3, fig. 20)

Etymology: gigant, Greece, gigantic.

Description: Spongy shell gigantic (Shell diameter 450μm, length of main spine 300μm, width of basal part 100μm) and subglobose. Four three-bladed main spines, having similar shape and size, but shorter than the shell diameter, arranged in the form of tetrahedron, By-spine is absent. No internal structureobserved.

Comparison: This new species distinguishes from all other descripted species of this genus by the characteristics of gigantic shell diameter and long main spines.

Horizon and locality: Upper Devonian Famennian; Hoh Xil region, Qinghai of China.

Archaeospongoprunum sinisterispinosum Wang sp. nov.

(Pl. 6, fig. 28)

Etymology: sinister, Latin, sinistral; spin, Latin, spine.

Description: Spongy shell elliptic in outline and medium in size. Shell pores small. The two three-bladed main spines of similar form and size are slender and slightly sinistral twist. Equal to the long diameter of shell in length. No by-spines observed.

Comparison: This new species resembles *A. patricki* Jud in the form and structure of shell body, but the latter has straight and untwist main spines or only one spine slightly sinistrally twist.

Horizon and locality: Middle Permian lower Maokouan. Hoh Xil region, Qinghai of China.

5 Composition, age and correlation of Early Mesozoic radiolarian faunas in Shuanghu of north Tibet and Hoh Xil of Qinghai

5.1 Early Mesozoic radiolarian faunas and their age in Shuanghu of north Tibet and Hoh Xil of Qinghai

5.1.1 Radiolarian fauna in sample W5418-2 from Hoh Xil of Qinghai

This radiolarian fauna contains 15 genera, 27 species (including 3 new species *Paroertlispongus longispinosa* Wang, *Parasepsagon hohxilensis* Wang and *Pseudostylosphaera qinghaiensis* Wang), belonging to 3 Orders, 5 Superfamilies and 6 Families. This fauna is characterized by rich species *Spongserrula rarauana* Dumitrica and associated with many other radiolarian fossils. *Sp. rarauana* has be recognized from Middle Triassic late Ladinian strata of Romania (Dumitrica 1982), Huangary (De Wever, 1984), Bosnia-Hercegovina (Dosztaly 1991, 1994, Kozur and Mostler, 1996), Turkey (Tekin, 1999), Thailand (Kamata et al., 2002). The list and range of the genera and species in this fauna see in Table 5.1 of Chapter 5.

We named this fauna as the *Spongoserrula rarauana* zone of late Ladinian.

According to radiolarian material of Hungary, Kozur and Mostler (1994) have erected the *Muelleritortis cochleata* zone. Then based on radiolarian data of Bosnia-Hercegovina, they (1996) have again subdivided this zone into 3 subzone, namely the *Pterospongus priscus* subzone, *Spongoserrula rarauana* and *S. fluegeli* subzone. Although there are no *Pt. priscus* and *S. fluegeli* in our sample, our fauna could be correlated to European by the linking of index zonal fossil *S. rarauana*.

5.1.2 Radiolarian fauna in sample Bb8311-1 from Hoh Xil of Qinghai

It contains 13 genera, 27 species (including 6 new species, *Paraoertlispongus longispinosa* Wang, *P. opiparus* Wang, *Betariella variospinosa* Wang, *Triassospongosphaera brevispinosa* Wang, *T. qinghaiensis* Wang and *Hozmadia pararotunda* Wang). They belong to 3 Orders, 4 Superfamilies and 7 Families. The list and range of the genera and species in this fauna see in Table 5.2 of Chapter 5.

In this fauna, *Annulotriassocampe multisegamentus* and many species of genus *Pseudostylosphaera* are characteristic fossils. The *Muelleritortis cochleata* zone and *Tritortis kretaensis* are considered as index fossils of late Ladinian and early Carnian, and the *Ladinocampe multiperforata* zone including two subzones, the *L. annuloperforata* subzone and the *L. vicentinensis* subzone is of middle—late Ladinian. They are all not found in our fauna. However, the range of important genera and spcies mostly are middle—late Ladinian in time, therefore, we temporarily name the *Annulotriassocampe multisegmentatus* zone as an independent zone. This zone could be correlated to the unnamed radiolarian zone of Europe Tethys proposed by Kozur and Mostler (1994, 1996).

5.1.3 Radiolarian fauna in sample Z144b1 from Hoh Xil of Qinghai

This radiolarian fauna contains 12 genera, 30 species (including 5 new species, *Pseudostylosphaera inaequispinosa* Wang, *P. paragracilis* Wang, *Tubotriassocyrtis fusiformis* Wang, *Tritortis robustispinosa* Wang and *Hindeosphaera bella* Wang), and is belonged to 3 Orders, 5 Superfamilies and 6 Families. The list and ranges of the genera and species in this fauna see in Table 5.3 of Chapter 5.

Many important genera and species of this fauna are discovered in late Ladinian—early Carnian strata in China and abroad. Two species *Muelleritortis cochleata* and *Tritortis kretaensis* are characteristic fossils of this fauna limited in late Ladinian. We termed this fauna as the *Muelleritortis cochleata* zone. It could be completely correlated to the same zone of Japan (Sugiyama, 1997) and Europe Tethys (Kozur and Mostler, 1994, 1996).

5.1.4 Radiolarian fauna in sample 8PFS6, 8PFS7 of Qinghai

This radiolarian fauna consists of 5 genera, 8 species. They were merged into 2 Orders, 3 Superfamilies and 4 Families, and the taxa and their ranges see in Table 5.4 of Chapter 5.

This radiolarian fauna is characterized by having the species *Tiborella florida* found in Middle Triassic late Anisian strata of Austria (Kozur et al., 1996), Slovinia (Ramovs and Gorican, 1995), the East Far of Russia (Bragin, 1991), Japan (Nakaseko and Nishimura, 1979), South China (Feng and Liu, 1993) and Yunnan of China (Feng et al., 2001). We called this fauna *Tiborella florda* zone, which could be correlated to the same zone of Japan, Europe Tethys and Yunnan of China, and to the *Shengia yini* (= *Archaeospongoprunum mesotriassicum*) assemblage of South China (Feng, 1992; Feng and Liu, 1993) as well.

5.1.5 Radiolarian fauna in sample D1339WF1 of Tibetan Jiaomuri

This radiolarian fauna includes 6 genera, 8 species. They are classified into 3 Orders, 4 Superfamilies and 5 Families. The list and range of the genera and species in this fauna see in Table 5.5 of Chapter 5.

Besides Tibet, the species *Eptingium nakasekoi* are also found in early Anisian strata of Huangary (Kozur and Mostler 1994), Slovinia (Ramovs and Gorican, 1995), Thailand (Kamata et al., 2002), Malaysia (Spiller, 2002), the Philippes (Cheng, 1989), Japan (Nakaseko and Nishimura, 1979; Sugiyama, 1997) and Yunnan of China (Feng et al., 2001).

Although many associated radiolarians of this species are spanning in middle—late Anisian in time. We designated this fauna as the *Eptingium nakasekoi* zone which could be correlated to the same zone of Japan (Sugigyama 1997), to the *Triassocampe dumitricai* zone of Yunnan, China (Feng et al., 2001), and roughly to the *Parasepsagon robustus* zone of Europe Tethys (Kozur and Mostler, 1994,

1996).

5.2 Early Mesozoic radiolarian zones in Shuanghu of north Tibet and Hoh Xil region of Qinghai and their correlation

In a world, in Shuanghu area of north Tibet and Hoh Xil region of Qinghai, the Early Mesozoic radiolarian faunas include 5 zones, spanning Anisian—Ladinian, and their global correlation framework see in Table 5.6 of Chapter 5.

6 Systematic classification and description of new species of Early Mesozoic radiolarians

Subphylum Radiolaria Müller, 1858
  Superorder Polycystina Ehrenberg, 1838, emend. Riedel, 1967
    Order Spumellaria Ehrenberg, 1875, emend. De Wever et al., 2001
      Superfamily Actinommacea Haeckel, 1862, emend. De Wever et al., 2001
        Family Pantanellidae Pessagno, 1977
          Subfamily Pantanellinae Pessagno, 1977
            Genus *Pantanellium* Pessagno, 1977
                Pantanellium multiporum Wang sp. nov., 23 (29, 30)
                P. sp. A, 24 (22, 24)
      Superfamily Sponguracea Ehrenberg, 1875, emend. De Wever et al., 2001
        Family Archeospongoprunidae Pessagno, 1973
          Subfamily Archeospongopruninae Pessagno, 1973
            Genus *Archeospongoprunum* Pessagno, 1973
                Archaeospongoprunum globosum Tekin et Mostler, 2005, 23 (33, 34)
        Family Pyramispongiidae Kozur et Mostler, 1978, emend. De Wever et al., 2001
          Genus *Paurinella* Kozur et Mostler, 1981
            Paurinella aequispinosa Kozur et Mostler, 1981, 19 (13, 14), 20 (23—25), 25 (31)
            P. latispinosa Kozur et Mostler, 1994, 25 (29, 30)
            P. sp. A, 19 (15)
          Genus *Triassospongosphaera* (Kozur et Mostler), 1981
            Triassospongosphaera brevispinosa Wang sp. nov., 22 (33)
            T. latispinosa (Kozur et Mostler), 1979, 22 (37)
            T. qinghaiensis Wang sp. nov., 22 (34—36)
            T. sp. A, 12 (8—11)
            T. triassica (Kozur et Mostler), 1979, 12 (12), 19 (17—19), 20 (11—22), 22 (38—40), 25 (27, 28)
        Family Oertlispongidae Kozur et Mostler, 1980

Genus *Paroertlispongus* Kozur et Mostler, 1981

    Paroertlispongus longispinosus Wang sp. nov. , 19 (1, 2, 21, 22), 22 (16—20)

    P. opiparus Wang sp. nov. , 22 (23, 24)

    P. sp. A, 22 (21)

Genus *Spongoserrula* Dumitrica, 1982

    Spongoserrula rarauana Dumitrica, 1982, 19 (10, 11), 20 (1—10)

Order Entactinaria Kozur et Mostler, 1982

Superfamily Eptingiacea Dumitrica, 1978

Familty Eptingiidae Dumitrica, 1978

Supfamily Eptingiinae Dumitrica, 1978

Genus *Eptingium* Dumitrica, 1978

    Eptingium manfredi Dumitrica, 1978, 19 (34—39), 23 (15, 20—26)

    E. nakasekoi Kozur et Mostler, 1994, 12 (16—23)

Genus *Cryptostephanidium* Dumitrica, 1978

    Cryptostephanidium cornigerum Dumitrica, 1978, 18 (22, 23, 25), 21 (33—36), 23 (13), 25 (41)

    C. longispinosum (Sashida), 1991, 23 (16, 28)

    C. sp. A, 21 (11, 37)

Genus *Spongostephanidium* Dumitrica, 1978

    Spongostephanidium spongiosum Dumitrica, 1978, 18 (27)

Genus *Triassistephanidium* Dumitrica, 1978

    Triassistephanidium laticornis Dumitrica, 1978, 18 (24, 26, 41)

Superfamily Hexastylacea Haecked, 1882, emend. Petrushevskaya, 1970

Family Hindeosphaeridae Kozur et Mostler, 1981

Genus *Hindeosphaera* Kozur et Mostler, 1979

    Hindeosphaera bella Wang sp. nov. , 24 (25)

    H. bispinosa Kozur et Mostler, 1979, 21 (20—32), 24 (27, 28, 31)

    H. spinulosa (Nakaseko et Nishimura), 1979, 19 (27), 23 (10, 12), 24 (26, 29, 30)

Genus *Muelleritortis* Kozur, 1988

    Muelleritortis cochleata (Nakaseko et Nishimura), 1979, 25 (38, 40)

    M. expansa Kozur et Mostler, 1988, 24 (33, 34)

    M. koeveskalensis Kozur, 1996, 24 (35), 25 (37, 39)

Genus *Parasepsagon* Dumitrica, Kozur et Mostler, 1980

    Parasepsagon firmum (Gorican), 1990, 18 (30, 31, 33, 35—39, 42—45), 21 (1—7, 14)

    P. hohxiliensis Wang sp. nov. , 18 (32, 34, 40, 46)

    P. praetetracanthus Kozur et Mostler, 1994, 12 (7)

    P. sp. A, 19 (9)

     P. sp. B, 21 (9, 10)

     P. variabilis (Nakaseko et Nishimura), 1979, 19 (33), 23 (14)

    Genus *Tritortis* Kozur, 1988

     Tritortis dispiralis (Bragin), 1986, 18 (28, 29), 25 (32, 33)

     T. kretaensis (Kozur et Krahl), 1984, 25 (35, 36, 42)

     T. robustispinosa Wang sp. nov., 25 (34)

    Genus *Pseudostylosphaera* Kozur et Mock, 1981

     Pseudostylosphaera compacta (Nakaseko et Nishimura), 1979, 12 (24, 26—30), 18 (20, 21), 19 (12, 24, 25, 47, 48), 21 (16), 23 (2, 4—7)

     P. coccostyla (Rüst), 1892, 18 (19)

     P. fragilis (Bragin), 1991, 24 (22)

     P. gracilis Kozur et Mock, 1981, 18 (11—13, 15—18), 21 (17, 18), 23 (17, 18), 24 (16—18)

     P. goestlingensis (Kozur et Mostler), 1979, 23 (11), 24 (21, 23)

     P. helicata (Nakaseko et Nishimura), 1979, 23 (9)

     P. imperspicua (Bragin), 1986, 24 (7—9)

     P. inaequispinosa Wang sp. nov., 24 (1—3, 6, 12, 13)

     P. japonica (Nakaseko et Nishimura), 1979, 12 (25), 18 (1—9), 23 (31), 24 (10, 11)

     P. longispinosa Kozur et Mostler, 1981, 19 (28, 49), 23 (32), 24 (4, 5)

     P. magnispinosa Yeh, 1989, 18 (14), 19 (23, 32), 23 (1, 3)

     P. nazarovi (Kozur et Mostler), 1979, 18 (3a, 10), 19 (29—31), 21 (15, 20, 21), 24 (20, 32)

     P. paragracilis Wang sp. nov., 24 (14, 15, 19)

     P. qinghaiensis Wang sp. nov., 21 (19)

     P. sp. A, 21 (8)

     P. sp. C, 21 (12, 13)

     P. sp. D, 23 (8)

  Family Multiarcusellidae Kozur et Mostler, 1979

   Subfamily Multiarcusellinae Kozur et Mostler, 1979

    Genus *Beturiella* Dumitrica, Kozur et Mostler, 1980

     Beturiella robusta Dumitrica, Kozur et Mosler, 1980, 22 (24, 31)

     B. sp. A, 22 (28, 30)

     B. variospinosa Wang sp. nov., 22 (25—27, 29, 32)

   Subfamily Austrisaturnalinae Kozur et Mostler, 1983

    Genus *Tiborella* Dumitrica, Kozur et Mostler, 1980

     Tiborella florida (Nakaseko et Nishimura), 1979, 12 (1—6), 19 (40—45)

Order Nassellaria Ehrenberg, 1875

  Family Poulpidae De Wever, 1981

Genus *Poulpus* De Wever, 1981
  Poulpus sp. A, 23 (35)
Genus *Hozmadia* Dumitrica, Kozur et Mostler, 1980
  Hozmadia pararotunda Wang sp. nov., 23 (19, 27)
Superfamily Acanthodesmiacea Hertwig, 1879
 Family Ruestcyrtiidae Kozur et Mostler, 1979
  Genus *Annulotriassocampe* Kozur, 1994
   Annulotriassocampe multisegmantatus Tekin, 1999, 22 (1—11), 25 (4—10)
   A. nova (Yao), 1980, 20 (26—28, 31)
   A. proprium (Blome), 1984, 22 (12)
   A. sulovensis (Kozur et Mock), 1981, 19 (5—7), 20 (30, 36, 38, 39), 22 (13), 25 (16, 17, 20, 23)
  Genus *Paratriassocampe* Kozur et Mostler, 1994
   Paratriassocampe brevis Kozur et Mostler, 1994, 22 (14, 15)
   P. gaetanii Kozur et Mostler, 1994, 25 (14, 15, 18)
  Genus *Striatotriassocampe* Kozur et Mostler, 1994
   Striatotriassocampe laeviannulata Kozur et Mostler, 1994, 25 (1—3)
   S. nodosoannulata Kozur et Mostler, 1994, 25 (11—13, 19)
  Genus *Triassocampe* Dumitrica, Kozur et Mostler, 1980
   Triassocampe coronata Bragin, 1991, 12 (14, 15)
   T. deweveri (Nakaseko et Nishimura), 1979, 20 (29, 32, 34, 35, 37), 25 (22, 24).
   T. nanpanensis (Feng), 1992, 12 (13)
   T. scalaris Dumitrica, Kozur et Mostler, 1980, 25 (21, 25)
   T. spp. Gorican et Buser, 1990, 20 (40, 41).
 Family Monicastericidae Kozur et Mostler, 1994, emend. Dumitrica, 2001.
  Genus *Tubotriassocyrtis* Kozur et Mostler, 1994
   Tubotriassocyrtis fusiformis Wang sp. nov., 25 (26)
   T. sp. A, 19 (16)
Superfamily Archaeodictyomitracea Pessagno, 1976
 Family Archaeodictyomitridae Pessagno, 1976
  Genus *Archaeodictyomitra* Pessagno, 1976, emend. Pessagno, 1977
   Archaeodictyomitra sp. A, 19 (8)
   A. sp. B, 20 (33)

***Pantanellium multiporum* Wang sp. nov.**
(Pl. 23, figs. 29, 30)

Etymology: multi, Latin, many; porus, Latin, pore.
Description: Cortical shell subsphaerical. Pored frames larger in size. 8 to 9 pores visible in every half

volume on shell surface. The two three-bladed main spines unequal in length. The three grooves alternate with three ridges. The secondary grooves visible on the ridgy basement. The short spine straight while long spine slightly sinistral twist at the end.

Comparison: This new species characterized by having numerous pored frames and slightly sinistral twist at distal end of long spine, which are distinguished from other species of this genus.

Horizon and locality: Middle Triassic upper Ladinian; Hoh Xil region, Qinghai of China.

Triassospongosphaera brevispinosa Wang sp. nov.
(Pl. 22, figs. 33, 33A)

Etymology: brev, Latin, short; spinosa, Latin, spine.

Description: Spongy shell subsphaerical. Shell body larger in size. The shell pores smaller and angular cone-shaped. Main spines over 40 in every half volume in number.

Comparison: This new species differs from *T. latispinosa* (Kozur et Mostler) in that the latter has smaller shell body and strong cone-shaped and less (ca. 14—18) main spines in every half volume.

Horizon and locality: Middle Triassic upper Ladinian; Hoh Xil region, Qinghai of China.

Triassospongosphaera qinghaiensis Wang sp. nov.
(Pl. 22, figs. 34—36)

Etymology: After the locality of the species, Qinghai Province.

Description: Spongy shell smaller and subsphaerical. Shell pores minor. The rod-shaped main spines, arranged radiantly, are about 1/3 of shell diameter in length. 16—18 main spines visible in every half volume. Each spine at the base is wider, then gradually decreased toward distal end and sharped at tip.

Comparison: This new species differs from *T. triassica* (Kozur et Mostler) in that the latter has a larger shell body and longer main spine, about over 1/2 of shell diameter in length.

Horizon and locality: Middle Triassic upper Ladinian; Hoh Xil region, Qinghai of China.

Paroertlispongus longispinosa Wang sp. nov.
(Pl. 19, figs. 1, 2, 21, 22; pl. 22, figs. 16—20)

Etymology: long, Latin, long; spinos, Latin, spine.

Description: Spongy shell smaller and subsphaerical. Shell pores also smaller. The two rod-shaped main spines straight and similar in form and in size. Its length about over two times of the shell diameter.

Comparison: This new species differs from *P. chinensis* (Feng) in that the latter has shorter main spines and the length of main spines similar to shell diameter.

Horizon and locality: Middle Triassic upper Ladinian; Hoh Xil region, Qinghai of China.

Paroertlispongus opiparus Wang sp. nov.
(Pl. 22, figs. 22, 23)

Etymology: opipar, Latin, beautiful.

Description: Spongy shell larger and subsphaerical. Shell pores smaller. The two strong rod-shaped main spines of different length. The long spine approximately twice as long as the short one, but similar in shell diameter.

Comparison: This new species differs from P. hermi (Lahm) in that the latter has weaker main spines.

Horizon and locality: Middle Triassic upper Ladinian; Hoh Xil region, Qinghai of China.

Hindeosphaera bella Wang sp. nov.
(Pl. 24, fig. 25)

Etymology: bell, Latin, beautiful.

Description: Cortical shell subsphaerical. Double layered, pored frames of inner layer smaller than those of outer layer, elevating with nodes at vertices. The two three-bladed main spines straight, but the long spine slender and longer than shell diameter, the short spine smaller than shell diameter. Longitudinal grooves wider than longitudinal ridges. No secondary grooves observed.

Comparison: This new species differs from H. spinulosa (Nakaseko et Nishimura) in that the latter has a strong long main spine, its length smaller than or similar to shell diameter.

Horizon and locality: Middle Triassic upper Ladinian; Litang area, Sichuan of China.

Parasepsagon hohxiliensis Wang sp. nov.
(Pl. 18, figs. 32, 34, 40, 46)

Plafkerium? firmum Gorican, 1990, pl. 6, fig. 6, non 3—5.

Etymology: This species is named after Hoh Xil region, Qinghai of China.

Description: Cortical shell subsphaerical, Double layered pored frames of inner layer smaller than those of outer layer, elevating with nodes at vertices. The four three-bladed main spines strong and straight, while three of them are similar in both form and size, and equal to shell diameter in length, the other spine is longer than shell diameter.

Comparison: This new species differs from P. variabilis (Nakaseko et Nishimura) in that the length of long and short main spines of the latter is all smaller than shell diameter, while their width is reduced from basement to distant end. Compared with P. firmum (Gorican) the latter has a smaller length of main spines, the long spine is similar to but one of short spines is smaller than shell diameter in length. In addition, the middle and upper pants of main spines in the latter are expanded, but the former does not.

Horizon and locality: Middle Triassic upper Ladinian; Slovinia and Hoh Xil region, Qinghai of China.

Tritortis robustispinosa Wang sp. nov.
(Pl. 25, fig. 34)

Etymology: robust, Latin, strong; spines, Latin, spine.

Description: Cortical shell small and subsphaerical. Double layered, pored frames of outer layer larger and elevated with nodes at vertices. Only two mostly strong three-bladed main spines preserved which are slightly sinistral twisted. Longitudinal ridges are narrower, but longitudinal grooves wider. The width of main spines is similar to but their length is longer than shell diameter.

Comparison: This new species differs from T. dispiralis (Bragin) in that the latter species has a larger shell body and slender main spines with smaller width.

Horizon and locality: Middle Triassic upper Ladinian; Litang area, Sichuan of China.

Pseudostylosphaera inaequispinosa Wang sp. nov.
(Pl. 24, figs. 1—3, 6, 12, 13)

Etymology: inaequi, Latin, unequal; spinos, Latin, spine.

Description: Cortical shell sphaerical or elliptical. Double layered, pored frames of inner layer small, while those of outer layer bigger. Vertices of pored frames elevated with nodes. The two three-bladed main spines medium in size and unequal in length. The long spines larger than, but the short ones smaller or similar to shell diameter in length. One spine straight, whereas the others slightly dextrally twisted.

Comparison: This new species differs from *P. nazarovi* (Kozur et Mostler) in that the latter has two slightly sinistral twisted main spines. It could be distinguished from *P. compacta* (Nakaseko et Nishimura) in that the latter has two straight and untwisted main spines.

Horizon and locality: Middle Triassic middle-upper Ladinian. Litang area, Sichuan of China.

Pseudostylosphaera paragracilis Wang sp. nov.
(Pl. 24, figs. 14, 15, 19)

Etymology: para, Greece, similar; gracilis, former species name.

Description: Cortical shell subsphaerical. Double layered, pored frames of inner layer smaller, whereas those of outer layer bigger. Vertices of pored frames elevated with nodes. The two three-bladed main spines stronger, and similar in length. The end of main spines is swelling. One spine straight, the other one slightly sinistrally twisted.

Comparison: This new species differs from *P. gracilis* Kozur et Mock in that the latter has two sinistrally twisted main spines.

Horizon and locality: Middle Triassic middle-upper Ladinian; Litang area, Sichuan of China.

Pseudostylosphaera qinghaiensis Wang sp. nov.
(Pl. 21, fig. 19)

Pseudostylosphaera sp. Sashida et al., 1993, fig. 7 (19).

Etymology: After the locality of the species, Qinghai Province.

Description: Cortical shell elliptical. Double-layered, vertices of pored frames elevated with nodes. The two three-bladed main spines, composed of three longitudinal ridges and three longitudinal grooves, strong. On the ridges no secondary grooves discovered. The length of two main spines is various, but longer than shell diameter. The spines are in the form of slightly dextrally twisted.

Comparison: This new species differs from *P. goestlingensis* (Kozur et Mostler) in that the latter has a bigger shell body and slender main spines which strongly dextrally twisted.

Horizon and locality: Middle Triassic upper Ladinian; Japan and Hoh Xil region, Qinghai of China.

Beturiella variospinosa Wang sp. nov.
(Pl. 22, figs. 25—27, 29, 32)

Etymology: vario, Latin, change; spinos, Latin, spine.

Description: Cortical shell subsphaerical, with small pored frames. The six three-bladed main spines composed of three narrower ridges and three wider grooves, smaller than shell diameter in length. One of them is strong, whereas the other five spines is smaller and similar in form and size. Their width is about 1/2 the width of big spine.

Comparison: This new species differs from *B. robusta* Dumitrica, Kozur et Mostler in that the latter has six strong three-bladed main spines.

Horizon and locality: Middle Triassic upper Ladinian; Hoh Xil region, Qinghai of China.

Hozmadia pararotunda Wang sp. nov.
(Pl. 23, figs. 19, 27)

Etymology: para, Greece, similar; rotund, Latin, circular.

Description: Cephalis small and subsphaerical. Double layered, pore frames of inner layer small, but those of outer layer bigger. With nodes at pored frames vertices. Top spine strong, three-bladed, the three three-bladed feet stronger, straight and untwisted. They all are larger than cephalic diameter in length.

Comparison: This new species differs from *H. rotunda* (Nakaseko et Nishimura) in that the latter has shorter and inwardly curved feets, and the length of top spine and 3 feets smaller than or equal to the diameter of cephalis.

Horizon and locality: Middle Triassic upper Ladinian; Hoh Xil region, Qinghai of China

Tubotriassocyrtis fusiformis Wang sp. nov.
(Pl. 25, fig. 26)

Etymology: fusiform, Latin, spindle-shaped.

Description: Shell body small and long spindle-shaped, composed of 12 segments. Cephalic segment smaller, conical and unpored with one apical horn. Thorax and abdominal segments shorter. Postabdominal segments consist of 8 ones. The width of shell body gradually increased from cephalic segment to 4th postabdominal one, then again gradually decreased from 5th one to endmost segment, in the form of antitrapziform. Circumferential ridges from the beginning of 4th postabdominal segment visible and with one row of pored ring on every ridge. The stricture between circumferential ridges shallower, smooth and poreless. A long and narrow tube smooth and unpored presents at the end of postabdominal segment.

Comparison: This new species resembles *Triassocampe fusiformis* Bragin in the form and structure, but the latter has no apical horn, deeper stricture between every segment and specially three rods of pored ring on every postabdominal segment.

Horizon and locality: Middle Triassic middle-upper Ladinian; Litang area, Sichuan of China.

7 Palaeogeographic implications of the Longmuco-Shuanghu suture zone

In the Longmuco-Shuanghu suture zone of northern Qinangtang, northern Tibet Plateau, there exist a number of the Late Palaeozoic—Early Mesozoic ophiolites and radiolarian bedded cherts rock formations, which contain globally distributed deep-water radiolarians. Such data not only provide us the new evidence of Tethys evolution, but also bring us some inspiration for studying the formation and closure age of Palaeotethys and the open time of Neotethys Ocean, and the boundary of the Southern and Northern continents as well.

7.1 Ophiolites and radiolarian faunas

There are Cambrian—Triassic ophiolites records in the Longmuco-Shuanghu suture zone of Qiangtang, based on the monograph of Li et al. (2016), *Geology in the Qiangtang Region*. The Late Devonian Frasnian, Early Permian early Longlinian, Middle Permian late Chihsian, Late Permian Changhsingian, Middle Triassic Anisian (e.g., Zhu et al., 2010; Li et al., 1997), and even possibly Late Triassic (Li et al., 1997) radiolarian faunas, so far, were recognized from the ophiolites and/or radio-larian bedded cherts of this suture zone.

The Late Palaeozoic—Early Mesozoic ophiolites in the Longmuco-Shuanghu suture zone are subdivided, by region, into Hongjishan, Jiaomuri and Shuanghu ophiolitic mélanges (see in Table 1.1 of Chapter 1), yielding five radiolarian faunas, i.e., *Helenifore robustum* fauna, *Pseudoalbaillella sakmarensis-P. lomentaria* fauna, *Pseudoalbaillella ishigai* fauna, *Neoalbaillella ornithoformis* fauna and *Eptingium nakasekoi* fauna (see in Table 7.1 of Chapter 7).

In the first radiolarian fauna, the most important elements include *Trilonche pittmani* (26-1), *T. davidi* (26-2), *T. echinata* (26-3), *T. elegans* (26-4), *T. tretactinia* (26-5), *Triaenosphaera sicarius* (26-6) and *Tetrentactinia spongacea* (26-7). Previously, they were mainly discovered in Middle Devonian Givetian—Late Devonian Famennian strata and commonly in Late Devonian Frasnian strata of East Australia, Western coast of America, Ural and Rudny Altai areas of Russia, Kazakhstan, Thailand, Malaysia, and Yunnan, Guangxi, Guizhou and Xinjiang of China. Although index zonal genus *Helenifore* of Late Devonian was not found in present fauna, but the main members of this fauna are of Late Devonian Frasnian in age, so we still named the present fauna the *Helenifore robustum* zone of Late Devonian Frasnian. It definitely can be correlated to the same zone in East part of Australia (Aitchison, 1988; Ishiga, 1988), the western coast of America (Boundy-Sanders and Murchey, 1999), Thailand (Sashida et al., 1993, 1998), Malaysia (Spiller, 2002), and Yunnan (Wang et al., 2000, 2003), Guizhou (Wang and Luo, 2006), Guangxi (Wang et al., 2003, 2012) and Xinjiang (Wang et al., 2013) of China.

The second radiolarian fauna is characterized by abundant pseudoalbaillellids. In this fauna, *Psedoalbaillella sakmarensis* (26-18—20) and *P. lomentaria* (26-12—14) are the dominant species. Besides the Qiangtang region of China, they have been found in Early Permian middle—late Wolfcampian strata in Ural and the Far East of Russia, America, Japan, Malaysia, Thailand, and Yunnan, Guizhou and Qinghai of China. We termed this fauna as the *P. sakmarensis-P. lomentaria* zone of Early Permian early Longlinian, corresponding to the middle—late Wolfcampian of American and Sakmarian of Russia. They could be correlated to the *P. lomentaria* zone of Japan (Ishiga, 1986, 1990), Thailand (Sashida et al., 1998, 2000; Wonganan and Caridroit, 2007), Malaysia (Spiller, 2002), western coast of America (Blome and Reed, 1992, 1995) and also to the *P. sakmarensis* zone of the far East Russia (Rudenko and Panasenko, 1997) and to the same zone of South China and Hoh Xil region, Qinghai of China.

The third fauna is characterized by having the species *Psedoalbaillella ishigai* (26-10, 11) which is associated with *Albaillella sinuata* (26-8), *Pseudotormentus kamigoriensis* (26-15, 16), *Latentifistula texana* (26-17) and *L. patagilaterala* (26-9). *Psedoalbaillella ishigai* is synonymous with *Psedoalbaillella* sp. C identified by Ishiga et al. (1982). In South China, this species is mainly discovered in Middle Permian late Chihsian strata and it is considered as an index zonal species. The present fauna is thus named *Ps. ishigai* zone. It can be correlated to the same zone of South China (Wang et al., 1998, 2006, 2007), the *P.* sp. C zone (Ishiga, 1986) and subsequent *Psedoalbaillella longtanensis* zone (Ishiga, 1990) of Nabeyama, Japan, the *P.* C zone of Leonardian of Oregon area, America (Blome and Reed, 1992) and to the *Psedoalbaillella corniculata* (= *Psedoalbaillella longtanensis*) zone of Artinskian of the Far East of Russia (Rudenko and Panasenko, 1997).

The fourth fauna is characterized by the abundant neoalbaillellids. In this fauna, *Neoalbaillella ornithoformis* (26-24, 28, 29) is a dominant species, thus this fauna is termed as *N. ornithoformis* zone which is the third radiolarian zone of Late Permian Changhsingian in South China (Wang et al., 2006, 2007). It is associated with particularly abundant radiolarians, including *Trilonche pseudocimelia* (26-21, 36, 37), *Archaeospongoprunum chiangdaoensis* (26-22), *Raciditor gracilis* (26-23), *Ishigaum trifistis* (26-25, 40), *I. craticula* (26-26), *I. obesum* (26-34), *Triplanospongos musashiensis* (26-27), *Neoalbaillella optima* (26-30), *N. pseudogrypa* (26-31), *N. gracilis* (26-32, 33), *Ormistonella robusta* (26-35), *Albaillella levis* (26-38), *A. lauta* (26-39), *A. triangularis* (26-44) and *Copicyntra akikawaensis* (26-41) etc. This zone can be entirely correlated to the same zone of Japan (Ishiga, 1990; Kuwahara et al., 1998), Malaysia (Sashida et al., 1995; Spiller, 2002), the Philippines (Cheng, 1989; Tumanda et al., 1990), Thailand (Sashida et al., 2000), the East Far of Russia (Rudenko et al., 1997), the western coast of America (Blome and Reed, 1992, 1995), and Yunnan, Guizhou and Guangxi of China (Wang et al., 2006, 2007).

The fifth fauna is characterized by having specific nasselarians in Mesozoic Period. The six genera and eight species, namely, *Eptingium nakasekoi* (26-42, 52), *Parasepsagon praetetracanthus* (26-43), *Pseudostylosphaera japonica* (26-45), *P. compacta* (26-46), *Tiborella florida* (26-50), and *Triassospongosphaera triassica* (26-51). These radiolarians are commonly found in the Middle Triassic early Anisian strata of all over the world. In this radiolarian fauna, the specific species *Eptingium nakasekoi* is mainly found in the early Anisian strata of Hungary (Kozur and Mostler, 1994), Slovinia (Ramovs and Gorica, 1995), Thailand (Kamata et al., 2002), Malaysia (Spiller, 2002), the Philippines (Cheng, 1989), Japan (Nakaseko and Nishimura, 1979; Sugiyama, 1997) and Yunnan of China (Feng et al., 2001). This species is known by Sugiyama (1997) as an index zonal fossil of Middle Triassic early Anisian. We also considered this fauna as the *Eptingium nakasekoi* zone which can correlated to the same zone of Japan (Sugiyama, 1997) and Hoh Xil region, Qinghai of China (Feng et al., 2007), also to the *Triassocampe dumitricai* zone of Chinese Yunnan (Feng et al., 2006), *Parasepsagon robustum* one of Europe Tethys (Kozur and Mostler, 1994), *Hozmadia* one of the East Far of Russia (Bragin, 1991) and *Triassocampe coronata* one of Malaysia (Spiller, 2002).

7.2 Palaeogeographic implications

The records of radiolarians-bearing ophilites and radiolarian bedded cherts, establishment of radiolarian zonation and its global correlation of radiolarian zones above, not only proved that there did exist a deep ocean basin or rift along the line of Longmuco-Shuanghu during the Late Palaeozoic – Early Mesozoic, but also clearly demonstrated the major evolution courses of Chinese north Tethys.

(1) The lowest radiolarian zones of Longmuco-Shuanghu and Longmuco-Yushu suture zones are all the Late Devonian, as early as early Late Devonian Frasnian in age (see in Table 7.1 of Chapter 7), demonstrating that the Palaeotethys formed in Late Devonian, even earlier.

(2) These two suture zones both lack the ophilites and radiolarian bedded chert formations and radiolarian fossil of the early Late Permian Wuchiapingian and Early Triassic Olenekian (see in Table 7.1 of Chapter 7), indicating that these two Palaeotethys basins or rifts had started to close since the latest Middle—earliest Late Permian Wuchiapingian, and completely disappeared in Late Permian or the end of the Permian, replaced by open shallow sea in Early Triassic. However, due to the stretching or sinking, the shallow sea crust rift again since the early Middle Triassic early Anisian, formed a new ocean, Neotethys Ocean.

The limestone member of the carbonate strata of the Hantanshan Group in northwest Xijinwulanco-Yishanco of Longmuco-Yushu suture zone, Hoh Xil, yields typical and abundant early Late Permian Wuchiapingian and Early Triassic Olenekian Tethys neritic biota fossils (Sha et al., 1992; Zhang and Zheng, 1994; Zhang and Sha, 1995; Sha, 1998; Sha et al., 2004). Wuchiapingian biota is composed of calcareous algae *Gymnocodium* cf. *exile*, *Permocalculus plumosus*, *Mizzia velebitana*, *Pseudovermiporella elliotti* ect., (Bao, 1995), foraminifer assemblage *Colaniella-Baisalina pulchra reitlingerae* (Luo, 1995), fusulinid zone of *Codonofusiella lui*, including *C.* cf. *pseudolui* (Zhang, 1995), and hydra, sponges, bryozoans, crinoid stems etc. (Sha, 1995a). It comes from the base of the limestone member of Hantanshan Group. Olenekian fauna consists of conodont assemblage of *Neospathodus waageni-N. timorensis* (Xu, 1995), bivalve assemblage of *Bakevellia costata-Leptochondria virgalensis-Entolium microtis* (Sha, 1995b; Sha and Grant-Macie, 1996), gastropods *Neritaria intexa*, *N. comensis*, *Naticopsis eyerichi* etc. (Zhu, 1995), and fragments of crinoid stems etc. (Sha, 1995a). This fauna comes from the lower limestone member of Hantanshan Group.

Such recordings that Wuchiapingian and Olenekian carbonate strata and neritic fossil biota are sandwiched in between the Middle Permian and Middle Triassic ophilite-radiolarite series and radiolarian fossil faunas in Longmuco-Yushu suture zone strongly supported that the conclusion above, Palaeoethys Ocean began to close during the latest Middle—earliest Late Permian, disappeared in Late Permian or the end-Permian, the Neotethys Ocean opened in the earliest middle Triassic. It is speculated that there exist Wuchiapingian and Olenekian carbonate rocks and neritic biota in Longmuco-Shuanghu suture zone, though it is remained to be found.

(3) The radiolarian zones in Changning-Shuangjiang-Menglian suture zone [i.e., Lancangjiang suture zone (Liu et al., 1993)] are more complete, compared with those of Longmuco-Shuanghu. Nevertheless, in this suture, the oldest radiolarian zone is Late Devonian Frasnian too, and there are no the latest Parmian—early Early Triassic Induan, and even possibly late Early Triassic Olenekian radiolarian and radiolarian-bearing ophiolites either (Table 7.1). Such records of strata and fossils imply that the major evolutionary process of Changning-Shuangjiang-Menglian ocean basin are same as that of Longmuco-Shuanghu and Longmuco-Yushu ocean basins above, Palaeotethys Ocean formed in

early Late Devonian, even earlier, but closed in Late Permian or by the end of the Permian, Neotethys opened in early Middle Triassic or the latest Early Triassic.

(4) Karakorum Pass-Longmuco-Yushu-Changning-Shuangjiang-Menglian suture zone built up the northern main suture zone or Palaeotethys Suture Zone (Huang and Chen, 1987). The similar or same radiolarian faunas (Table 7.1) demonstrate that these three Late Palaeozoic-Early Mesozoic deep ocean basins or rifts, Longmuco-Shuanghu, Longmuco-Yushu and Changning-Shuangjiang-Menglian (Lancangjiang) basins or rifts were interlinked, forming the vast Palaeotethys and Neotethys oceans together with the shallow Tethys. To the west, Longmuco-Shuanghu suture zone ends and connects with the Longmuco-Yushu suture zone in Longmuco area (see in Fig. 1 of Praface). Such geographic distribution pattern of suture zones obviously also shows that the west end of Longmuco-Shuanghu suture zone connects with the northern main suture zone, and the Longmoco-Shuanghu-Lancangjiang suture zone has been accepted as the boundary separating the Gondwana and Eurasia by some authors (e.g., Metacalfe, 2002, 2013; Li et al., 2009).

8 Conclusions

Radiolarian-bearing ophilites and radiolarian bedded cherts are the most reliable sign and radiolarian fossil is the essential evidence for the recognition of ocean basin or rift. The suture zones of Longmuco-Shuanghu and Longmuco-Yushu of Qiangtang region do record not only the Late Palaeozoic-Early Mesozoic ophilites-radiolarite series, but also full of variety and globally distributed radiolarian faunas, demonstrating that there is a rare archives recording the evolution, including the establishment and disappearance age of Plaeotethys Ocean and the open time of Neotethys Ocean in the north Tethys of China.

25 families, 53 genera and 155 species (including 25 new species) of Late Devonian—Middle Triassic radiolarians have been recognized from there so far. The sequence of the ophilites and radiolarian bedded cherts formations, particularly the 14 radiolarian fossil zones recorded in two suture zones suggest that the Plaeotethys Ocean situated in northern Palaeotethys of China formed in Late Devonian, even earlier. However, with the subduction of oceanic crust of Palaeotethys and northward drift of Tanggula Block, it began to close during latest Middle Permian—earliest Late Permian (e.g., Zhang 1991; Sha et al., 1992; Bian et al., 1993, 1996, 1997; The Comprehensive Scientific Expedition to the Hoh Xil region, 1994; Zhang and Zheng, 1994; Wang, 1995; Zhang, 1995; Sha, 1998, 2009, 2018; Sha and Fürich, 1999; Sha et al., 2004) or Late Permian (e.g., Zhang, 1993), and completely disappeared by the end of the Permian, entered the important turning point of Tethys evolution (Huang and Chen, 1987). It was replaced by an open shallow sea dominated by carbonate facies, but it was still connected with the Tethys shallow sea to the south, formed a vast and vibrant blue sea, Tethys sea (Sha et al., 1992, 2004; Sha and Fürich, 1999; Sha, 1998, 2009, 2018). Nevertheless, the seabed of north Tethys rift and disintegrated in the early Middle Triassic early Anisian, caused the Neotethys Ocean be opened there.

Although many Tethys researchers concluded that the Neo-Tethys Ocean opened in Late Palaeozoic Middle Permian even earlier (e.g., Muttoni et al., 2009; Zanchi and Gaetani, 2011; Angiolini et al.,

2013, 2015; Metcalfe, 2013; Berra and Angiolini, 2014; Zhang and Wang, 2019), there is no radiolarian evidence supporting such conclusion, so far.

In the areas south of the northern main suture zone, i. e., Karakorum Pass-Longmuco-Yushu-Jinsha River-Changning-Shuangjiang-Mengliansuture zone (including Longmuco-Shuanghu suture zone), there are two suture zones recording the evolution of south Tethys of China, i. e., Yindus-Yarlung suture zone and its branch Bangongco-Dinqing-Nu River suture zone (Huang and Chen, 1987). In the end of last century and the beginning of this century, Yang et al. (2000) and Wang et al. (2002a) found the Middle Triassic Ladinian radiolarian assemblage of *Pseudostylophaera nazarovi* from the silica rocks of ophiolitic mélange of Yindus-Yarlung suture zone in Zedang, southern Tibet. Wang et al. (2000b) recognized the Late Triassic Carnian radiolarian fauna of *Capnuchospharera triassica* from the siliceous rocks of ophiolitic mélange of Bangongco-Dinqing-Nu River suture zone in Dinqing, northern Tibet. These two critical findings respectively demonstrated that the open time of Yindus-Yarlung ocean basin was late Middle Triassic Ladinian, and Bangongco-Dinqing-Nu River one was early Late Triassic Carnian (Sha, 2018).

In a word, in the south Tethys of China south of northern main suture zone, there is no radiolarian fossil record earlier than Middle Triassic so far, thus the Neotethys Ocean there initiated in Middle Triassic, rather than Permian in age. Did the north and south Neotethys oceans of China formed at the same time or successfully born? The answer to such great geographic mystery will be remained the discovery of more radiolarian fossils, particularly the precise study of radiolarian taxonomy.

The similarity or commonality of Late Palaeozoic—Early Mesozoic radiolarian faunas in Longmuco-Shuanghu, Longmuco-Yushu and Changning-Shuangjiang-Menglian suture zones and, the geographic distribution pattern of suture zones that the west end of Longmuco-Shuanghu suture zone ends and connects with the Longmuco-Yushu suture zone in Longmuco area (see in Fig. 1 of Introduction), all obviously show that the Longmuco-Shuanghu suture zone communicates with the northern main suture zone (Karakorum Pass-Longmuco-Yushu-Changning-Shuangjiang-Menglian suture zone). Nevertheless, whether or not and when the Longmuco-Shuanghu suture zone extended to southwest until Lancangjiang (i. e., Changning-Shuangjiang-Menglian suture zone (Liu et al., 1993)) (e. g., Metacalfe, 2002, 2013; Li et al., 2009), are remained to be confirmed by the Late Palaeozoic—Early Mesozoic radiolarian fossils from the north *Lancang* River belt, the transtional zone linking Longmuco-Shuanghu and *Lancangjiang* suture zones.

图 版 说 明
(PLATE EXPLANATIONS)

图版 1（1—41），图版 2（1—45）（8P$_2$W22-2）；图版 3（1—5，7—17，19—30）（8P$_2$W16-2），图版 3（6，18）（W7008），图版 3（31—50）（ZD305-90b1）；图版 4（1—45），图版 5（1—54），图版 6（1—39）（W5032），图版 6（40—50），图版 7（1—46）（W5031）；图版 8（1—50）（8PFS5014）；图版 9（1—43）（CDP14WF1）；图版 10（1—48）（D7052WF1）；图版 11（1—42）（D7052WF3）；图版 12（1—30）（D1339WF1），图版 12（31—43）（D9002WF1）；图版 13（1—48），图版 14（1—65）（D9002WF3-2）；图版 15（1—24），图版 16（1—21），图版 17（1—20），图版 18（1—46），图版 19（1—45）（8PFS6），图版 19（46—49）（8PFS7）；图版 20（1—41），图版 21（1—37）（W5418—2）；图版 22（1—40），图版 23（1—35）（Bb8311-1）；图版 24（1—35），图版 25（1—42）（Z 上部 Z114b1）；图版 26（1—52）（这一图版为青海可可西里地区 8P$_2$W22-2，8P$_2$W16-2，W7008，W5032，W5031，8PFS5014，W5418-2，8PFS6，8PFS7，Bb8311-1，西藏双湖地区 CDP14WF1，D7052WF1，D7052WF3，D9002WF1，D9002WF3-2，D1339WF1 和四川理塘地区 Z 上部 Z114b1，ZD305-90b1 样品的重要放射虫化石照片）。

所有放射虫图影均为电镜扫描照片，标本保存在中国科学院南京地质古生物研究所。图版中比例尺长度（scale bar）= 100 μm

图 版 1

1，10. *Stigmosphaerostylus variospina*（Won）

2，5，6. *Spongentactinia* sp. B

3，4，9，29—31，33—35. *Stigmosphaerostylus diversitus*（Nazarov）

7，8. *Spongentactinia* sp. C

11，15，16，20，22，24，25. *Astroentactinia biaciculata* Nazarov

12—14，17—19. *Astroentactinia stellata* Nazarov

21，23，26—28，36—38. *Astroentactinia paronae*（Hinde）

32. *Stigmosphaerostylus proceraspina*（Aitchison）

39—41. *Stigmosphaerostylus spiciocus* Wang sp. nov.；40. 正模（Holotype）

图 版 2

1. *Holoeciscus elongata* Kiessling et Tragelehn

2，3. *Archocyrtium castuligerum* Deflandre

4，6. *Archocyrtium diductum* Deflandre

5. *Archocyrtium strictum* Deflandre

7. *Archocyrtium ludicrum* Deflandre

8，9. *Archocyrtium ferreum* Braun

10. *Pylentonema* sp. A Schwartzapfel et Holdsworth

11，13—15，25. *Astroentactinia paronae*（Hinde）

12，18，23. *Tetrentactinia spongacea* Foreman

16. *Astroentactinia stellata* Nazarov

17. *Polyentactinia aranea* Gourmelon

· 177 ·

19，20，22. *Triaenosphaera sicarius* Deflandre

21. *Triaenosphaera hebes* Won

24. *Palaeoscenidium cladophorum* Deflandre

26，32—36，38. *Spongentactinia exilispina* (Foreman)

27—31，43，45. *Stigmosphaerostylus micula* (Foreman)

37，39—42，44. *Astroentactinia biaciculata* Nazarov

图 版 3

1. *Archocyrtium wonae* Cheng

2. *Archocyrtium diductum* Deflandre

3，4. *Archocyrtium* sp. A

5，13—15，19，22. *Stigmosphaerostylus variospina* (Won)

6—11. *Spongentactinia* sp. A

12. *Spongentactinia* sp. C

16，25. *Spongentactinia indisserta* Nazarov

17，21. *Triaenosphaera hebes* Won

18，23. *Spongentactinia spongites* (Foreman)

20. *Tetrentactinia gigantia* Wang sp. nov.；20. 正模 (Holotype)

24. *Stigmosphaerostylus pantotolma* (Braun)

26，28—30. *Astroentactinia multispinosa* Won

27. *Stigmosphaerostylus micula* (Foreman)

31—40. *Pseudoalbaillella litangensis* Wang sp. nov.；35. 正模 (Holotype)

41，45，46，50. *Pseudoalbaillella scalprata* Holdsworth et Jones

42—44，47—49. *Pseudoalbaillella globosa* Ishiga et Imoto

图 版 4

1—10. *Pseudoalbaillella longtanensis* Sheng et Wang

11—16. *Pseudoalbaillella fusiformis* (Holdsworth et Jones)

17—30. *Pseudoalbaillella globosa* Ishiga et Imoto

31—37. *Pseudoalbaillella scalprata* Holdsworth et Jones

38—45. *Latentibifistula triacanthophora* Nazarov et Ormiston

图 版 5

1，13，15，16，18，19. *Stigmosphaerostylus modestus* (Sashida et Tonishi)

2—11，14，17. *Stigmosphaerostylus itsukaichiensis* (Sashida et Tonishi)

12，20—22. *Stigmosphaerostylus cruciformis* Wang sp. nov.；21. 正模 (Holotype)

23，24，29，30. *Stigmosphaerostylus gracilentus* Wang sp. nov.；23. 正模 (Holotype)

25—28，31. *Stigmosphaerostylus vetulus* Wang sp. nov.；26. 正模 (Holotype)

32—42. *Raciditor phlogidea* (Wang)

43. *Raciditor oblatum* Wang sp. nov.；43. 正模 (Holotype)

44，45，53，54. *Pseudotormentus kamigoriensis* De Wever et Caridroit

46—52. *Quadricaulis femoris* Caridroit et De Wever

图 版 6

1，3，5—10. *Raciditor gracilis* (De Wever et Caridroit)
2，4. *Raciditor phlogidea* (Wang)
11—19. *Raciditor oblatum* Wang sp. nov.
20，21. *Quinqueremis robusta* Nazarov et Ormiston
22. *Stigmosphaerostylus ichikawai* (Caridroit et De Wever)
23，24，26，27. *Stigmosphaerostylus* sp. B
25，32. *Stigmosphaerostylus* sp. A
28. *Archaeospongoprunum sinisterispinosum* Wang sp. nov.；28. 正模（Holotype）
29—31. *Copiellintra diploacantha* Nazarov et Ormiston
33. *Copicyntra akikawaensis* Sashida et Tonishi
34，35，46，47. *Stigmosphaerostylus* sp. C
36—38. *Stigmosphaerostylus vetulus* Wang. sp. nov.
39，44. *Hegleria mammilla* (Sheng et Wang)
40，45—47. *Copicyntra cuspidata* Nazarov et Ormiston
41—43，48—50. *Trilonche* sp. A.

图 版 7

1，2. *Pseudoalbaillella* sp. A
3，11—14. *Albaillella xiadongensis* Wang
4—10. *Pseudoalbaillella rhombothoracata* Ishiga et Imoto
15—18. *Albaillella sinuata* Ishiga et Watase
19，20，22，25—28. *Latentifistula patagilaterala* Nazarov et Ormiston
21. *Latentifistula* sp. A
23，24. *Latentifistula texana* Nazarov et Ormiston
29. *Stigmosphaerostylus* sp. E
30—34. *Stigmosphaerostylus cruciformis* Wang sp. nov.
35—38. *Stigmosphaerostylus* sp. A
39—43. *Pseudotormentus kamigoriensis* De Wever et Caridroit
44. *Raciditor oblatum* Wang sp. nov.
45. *Trilonche* sp. B
46. *Quadriremis minima* Nazarov et Ormiston

图 版 8

1—5. *Pseudoalbaillella elongata* Ishiga et Imoto
6—17，35. *Pseudoalbaillella rhombothoracata* Ishiga et Imoto
18—32. *Pseudoalbaillella sakmarensis* (Kozur)
33，34，36—40. *Pseudoalbaillella simplex* Ishiga et Imoto
41，42，45. *Latentibifistula asperspongiosa* Sashida et Tonishi
43，44. *Pseudotormentus kamigoriensis* De Wever et Caridroit
46—50. *Quinqueremis arundinea* Nazarov et Ormiston

图 版 9

1. *Spongentactinia* sp. D
2—4. *Trilonche pittmani* Hinde
5. *Trilonche minax* (Hinde)
6，7，9，24，29，32，33. *Trilonche davidi* (Hinde)
8，10—16. *Trilonche echinatum* (Hinde)
17，35. *Trilonche elegans* (Hinde)
18. *Tetrentactinia spongacea* Foreman
19—21，25，30，41，42. *Triaenosphaera sicarius* Deflandre
22，23. *Triaenosphaera robustispina* Wang sp. nov.；22. 正模（Holotype）
26—28. *Stigmosphaerostylus variospina* (Won)
31，34，36—40，43. *Trilonche tretactinia* (Foreman)

图 版 10

1，4，5. *Latentifistula texana* Nazarov et Ormiston
2，3. *Latentifistula patagilaterala* Nazarov et Ormiston
6. *Raciditor inflata* (Sashida et Tonishi)
7. *Raciditor gracilis* (De Wever et Caridroit)
8. *Latentifistula* sp. B
9—12. *Albaillella sinuata* Ishiga et Watase
13，14. *Pseudotormentus kamigoriensis* De Wever et Caridroit
15，16，18，20—22，24，27，30，31，41. *Pseudoalbaillella ishigai* Wang
17，19，23，25，26，33—36，38，40，42. *Pseudoalbaillella monopteryla* Wang sp. nov.；40，正模（Holotype）
28，29，32，37，39，43—48. *Pseudoalbaillella nonpteryla* Wang sp. nov.；29. 正模（Holotype）

图 版 11

1—4. *Latentifistula conica* Wang sp. nov.；4. 正模（Holotype）
5，6，19—39，41. *Pseudoalbaillella sakmarensis* (Kozur)
7—18. *Pseudoalbaillella lomentaria* Ishiga et Imoto
40，42. *Radiolarids* gen. et sp. indet.

图 版 12

1—6. *Tiborella florida* (Nakaseko et Nishimura)
7. *Parasepsagon praetetracanthus* Kozur et Mostler
8—11. *Triassospongosphaera* sp. A
12. *Triassospongosphaera triassica* (Kozur et Mostler)
13. *Triassocampe nanpanensis* (Feng)
14，15. *Triassocampe coronata* Bragin
16—23. *Eptingium nakasekoi* Kozur et Mostler
24，26—30. *Pseudostylosphaera compacta* (Nakaseko et Nishimura)
25. *Pseudostylosphaera japonica* (Nakaseko et Nishimura)
31—34. *Neoalbaillella ornithoformis* Takemura et Nakaseko
35，36，39. *Triplanospongos musashiensis* Sashida et Tonishi

37. *Trilonche pseudocimelia* (Sashida et Tonishi)

38. *Archaeospongoprunum chiangdaoensis* (Sashida)

40，42. *Ishigaum trifistis* De Wever et Caridroit

41. *Raciditor gracilis* (De Wever et Caridroit)

43. *Trilonche* sp. C

图 版 13

1—5，7，8，11—13，20—22，25—27. *Neoalbaillella ornithoformis* Takemura et Nakaseko

6，9，23，28—33. *Neoalbaillella* sp. A.

10，24. *Neoalbaillella optima* Ishiga，Kito et Imoto

14. *Neoalbaillella pseudogrypa* Sashida et Tonishi

15—19. *Neoalbaillella gracilis* Takemura et Nakaseko

34. *Ishigaum trifistis* De Wever et Caridroit

35. 45—48. *Triplanospongos musashiensis* Sashida et Tonishi

36. *Ishigaum obesum* De Wever et Caridroit

37. *Cauletella manica* (De Wever et Caridroit)

38—43. *Copicyntra akikawaensis* Sashida et Tonishi

44. *Trilonche* sp. A.

图 版 14

1—3，5，6. *Triplanospongos musashiensis* Sashida et Tonishi

4，7—11，19，26，35. *Ishigaum trifistis* De Wever et Caridroit

12，16，21，22，27—29，34，36，38. *Ishigaum craticula* Shang，Caridroit et Wang

13，14. *Trilonche pseudocimelia* (Sashida et Tonishi)

15，18，23，41. *Ishigaum obesum* De Wever et Caridroit

17，24，25，37，40. *Ormistonella robusta* De Wever et Caridroit

20，39. *Raciditor inflata* (Sashida et Tonishi)

30. *Ishigaum* sp. A. Wang et Li

31—33. *Raciditor gracilis* (De Wever et Caridroit)

42—49，51，55—57，61—65. *Albaillella triangularis* Ishiga，Kito et Imoto

50，52—54，59，60. *Albaillella levis* Ishiga，Kito et Imoto

58. *Albaillella lauta* Kuwahara

图 版 15

1—9. *Pseudoalbaillella sakmarensis* (Kozur)

10—12，18，20—23. *Pseudoalbaillella scalprata* Holdsworth et Jones

13—17，19，24. *Pseudoalbaillella postscalprata* Ishiga

图 版 16

1，2. *Pseudoalbaillella elongata* Ishiga et Imoto

3—9. *Pseudoalbaillella lomentaria* Ishiga et Imoto

10. *Raciditor* sp. A

11，15—17，20. *Raciditor oblatum* Wang sp. nov.

12. *Quadricaulis femoris* Caridroit et De Wever

13. *Latentifistula texana* Nazarov et Ormiston

14，18. *Latentifistula conica* Wang sp. nov.

19. *Quinquiremis robusta* Nazarov et Ormiston

21. *Stigmosphaerostylus* sp. D

图　版　17

1，2. *Albaillella indensis* Won

3. *Albaillella undulata* Deflandre

4. *Stigmosphaerostylus variospina*（Won）

5. *Archocyrtium lagabriellei* Gourmelon

6，7. *Pylentonema mira* Cheng

8—15，17. *Stigmosphaerostylus vulgaris*（Won）

16. *Polyentactinia* sp. A

18—20. *Astroentactinia biaciculata* Nazarov

图　版　18

1—9. *Pseudostylosphaera japonica*（Nakaseko et Nishimura）

3a，10. *Pseudostylosphaera nazarovi*（Kozur et Mostler）

11—13，15—18. *Pseudostylosphaera gracilis* Kozur et Mock

14. *Pseudostylosphaera magnispinosa* Yeh

19. *Pseudostylosphaera coccostyla*（Rüst）

20，21. *Pseudostylosphaera compacta*（Nakaseko et Nishimura）

22，23，25. *Cryptostephanidium cornigerum* Dumitrica

24，26，41. *Triassistephanidium laticornis* Dumitrica

27. *Spongostephanidium spongiosum* Dumitrica

28，29. *Tritortis dispiralis*（Bragin）

30，31，33，35—39，42—45. *Parasepsagon firmum*（Gorican）

32，34，40，46. *Parasepsagon hohxiliensis* Wang sp. nov.；32. 正模（Holotype）

图　版　19

1，2，21，22. *Paroertlispongus longispinosa* Wang sp. nov.

3，4. *Cenosphaera* sp. A

5—7. *Annulotriassocampe sulovensis*（Kozur et Mock）

8. *Archaeodictyomitra* sp. A

9. *Parasepsagon* sp. A

10，11. *Spongoserrula rarauana* Dumitrica

12，24，25，47，48. *Pseudostylosphaera compacta*（Nakaseko et Nishimura）

13，14. *Paurinella aequispinosa* Kozur et Mostler

15. *Paurinella* sp. A

16. *Tubotriassocyrtis* sp. A

17—19. *Triassospongosphaera triassica*（Kozur et Mostler）

20. *Quinqueremis* sp. A

23，32. *Pseudostylosphaera magnispinosa* Yeh
26. *Pseudostylosphaera* sp. B
27. *Hindeosphaera spinulosa* （Nakaseko et Nishimura）
28，49. *Pseudostylosphaera longispinosa* Kozur et Mostler
29—31. *Pseudostylosphaera nazarovi* （Kozur et Mostler）
33. *Parasepsagon variabilis* （Nakaseko et Nishimura）
34—39. *Eptingium manfredi* Dumitrica
40—45. *Tiborella florida* （Nakaseko et Nishimura）
46. *Triassocampe* sp. A

图 版 20

1—10. *Spongoserrula rarauana* Dumitrica
11—22. *Triassospongosphaera triassica* （Kozur et Mostler）
23—25. *Paurinella aequispinosa* Kozur et Mostler
26—28，31. *Annulotriassocampe nova* （Yao）
29，32，34，35，37. *Triassocampe deweveri* （Nakaseko et Nishimura）
30，36，38，39. *Annulotriassocampe sulovensis* （Kozur et Mock）
33. *Archaeodictyomitra* sp. B
40，41. *Triassocampe* spp. Gorican et Buser

图 版 21

1—7，14. *Parasepsagon firmum* （Gorican）
8. *Pseudostylosphaera* sp. A
9，10. *Parasepsagon* sp. B
11，37. *Cryptostephanidium* sp. A
12，13. *Pseudostylosphaera* sp. C
15. *Pseudostylosphaera nazarovi* （Kozur et Mostler）
16. *Pseudostylosphaera compacta* （Nakaseko et Nishimura）
17，18. *Pseudostylosphaera gracilis* Kozur et Mock
19. *Pseudostylosphaera qinghaiensis* Wang sp. nov.；19. 正模（Holotype）
20—32. *Hindeosphaera bispinosa* Kozur et Mostler
33—36. *Cryptostephanidium cornigerum* Dumitrica

图 版 22

1—11. *Annulotriassocampe multisegmantatus* Tekin
12. *Annulotriassocampe proprium* （Blome）
13. *Annulotriassocampe sulovensis* （Kozur et Mock）
14，15. *Paratriassocampe brevis* Kozur et Mostler
16—20. *Paroertlispongus longispinosus* Wang sp. nov.；16. 正模（Holotype）
21. *Paroertlispongus* sp. A
22，23. *Paroertlispongus opiparus* Wang sp. nov.；22. 正模（Holotype）
24，31. *Beturiella robusta* Dumitrica，Kozur et Mostler
25—27，29，32. *Beturiella variospinosa* Wang sp. nov.；26. 正模（Holotype）

28，30. *Beturiella* sp. A

33. *Triassospongosphaera brevispinosa* Wang sp. nov.；33. 正模（Holotype）

34—36. *Triassospongosphaera qinghaiensis* Wang sp. nov.；35. 正模（Holotype）

37. *Triassospongosphaera latispinosa*（Kozur et Mostler）

38—40. *Triassospongosphaera triassica*（Kozur et Mostler）

图　版　23

1，3. *Pseudostylosphaera magnispinosa* Yeh

2，4—7. *Pseudostylosphaera compacta*（Nakaseko et Nishimura）

8. *Pseudostylosphaera* sp. D

9. *Pseudostylosphaera helicata*（Nakaseko et Nishimura）

10，12. *Hindeosphaera spinulosa*（Nakaseko et Nishimura）

11. *Pseudostylosphaera goestlingensis*（Kozur et Mostler）

13. *Cryptostephanidium cornigerum* Dumitrica

14. *Parasepsagon variabilis*（Nakaseko et Nishimura）

15，20—26. *Eptingium manfredi* Dumitrica

16，28. *Cryptostephanidium longispinosum*（Sashida）

17，18. *Pseudostylosphaera gracilis* Kozur et Mock

19，27. *Hozmadia pararotunda* Wang sp. nov.；27. 正模（Holotype）

29，30. *Pantanellium multiporum* Wang sp. nov.

31. *Pseudostylosphaera japonica*（Nakaseko et Nishimura）

32. *Pseudostylosphaera longispinosa* Kozur et Mostler

33，34. *Archaeospongoprunum globosum* Tekin et Mostler

35. *Poulpus* sp. A

图　版　24

1—3，6，12，13. *Pseudostylosphaera inaequispinosa* Wang sp. nov.；12. 正模（Holotype）

4，5. *Pseudostylosphaera longispinosa* Kozur et Mostler

7—9. *Pseudostylosphaera imperspicua*（Bragin）

10，11. *Pseudostylosphaera japonica*（Nakaseko et Nishimura）

14，15，19. *Pseudostylosphaera paragracilis* Wang sp. nov.；15. 正模（Holotype）

16—18. *Pseudostylosphaera gracilis* Kozur et Mock

20，32. *Pseudostylosphaera nazarovi*（Kozur et Mostler）

21，23. *Pseudostylosphaera goestlingensis*（Kozur et Mostler）

26，29，30. *Hindeosphaera spinulosa*（Nakaseko et Nishimura）

22，24. *Pantanellium* sp. A

25. *Hindeosphaera bella* Wang sp. nov.；25. 正模（Holotype）

27，28，31. *Hindeosphaera bispinosa* Kozur et Mostler

33，34. *Muelleritortis expansa* Kozur et Mostler

35. *Muelleritortis koeveskalensis* Kozur

图　版　25

1—3. *Striatotriassocampe laeviannulata* Kozur et Mostler

4—10. *Annulotriassocampe multisegmantatus* Tekin
11—13, 19. *Striatotriassocampe nodosoannulata* Kozur et Mostler
14, 15, 18. *Paratriassocampe gaetanii* Kozur et Mostler
16, 17, 20, 23. *Annulotriassocampe sulovensis*（Kozur et Mock）
21, 25. *Triassocampe scalaris* Dumitrica，Kozur et Mostler
22, 24. *Triassocampe deweveri*（Nakaseko et Nishimura）
26. *Tubotriassocyrtis fusiformis* Wang sp. nov.；26. 正模（Holotype）
27, 28. *Triassospongosphaera triassica*（Kozur et Mostler）
29, 30. *Paurinella latispinosa* Kozur et Mostler
31. *Paurinella aequispinosa* Kozur et Mostler
32, 33. *Tritortis dispiralis*（Bragin）
34. *Tritortis robustispinosa* Wang sp. nov.；34. 正模（Holotype）
35, 36, 42. *Tritortis kretaensis*（Kozur et Krahl）
37, 39. *Muelleritortis koeveskalensis* Kozur
38, 40. *Muelleritortis cochleata*（Nakaseko et Nishimura）
41. *Cryptostephanidium cornigerum* Dumitrica

图 版 26

1. *Trilonche pittmani* Hinde
2. *Trilonche davidi*（Hinde）
3. *Trilonche echinata*（Hinde）
4. *Trilonche elegans*（Hinde）
5. *Trilonche tretactinia*（Foreman）
6. *Triaenosphaera sicarius* Deflandre
7. *Tetrentactinia spongacea* Foreman
8. *Albaillella sinuata* Ishiga et Watase
9. *Latentifistula patagilaterala* Nazarov et Ormiston
10, 11. *Pseudoalbaillella ishigai* Wang
12—14. *Pseudoalbaillella lomentaria* Ishiga et Imoto
15, 16. *Pseudotormentus kamigoriensis* De Wever et Caridroit
17. *Latentifistula texana* Nazarov et Ormiston
18—20. *Pseudoalbaillella sakmarensis*（Kozur）
21, 36, 37. *Trilonche pseudocimelia* Sashida et Tonishi
22. *Archaeospongoprunum chiangdaoensis*（Sashida）
23. *Raciditor gracilis*（De Wever et Caridroit）
24, 28, 29. *Neoalbaillella ornithoformis* Takemura et Nakaseko
25, 40. *Ishigaum trifistis* De Wever et Caridroit
26. *Ishigaum craticula* Shang，Caridroit et Wang
27. *Triplanospongos musashiensis* Sashida et Tonishi
30. *Neoalbaillella optima* Ishiga，Kito et Imoto
31. *Neoalbaillella pseudogrypa* Sashida et Tonishi
32, 33. *Neoalbaillella gracilis* Takemura et Nakaseko
34. *Ishigaum obesum* De Wever et Caridroit

35. *Ormistonella robusta* De Wever et Caridroit
38. *Albaillella levis* Ishiga, Kito et Imoto
39. *Albaillella lauta* Kuwahara
41. *Copicyntra akikawaensis* Sashida et Tonishi
42, 52. *Eptingium nakasekoi* Kozur et Mostler
43. *Parasepsagon praetracanthus* Kozur et Mostler
44. *Albaillella triangularis* Ishiga, Kito et Imoto
45. *Pseudosphaerostylus japonica* (Nakaseko et Nishimura)
46. *Pseudosphaerostylus compacta* (Nakaseko et Nishimura)
47, 48. *Triassocampe coronata* Bragin
49. *Triassocampe nanpanensis* (Feng)
50. *Tiborella florida* (Nakaseko et Nishimura)
51. *Triassospongosphaera triassica* (Kozur et Mostler)

图　　版
(PLATES)

图版 1

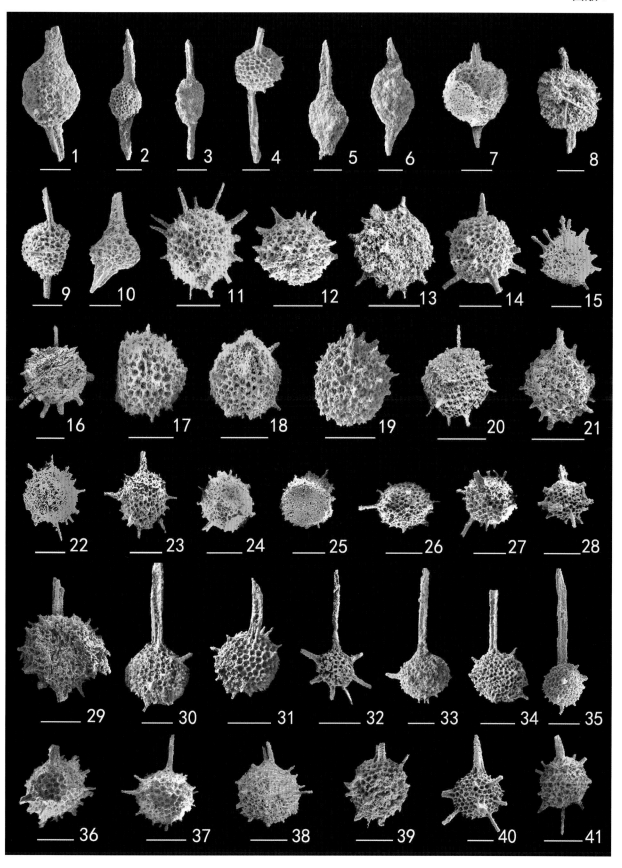

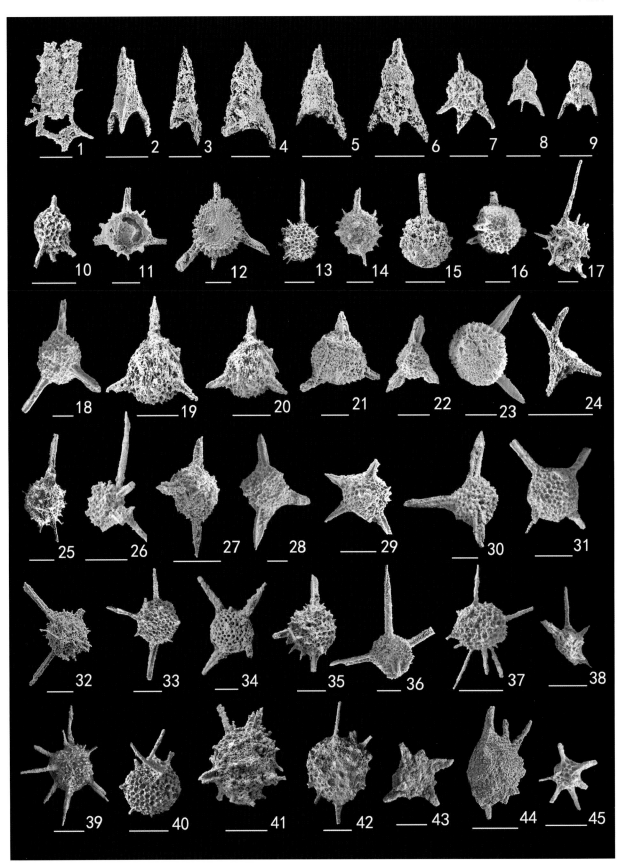

图版2

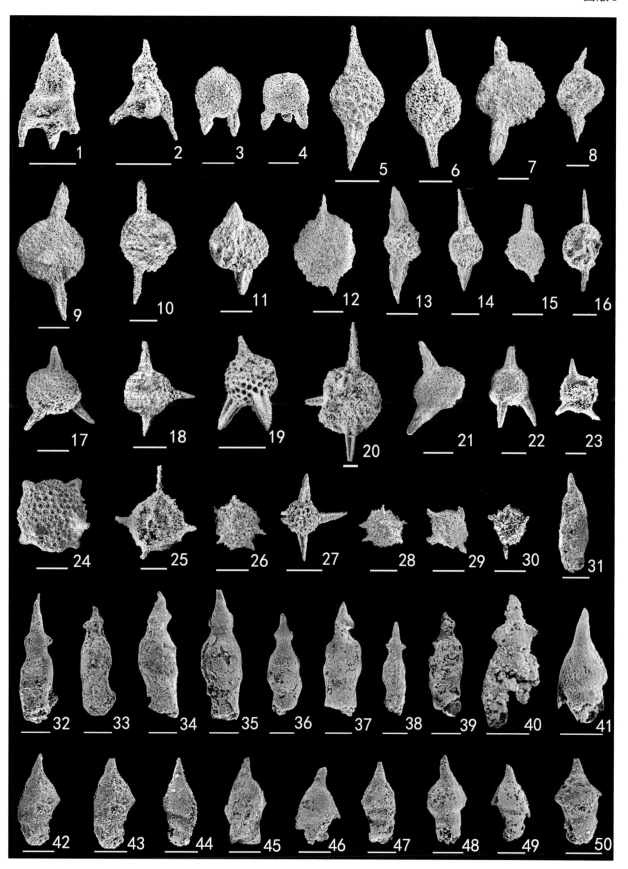

图版 3

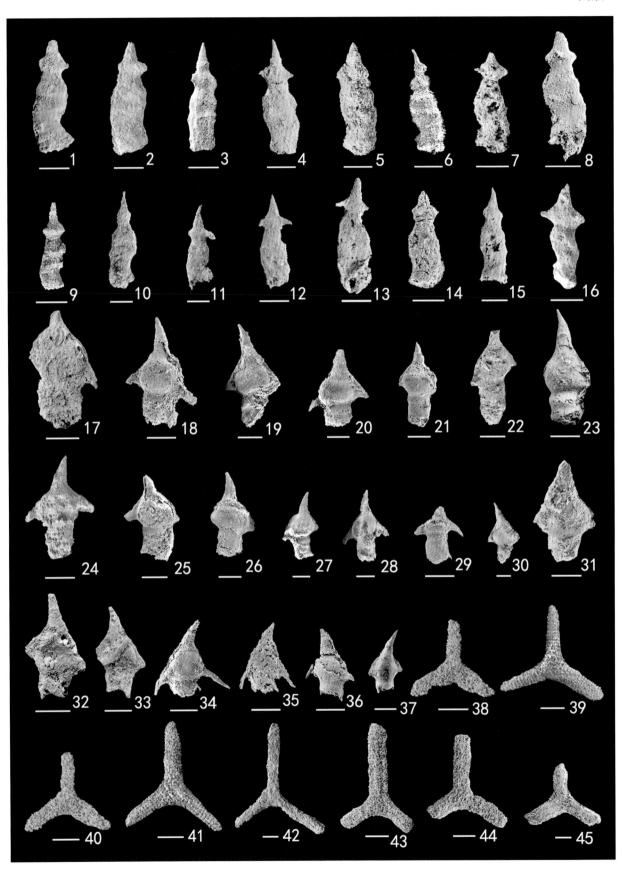

图版 5

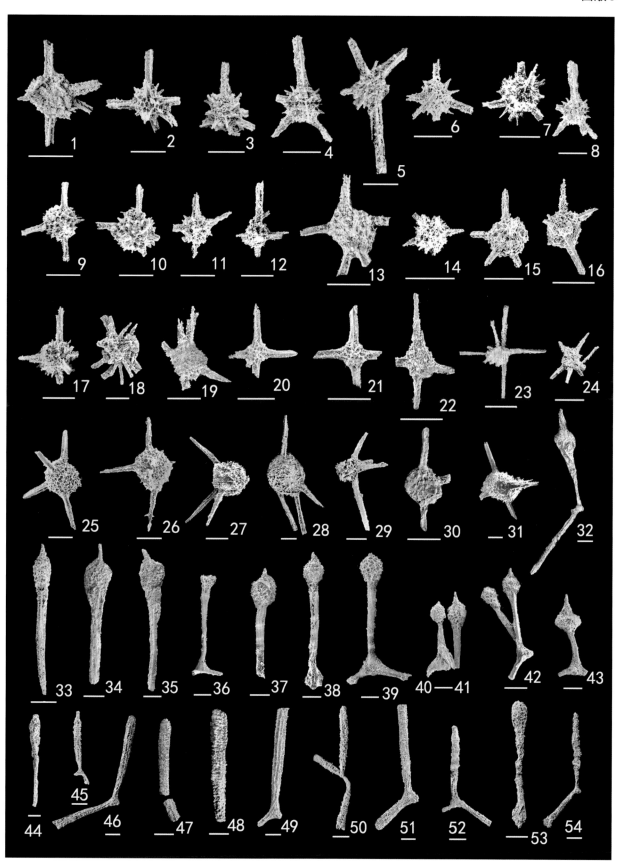

图版 6

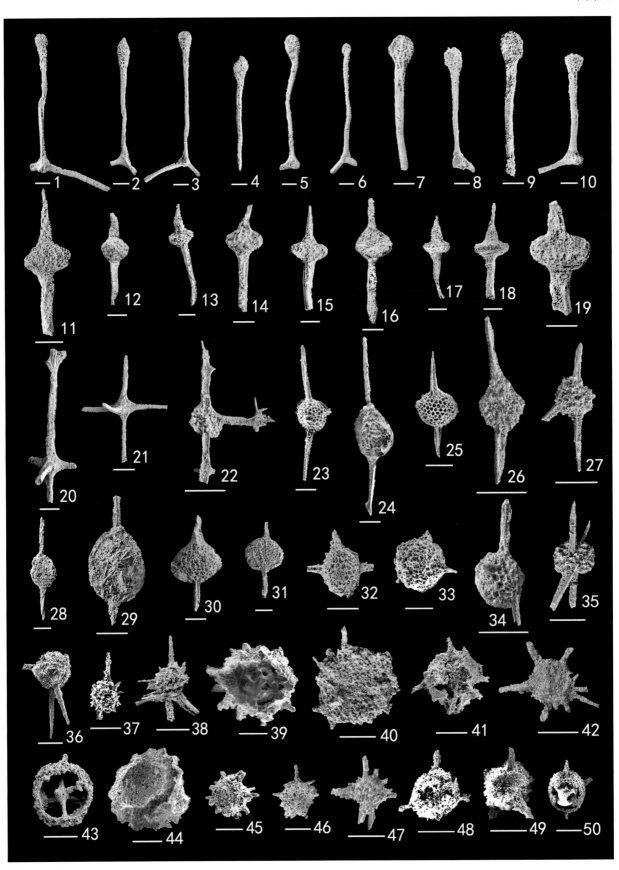

图版 7

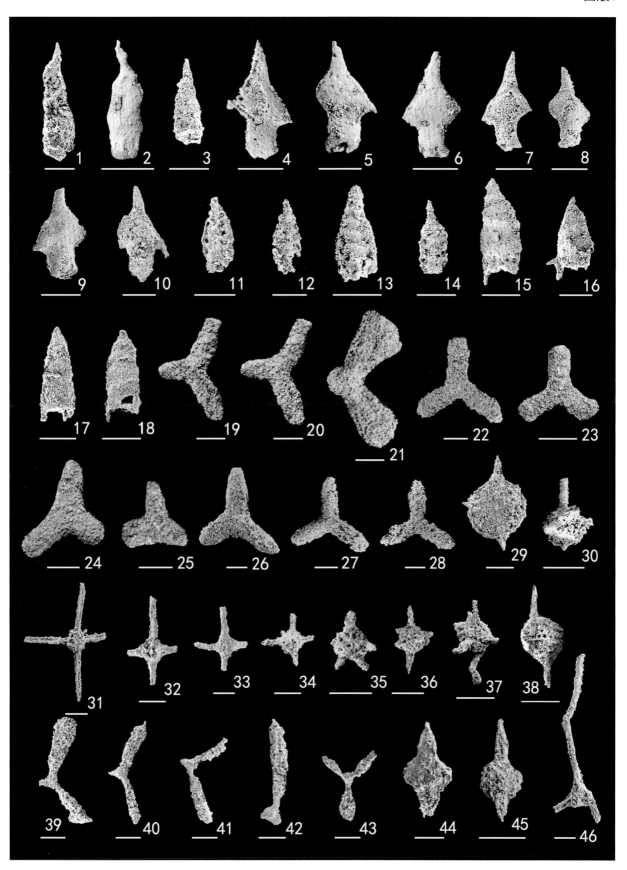

图版 8

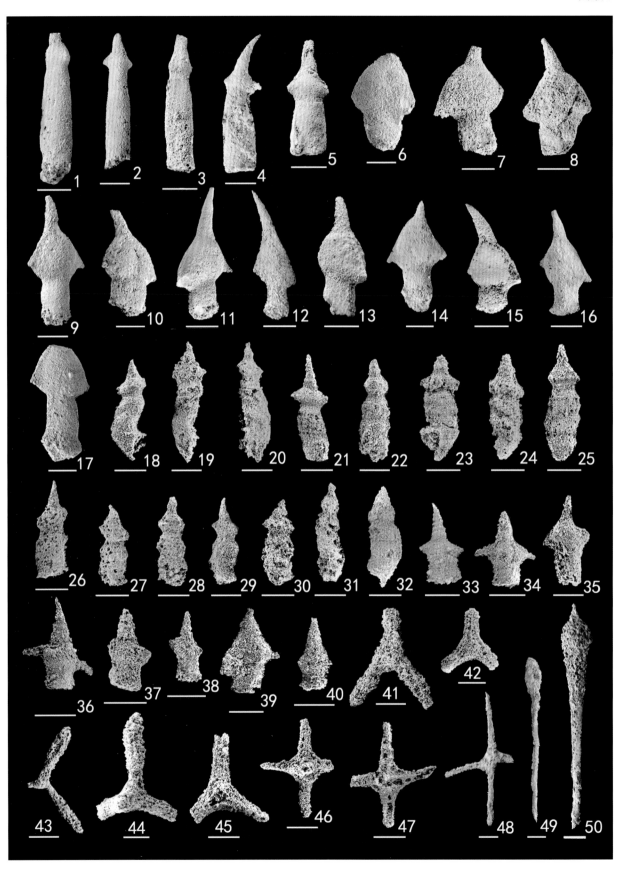

图版 9

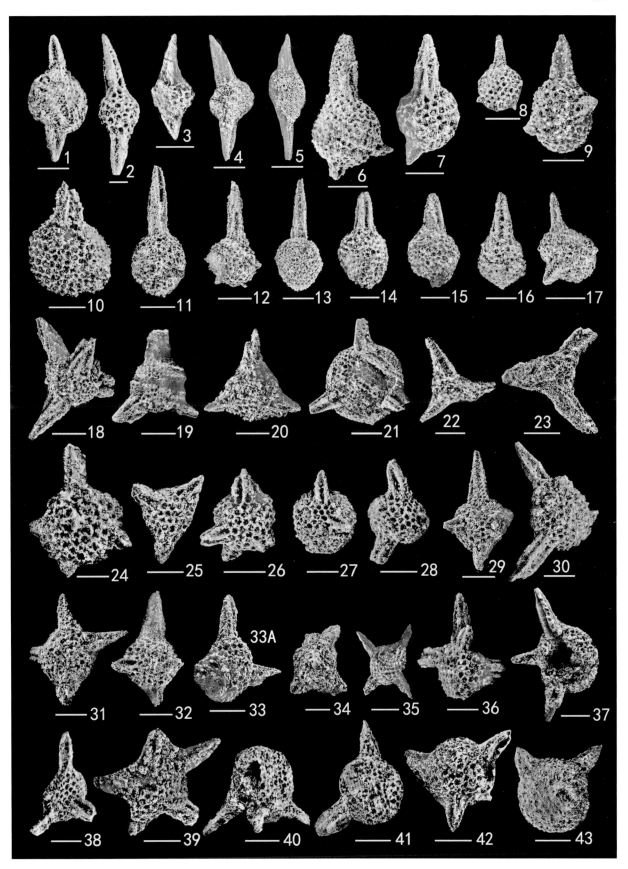

图版 10

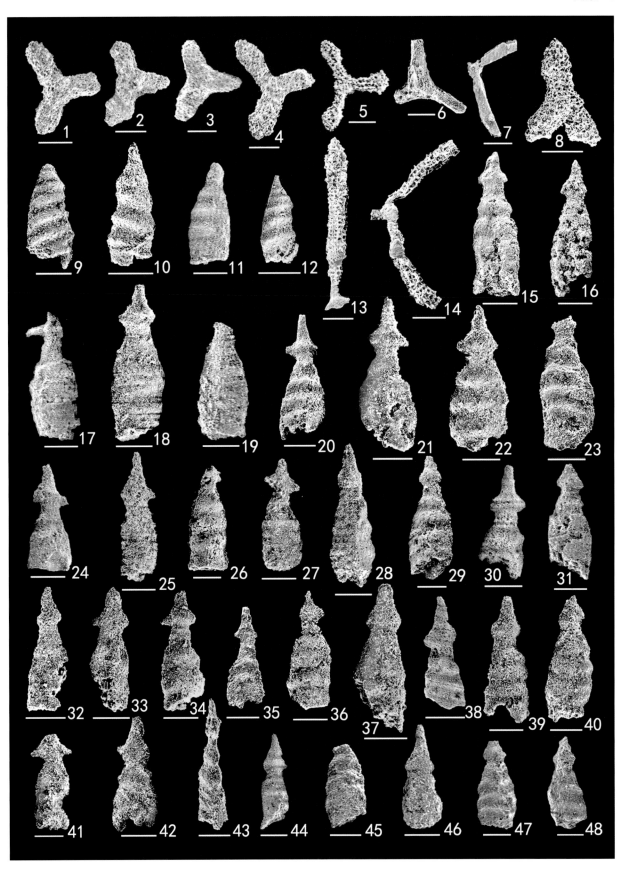

图版 11

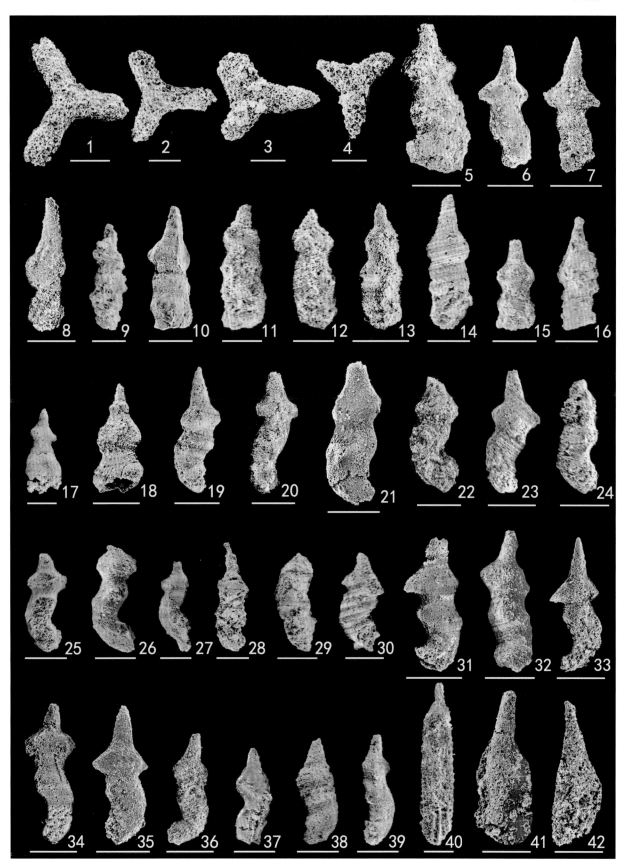

图版 12

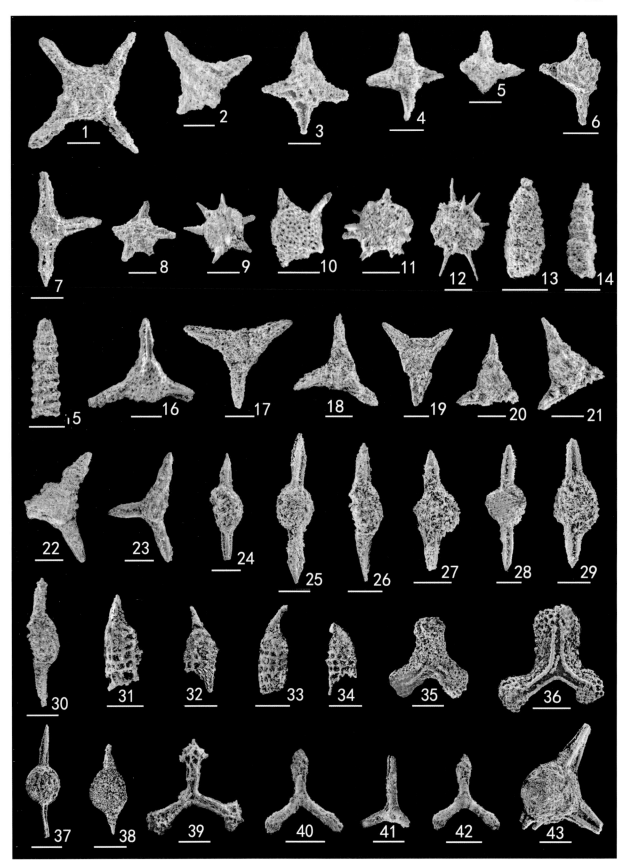

图版 13

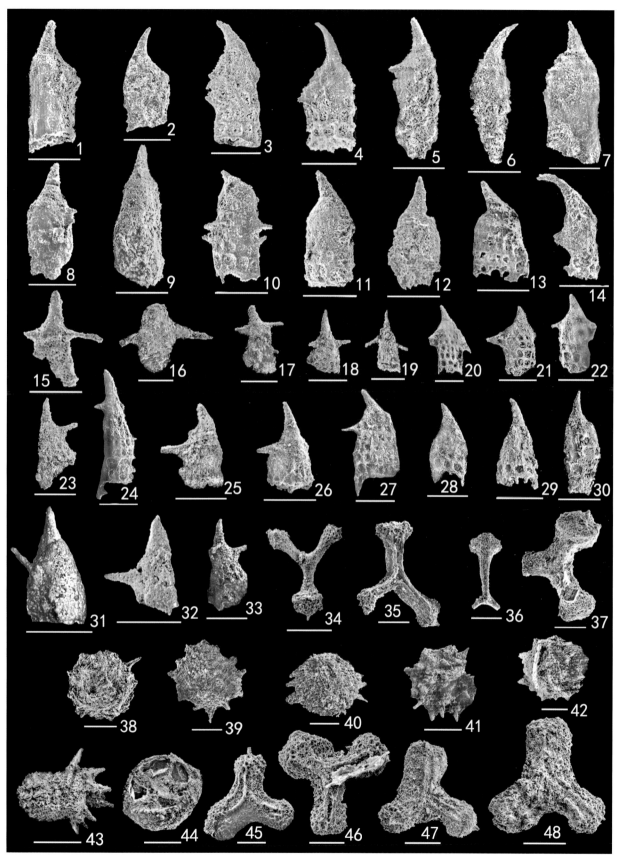

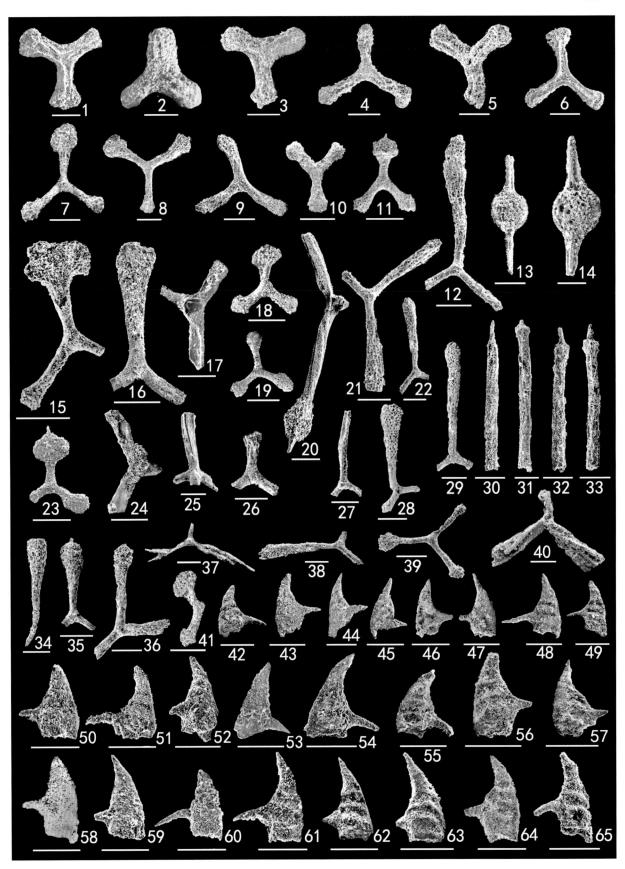

图版 14

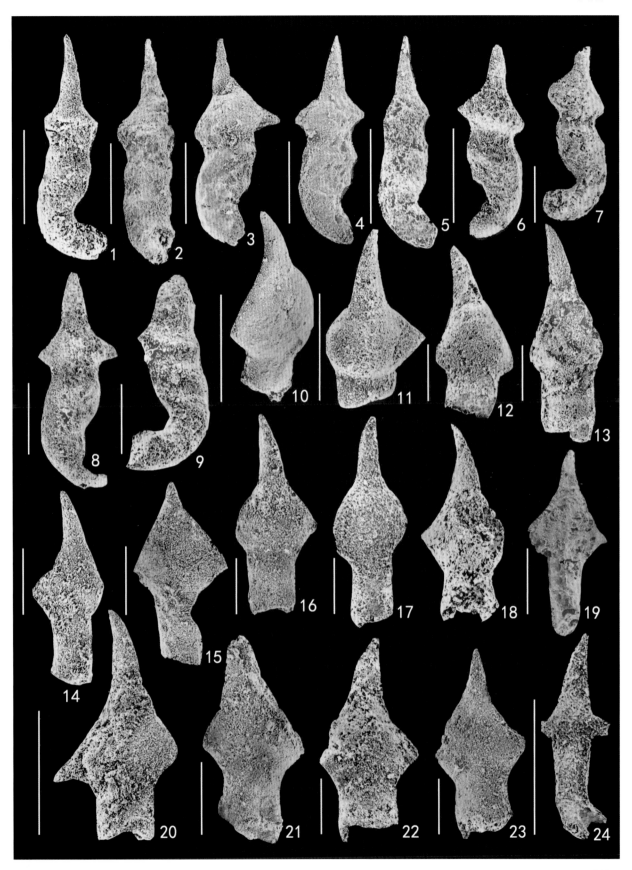

图版 15

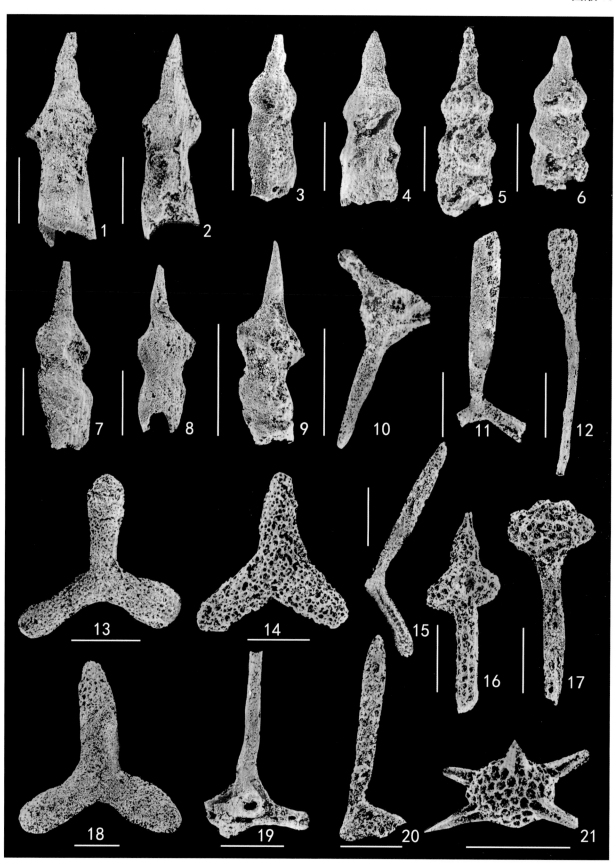

图版 16

图版 17

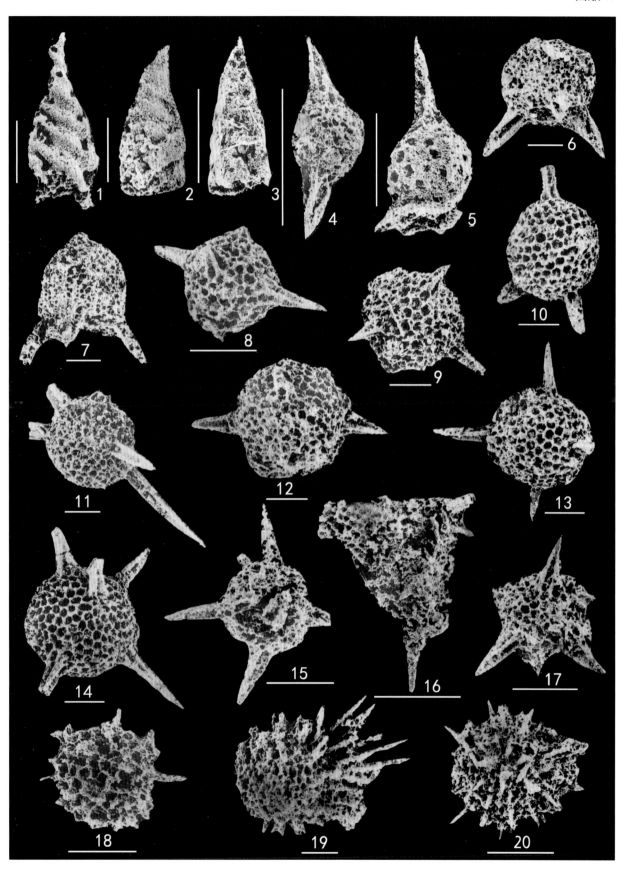

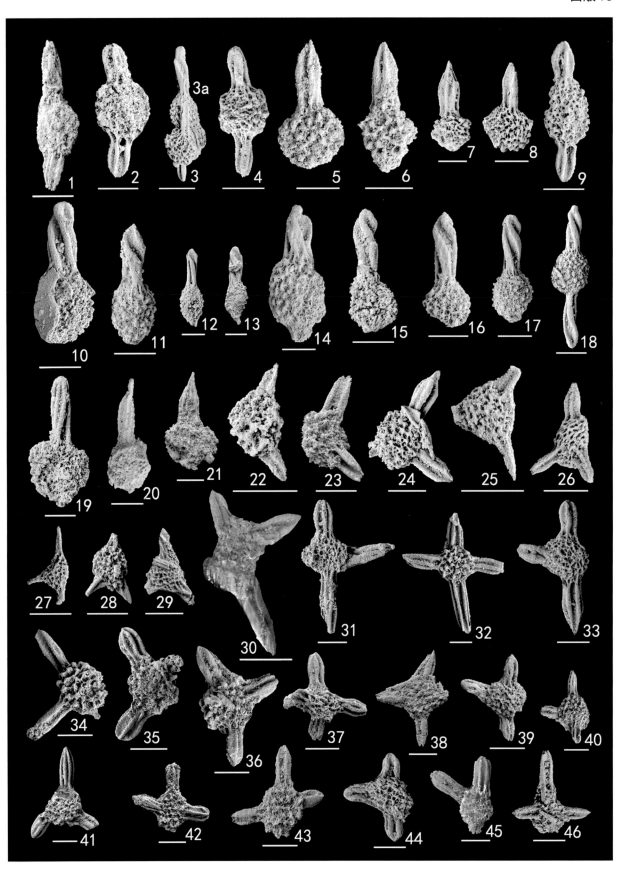

图版 18

图版 19

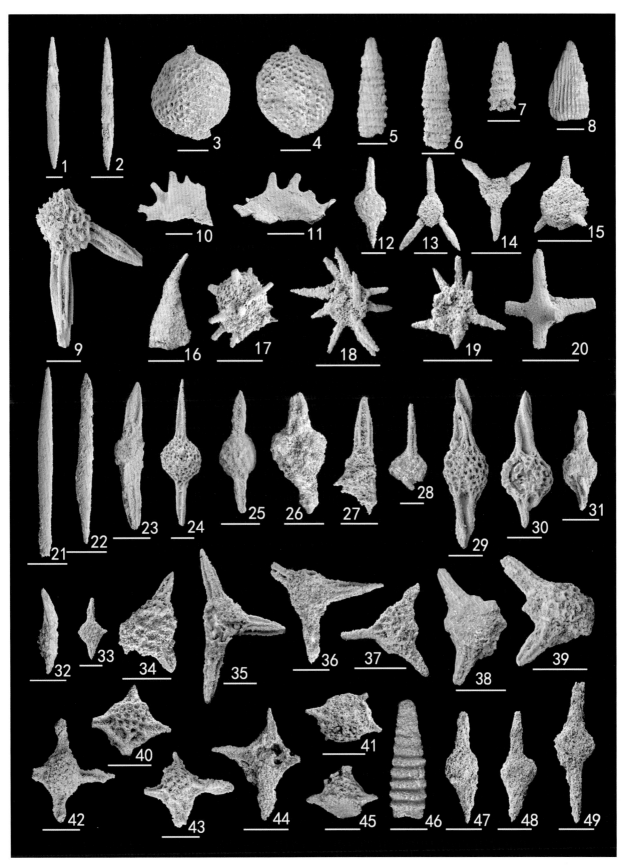

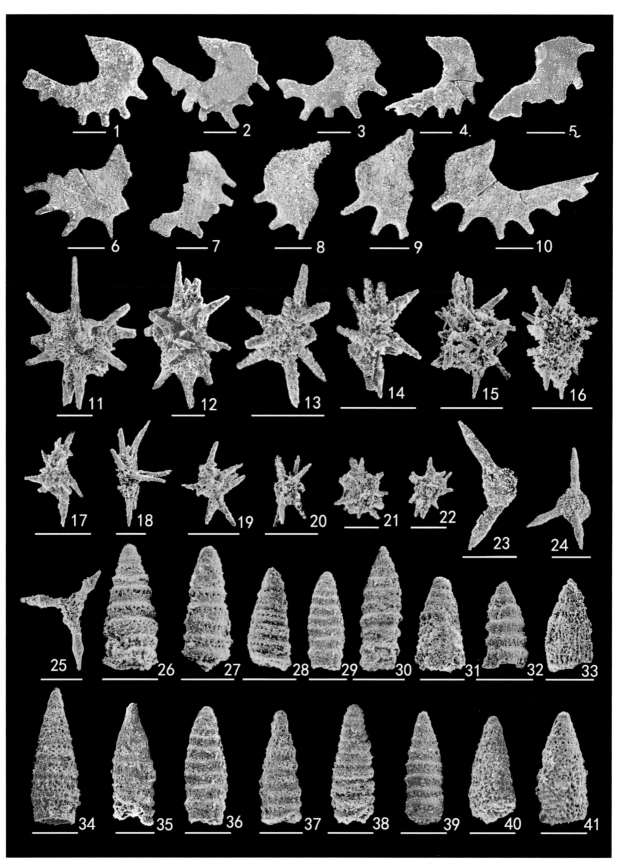

图版20

图版21

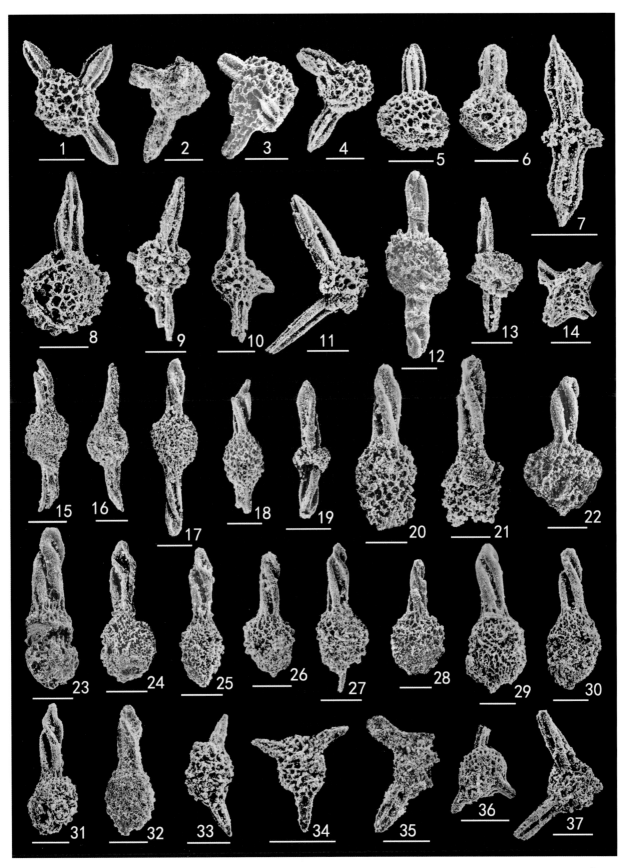

图版 22

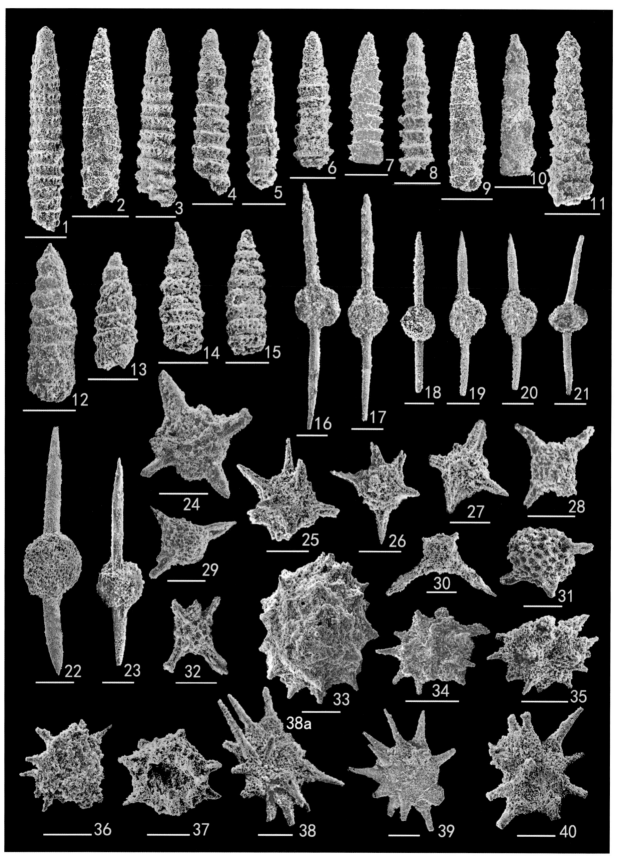

图版 23

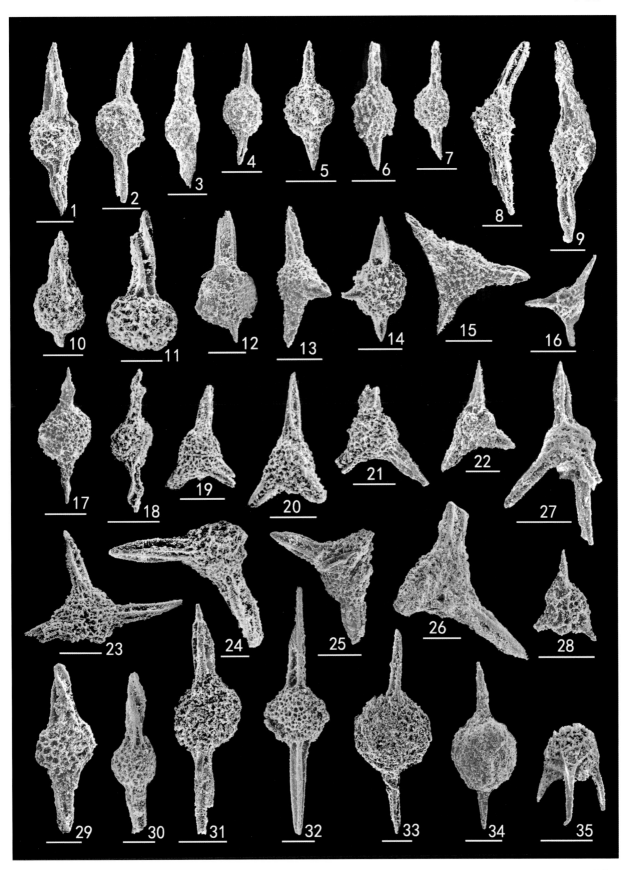

图版 24

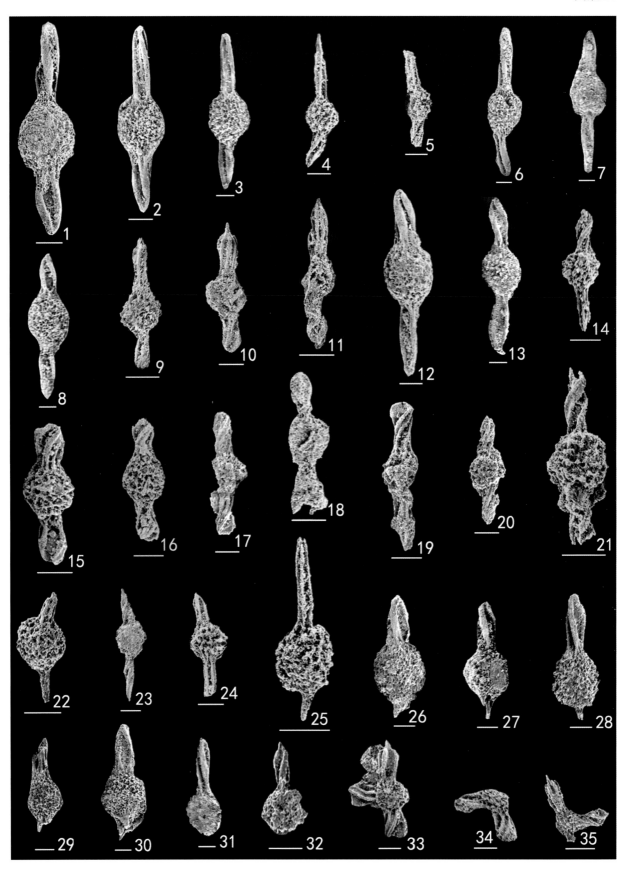

图版 25

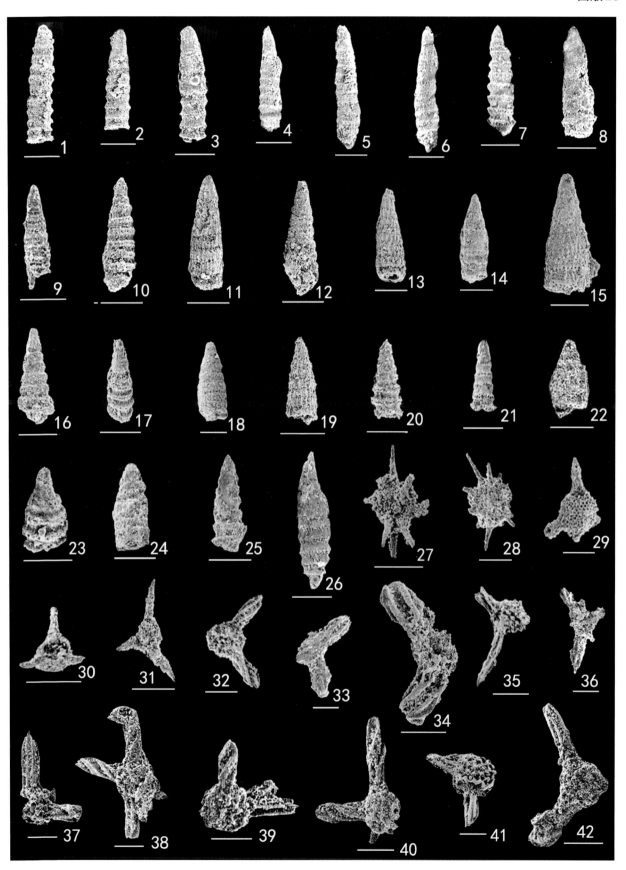

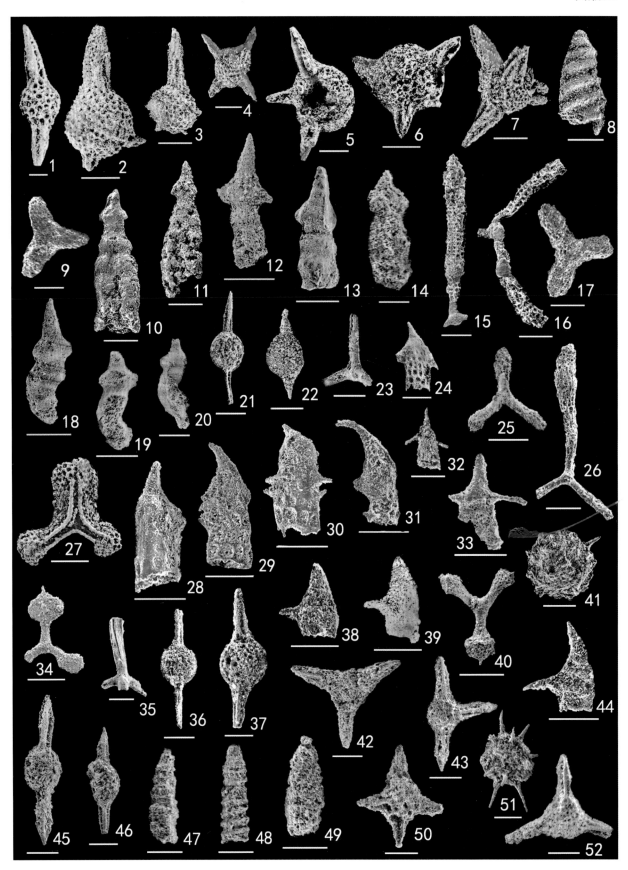

图版 26